Numerical Analysis for Engineers

Methods and Applications

SECOND EDITION

TEXTBOOKS in MATHEMATICS

Series Editors: Al Boggess and Ken Rosen

TEXTBOOKS in MATHEMATICS

Numerical Analysis for Engineers

Methods and Applications

SECOND EDITION

Bilal M. Ayyub

University of Maryland
College Park, USA

Richard H. McCuen

University of Maryland
College Park, USA

CRC Press
Taylor & Francis Group
Boca Raton   London   New York

CRC Press is an imprint of the
Taylor & Francis Group an **informa** business

A CHAPMAN & HALL BOOK

CRC Press
Taylor & Francis Group
6000 Broken Sound Parkway NW, Suite 300
Boca Raton, FL 33487-2742

© 2016 by Taylor & Francis Group, LLC
CRC Press is an imprint of Taylor & Francis Group, an Informa business

No claim to original U.S. Government works

Printed on acid-free paper
Version Date: 20150818

International Standard Book Number-13: 978-1-4822-5035-0 (Hardback)

Visit the Taylor & Francis Web site at
http://www.taylorandfrancis.com

and the CRC Press Web site at
http://www.crcpress.com

To my wife, Deena.

—Bilal M. Ayyub

Contents

Preface

In preparing this book, we strove to achieve a set of educational objectives that include (1) introducing numerical methods to engineering students and practicing engineers, (2) emphasizing the practical aspects of the use of these methods, and (3) establishing the limitations, advantages, and disadvantages of the methods. Although the book was developed with emphasis on engineering and technological problems, the numerical methods can also be used to solve problems in other fields of science.

The book demonstrates the power of numerical methods in the context of solving complex engineering and scientific problems. Studies show that engineering practice will continue to tackle more complex design problems, thus necessitating increased reliance on numerical methods because of the lack of analytical solution methodologies. Therefore, this book is intended to better prepare future engineers, as well as assist practicing engineers, in understanding the fundamentals of numerical methods, especially their applications, limitations, and potentials.

STRUCTURE, FORMAT, AND MAIN FEATURES

We have developed this book with a dual use in mind: as a self-learning guidebook and as a required textbook for a course. In either case, the text has been designed to achieve important educational objectives.

The 11 chapters of the book cover the following subjects: (1) introduction to the text, (2) matrix analysis, (3) introduction to numerical methods, (4) the identification of the roots of equations, (5) the solution of linear simultaneous equations, (6) the interpolation of values of a dependent variable for a given set of discrete measurements, (7) approximation of the differential or integral of an unknown function for a given set of discrete measurements from the function, (8) solutions to differential equations, (9) describing data graphically and computing important characteristics of sample measurements and making basic statistical computations, (10) fitting a curve or a model to data, and (11) numerical optimization. The book was designed for an introductory course in numerical analysis with emphasis on applications. In developing the book, a set of educational outcomes as detailed in Chapter 1 motivated the structure and content of its text. Ultimately, serious readers will find the material very useful in engineering problem solving and decision making.

Each chapter of the book contains many computational examples, as well as a section on applications that contain additional engineering examples. Also, each chapter includes a set of exercise problems that cover the materials of the chapter. The problems were designed carefully to meet the needs of instructors in assigning homework and the readers in practicing the fundamental concepts.

The book can be covered in one semester or two semesters in conjunction with a programming language. The chapter sequence can be followed as a recommended sequence. However, if needed, instructors can choose a subset of the chapters of the book for courses that do not

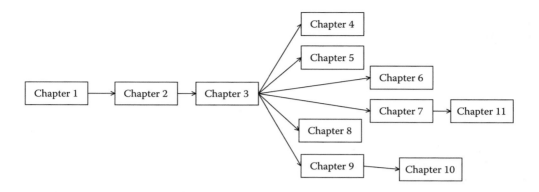

FIGURE P.1   Logical sequence of chapters in the book in terms of their interdependencies.

permit complete coverage of the materials or require coverage that cannot follow the presented order. After completing Chapters 1, 2, and 3, the readers will have sufficient background to follow and understand the materials in Chapters 4 through 8 in any sequence. However, before reading Chapter 10, Chapter 9 should be studied. Figure P.1 illustrates the logical sequence of these chapters.

Acknowledgments

This book was developed over several years and draws on our experiences in teaching courses on numerical methods and statistics. The first edition of this book was used repeatedly in several courses at the University of Maryland for about two decades and has been used at many institutions worldwide. This usage has proved to be a very valuable tool in establishing the second edition's contents and final format and structure.

We would like to acknowledge all the students who participated in the testing of the first edition of the book for their comments, input, and suggestions. The students who took courses on computational methods in civil engineering in the semesters of 1992 and 1994 contributed to this endeavor. Their feedback was very helpful and greatly contributed to the final product. Also, we would like to acknowledge the instructors who used the book during this testing period, Drs. P. Johnson and R. Muhanna. The assistance of Ru-Jen Chao, Dr. Maguid Hassan, and Dr. Ibrahim Assakkaf in critically reviewing the problems at the ends of the chapters and developing some of the example problems and problem solutions is gratefully acknowledged. The reviewers' comments that were provided to us by the publisher were used to improve the book to meet the needs of instructors and enhance the education process. We acknowledge the publisher and the reviewers for these comments.

We would like to invite users of this book, including both readers and instructors, to send us any comments on the book by e-mail at ba@umd.edu. These comments will be used in developing future editions of the book.

Bilal M. Ayyub
Richard H. McCuen

Authors

Bilal M. Ayyub, PhD, is a professor of civil and environmental engineering at the University of Maryland, College Park, and the director of the Center for Technology and Systems Management at the A. James Clark School of Engineering. Dr. Ayyub has been at the University of Maryland since 1983. He is a fellow of the American Society of Civil Engineers, the American Society of Mechanical Engineers, the Society of Naval Architects and Marine Engineers, and the Society for Risk Analysis, and is a senior member of the Institute of Electrical and Electronics Engineers (IEEE). Dr. Ayyub has completed many research and development projects for many governmental and private entities and is a multiple recipient of the ASNE Jimmie Hamilton Award for the best papers in the *Naval Engineers Journal* in 1985, 1992, 2000, and 2003. Also, he received the ASCE Outstanding Research Oriented Paper award for 1987 for his paper in the *Journal of Water Resources Planning and Management*, the ASCE Edmund Friedman Award in 1989, the ASCE Walter Huber Research Prize in 1997, and the K. S. Fu Award of the North American Fuzzy Information Processing Society (NAFIPS) in 1995. He received the Department of the Army Public Service Award in 2007. Dr. Ayyub is the author or coauthor of more than 600 publications in journals, conference proceedings, and reports. Among his publications are many books, including the following selected textbooks used by universities worldwide: *Uncertainty Modeling and Analysis for Engineers and Scientists* (Chapman & Hall/CRC, 2006; with G. Klir); *Risk Analysis in Engineering and Economics*, second edition (Chapman & Hall/CRC, 2003); *Elicitation of Expert Opinions for Uncertainty and Risks* (CRC Press, 2002); *Probability, Statistics and Reliability for Engineers and Scientists*, third edition (Chapman & Hall/CRC, 2011; with R. H. McCuen); and *Numerical Methods for Engineers* (Prentice Hall, 1996; with R. H. McCuen). He is the editor-in-chief of the ASCE-ASME *Journal of Risk and Uncertainty in Engineering Systems*.

Richard H. McCuen, PhD, is the Ben Dyer Professor of civil and environmental engineering at the University of Maryland, College Park. Dr. McCuen earned degrees from Carnegie Mellon University and the Georgia Institute of Technology. He received the Icko Iben Award from the American Water Resource Association and was co-recipient of the 1988 outstanding research award from the American Society of Civil Engineers Water Resources Planning and Management Division. He received the 2015 VenTe Chow Award from the American Society of Civil Engineers. Topics in statistical hydrology and storm water management are his primary research interest. Dr. McCuen is the author of 26 books and over 250 professional papers, including *Modeling Hydrologic Change* (CRC Press, 2002); *Hydrologic Analysis and Design*, third edition (Prentice Hall, 2005); *The Elements of Academic Research* (ASCE Press, 1996); *Estimating Debris Volumes for Flood Control* (Lighthouse Publications, 1996; with T. V. Hromadka); and *Dynamic Communication for Engineers* (ASCE Press, 1993; with P. Johnson and C. Davis).

Introduction

1.1 NUMERICAL ANALYSIS IN ENGINEERING

The advancement of computer technology and its availability to practicing engineers has affected engineering education in many aspects. To prepare engineering students and practicing engineers for current and future challenges, an introduction to and sometimes emphasis on computer-based methods are needed. Currently, practical engineering problems demand the use of computers to obtain solutions in a timely manner and with acceptable accuracy.

The solutions of engineering problems can be obtained using analytical methods or numerical methods. The complexity of many practical engineering problems makes it necessary to use numerical methods; that is, very often analytical solutions cannot be obtained. For example, engineering students learn analytical differentiation and integration of mathematical functions within the first two years of their college education. However, analytical differentiation and integration provide a closed-form derivative and integral, respectively, only for simple functions. As the problems take on greater reality and the complexity of these functions increases, our ability to use analytical differentiation or integration diminishes until the point where analytical solutions are not possible. Also, it is common in engineering to require the use of functions that cannot be expressed in closed form for which derivatives or integrals are needed. In such cases, analytical methods of differentiation and integration cannot be used. Therefore, there is a need for another class of solution methods. Numerical methods belong to this second class of solution tools that can be used where analytical methods are not capable of or practical for providing solutions.

The objective of this book is to introduce numerical methods to students of engineering and science and practicing engineers. The practical aspects of the use of these methods are emphasized throughout the book, with sections in each chapter where practical applications are provided. Although the book was developed with emphasis on engineering and technological problems, numerical methods can be used to solve problems in many other fields. Numerical methods are provided with varying levels of detail and emphasis. A critical presentation of these methods is provided to enhance the readers' understanding.

The numerical methods can be solved by hand or with a computer. The importance of numerical methods in engineering can be effectively demonstrated in cases dealing with complex problems for which analytical solutions cannot be obtained or hand calculations cannot be made. In this book, we use common engineering problems to demonstrate the computational procedures. The examples were intentionally selected with traceable solutions so that readers can reproduce them. It is helpful, but not necessary, for readers to be familiar with the fundamentals of a computer language. We present algorithms in the book as pseudocodes for selected methods. Also, programming considerations for selected methods are provided.

The use of any computational method, analytical or numerical, without the proper understanding of the limitations and shortcomings can have serious consequences. Before using them, users should become well versed with the numerical methods in terms of both the computational details and their limitations, shortcomings, and strengths. Numerical methods should not be used as black boxes with input and output or as numerical recipes.

1.1.1 DECISION MAKING IN ENGINEERING

Engineering is a profession that continually places its members in decision-making positions. To make the best decisions, engineers must be aware of and fully understand alternative solution procedures. This need has become more important in professional practice as the problems that engineers must address have become more important to society and as the design methods have become more complex. The term *best* decisions is multifaceted. Engineers need accurate solutions that are both unbiased and have high precision. The solution must be cost effective and have minimal environmental consequences. The adopted solution should improve the community and be esthetically appealing. These are just some of the criteria that engineers use in making design decisions.

When engineers are given a design problem, they typically approach the problem using a systematic procedure. One formulation of the process is as follows: (1) identify the problem, (2) state the objectives, (3) develop alternative solutions, (4) evaluate the alternatives, and (5) implement the best alternative.

Engineering decision problems can be classified into single- and multiple-objective problems. Example objectives are minimizing the total expected cost, maximizing safety, maximizing the total expected utility value, and maximizing the total expected profit. Decision analysis requires the definition of these objectives. For cases of multiple objectives, the objectives need to be stated in the same units, and weights that reflect the importance of the objectives and that can be used to combine the objectives need to be assigned. Then the problem can be formulated in a suitable form for attaining the decision objectives. The development of alternative solutions comes next, which is an especially critical stage of the process. The team of engineers must have a sound technical background and a broad understanding of alternative design methods. To properly evaluate each alternative, a complete understanding of the technical basis of the design method is required. Then alternatives need to be evaluated, and selection can be made.

Computers have increased the number of alternative solution procedures that are available to engineers. Whereas the engineers of past generations were limited to graphical, analytical, and simple empirical methods, engineers can now consider numerical and simulation methods in their design work. The increased complexity of the numerical and simulation methods is believed to improve the accuracy of the solutions, and since these methods are easily implemented with a computer, the increased effort over the less detailed methods is minimal. Thus the increase in design effort is more than offset by the expected improvement in accuracy. However, to achieve this benefit, the engineer must fully understand the more complex methods so that they can be properly applied.

1.1.2 EXPECTED EDUCATIONAL OUTCOMES

While a textbook can be the principal medium for someone to individually gain an understanding of a body of knowledge, books such as this one are typically used as part of a course on numerical methods. This book has been designed to be used in either capacity—as a self-learning guidebook or as a required textbook for a course. In either case, the text has been designed to produce the following educational outcomes:

1. The reader of the book should be able to numerically solve problems in (a) the identification of roots of equations, (b) the solution of linear simultaneous equations, (c) the interpolation of values of a dependent variable given a set of discrete measurements, (d) approximating the differential or integral of an unknown function given a set of discrete measurements from the function, (e) finding solutions to differential equations, (f) describing data graphically and computing important characteristics of sample measurements, (g) making basic statistical computations, and (h) fitting a curve or a model to data. Also, the reader will learn the fundamentals of matrix analysis.
2. The reader should be able to select from alternative methods the one method that is most appropriate for a specific problem and state reasons for the selection.
3. The reader should be able to formulate algorithms to solve problems numerically.
4. The reader should understand the limitations of each numerical method, especially the conditions under which they may fail to converge to a solution.
5. The reader should recognize engineering applications of the various numerical methods.

In developing the topics presented in the book, these expected educational outcomes motivated the structure and content of the text.

Ultimately, the serious reader will find the material very useful in engineering problem solving and decision making.

1.2 ANALYTICAL VERSUS NUMERICAL ANALYSIS

In the previous section, problem-solving methods were separated into two classes: analytical and numerical. These two approaches are distinguished on the basis of their algorithms. Analytical calculus forms the basis for analytical problem solving, while finite-difference arithmetic forms the basis of numerical methods. As an example, finding the minimum or maximum of a simple function is a common need in engineering. For the analytical approach, the derivative is computed analytically and set equal to zero, and the resulting equation is solved for the value of the unknown. If the minimum of the function $y = x^2 - 3x + 2$ was of interest, it is easily obtained analytically by computing the derivative as $2x - 3$. Therefore, the minimum solution occurs at $x = 1.5$, which is obtained by setting the derivative to zero and solving for x. One numerical solution is to iterate over a range of x values at a constant increment Δx and select the value of x for which y is smallest. For example, if we define the interval from 1 to 2 and use an increment of 0.2, the following tabulation indicates that the minimum value of y falls in the interval $1.4 < x < 1.6$:

x	1.0	1.2	1.4	1.6	1.8	2.0
y	0	−0.16	−0.24	−0.24	−0.16	0

To improve the accuracy of the solution, the search interval of x could be established as $1.4 < x < 1.6$, the increment decreased to 0.04, and the search repeated. For this example, the analytical approach is clearly superior. But what if the function is complex and an analytical derivative cannot be found? Then a numerical solution is necessary.

Both classes have advantages and disadvantages. Analytical techniques provide a direct solution and will result in an exact solution, if one exists. Analytical methods usually require less time to find a solution. However, this occurs partly because analytical methods are practical only for functions that have a simple, closed-form mathematical structure. Also, the analytical solution procedure becomes considerably more complex when constraints are involved.

Numerical analysis can be used with functions with a moderately complex structure, and it is easy to include constraints on the unknowns in the solution. However, numerical methods most often require a considerable number of iterations in order to approach the true solution. Furthermore, the solution usually is not exact, and it is also necessary to provide initial estimates of the unknowns. Since the solution is not exact, error analysis and error estimation are often necessary.

1.3 TAYLOR SERIES EXPANSION

A Taylor series is commonly used as a basis of approximation in numerical analysis. The objective of this section is to provide a review of the Taylor series.

A Taylor series is the sum of functions based on continually increasing derivatives. For a function $f(x)$ that depends on only one independent variable x, the value of the function at point $x_0 + h$ can be approximated by the following Taylor series:

$$f(x_0 + h) = f(x_0) + hf^{(1)}(x_0) + \frac{h^2}{2!} f^{(2)}(x_0) + \frac{h^3}{3!} f^{(3)}(x_0)$$

(1.1)

$$+ \cdots + \frac{(h^n)}{n!} f^n(x_0) + R_{n+1}$$

in which

x_0 is some base value (or starting value) of the independent variable x

h is the distance between x_0 and the point x at which the value of the function is needed—that is, $h = x - x_0$

$n!$ is the factorial of n (i.e., $n! = n(n-1)(n-2) \ldots 1$)

R_{n+1} is the remainder of the Taylor series expansion

The superscript (n) on the function $f^{(n)}$ indicates the nth derivative of the function $f(x)$

The derivatives of $f(x)$ in Equation 1.1 are evaluated at x_0. Equation 1.1 can also be written in the following form:

(1.2)
$$f(x_0 + h) = \sum_{k=0}^{\infty} \frac{h^k}{k!} f^{(k)}(x_0)$$

where $0! \equiv 1$. Equations 1.1 and 1.2 are based on the assumption that continuous derivatives exist in an interval that includes the points x and x_0.

Equations 1.1 and 1.2 can be used with a finite number of terms to produce approximations for the evaluated function at $(x_0 + h)$. The order of the approximation is defined by the order of the highest derivative that is included in the approximation. For example, the first-order approximation is defined by the terms of the Taylor series expansion up to the first derivative as follows:

(1.3)
$$f(x_0 + h) \approx f(x_0) + hf^{(1)}(x_0)$$

Similarly, the second- and third-order approximations are given, respectively, by

(1.4)
$$f(x_0 + h) \approx f(x_0) + hf^{(1)}(x_0) + \frac{h^2}{2!} f^{(2)}(x_0)$$

and

$$(1.5) \qquad f(x_0+h) \approx f(x_0)+hf^{(1)}(x_0)+\frac{h^2}{2!}f^{(2)}(x_0)+\frac{h^3}{3!}f^{(3)}(x_0)$$

The accuracy of the approximation improves as the order of the approximation is increased.

The importance of the individual terms of the Taylor series depends on the nature of the function and the distance h. The higher order terms become more important as the nonlinearity of the function increases and the difference $x - x_0$ increases. If the function is linear, then only the term with the first derivative is necessary. As the separation distance h increases, the importance of the nonlinear terms usually increases. Thus the error increases as fewer terms of the Taylor series are included.

Example 1.1: Nonlinear Polynomials

Taylor series are used to evaluate a function that is of practical interest, but with an unknown expression. If the expression of the function were known, then the Taylor series approximation would not be necessary. In this example, the true function will be assumed to be known in order to demonstrate the computational aspects of the Taylor series and the sizes of the remainder terms in the series. For this purpose, the following fourth-order polynomial is used:

$$(1.6) \qquad f(x)=\frac{1}{8}x^4-\frac{1}{6}x^3+\frac{1}{2}x^2-\frac{1}{2}x+2$$

$$(1.7) \qquad f^{(1)}(x) = 0.5x^3 - 0.5x^2 + x - 0.5$$

$$(1.8) \qquad f^{(2)}(x) = 1.5x^2 - x + 1$$

$$(1.9) \qquad f^{(3)}(x) = 3x - 1$$

$$(1.10) \qquad f^{(4)}(x) = 3$$

$$(1.11) \qquad f^{(5)}(x) = 0$$

For an assumed base point (x_0) of 1.0, the value of the function and its derivatives are

$$(1.12) \qquad f(x) = 1.95833$$

$$(1.13) \qquad f^{(1)}(x) = 0.5$$

$$(1.14) \qquad f^{(2)}(x) = 1.5$$

(1.15) $f^{(3)}(x) = 2$

(1.16) $f^{(4)}(x) = 3$

(1.17) $f^{(5)}(x) = 0$

Using a separation distance h of 0.5, the values of the individual terms of the Taylor series are shown in Table 1.1a. For $x = 1.5$, the linear term is dominant. For x from 2 to 3, the second term is dominant, with less emphasis given to the linear term and more emphasis given to the cubic term. For $x = 3.5$, the cubic term is the most important, while for $x = 4$, the fourth-order term is most important. These results illustrate the interaction between the importance of the terms and both the nonlinearity of the function and the separation distance. The cumulative values for the terms of the Taylor series are given in Table 1.1b, and the differences between the cumulative values and the true values are given in Table 1.1c. The differences show the need to consider the higher order terms as the separation distance increases. This is clearly evident from Figure 1.1, which shows the cumulative values from Table 1.1b as a function of separation distance h.

TABLE 1.1   a. Values of the Individual Terms of the Taylor Series Expansion of Equations 1.1 and 1.2 for the Quadratic Function of Equation 1.6; b. Cumulative Function of the Taylor Series Expansion; c. Difference between the Cumulative Function and the True Value of the Function

Part	x	h	$hf^{(1)}(x_0)$	$\dfrac{h^2}{2!}f^{(2)}(x_0)$	$\dfrac{h^3}{3!}f^{(3)}(x_0)$	$\dfrac{h^4}{4!}f^{(4)}(x_0)$
a. Individual	1.5	0.5	0.25	0.1875	0.04167	0.0078125
terms	2.0	1.0	0.50	0.7500	0.33333	0.1250000
	2.5	1.5	0.75	1.6875	1.12500	0.6328125
	3.0	2.0	1.00	3.0000	2.66667	2.0000000
	3.5	2.5	1.25	4.6875	5.20833	4.8828125
	4.0	3.0	1.50	6.7500	9.00000	10.1250000
b. Cumulative	1.5		2.20833	2.39583	2.43750	2.44531
function	2.0		2.45833	3.20833	3.54167	3.66667
	2.5		2.70833	4.39538	5.52083	6.15365
	3.0		2.95833	5.95833	8.62500	10.62500
	3.5		3.20833	7.89583	13.10417	17.98698
	4.0		3.45833	10.20833	19.20833	29.33333
c. Errors	1.5		−0.23698	−0.04948	−0.00781	0.00000
	2.0		−1.20833	−0.45833	−0.12500	0.00000
	2.5		−3.44531	−1.75781	−0.63281	0.00000
	3.0		−7.66667	−4.66667	−2.00000	0.00000
	3.5		−14.77865	−10.09115	−4.88281	0.00000
	4.0		−25.87500	−19.12500	−10.12500	0.00000

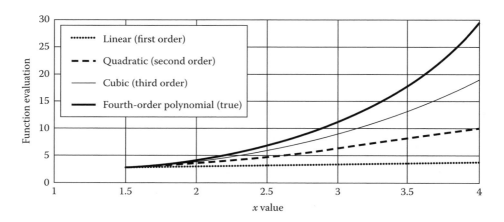

FIGURE 1.1   Taylor series approximations of a fourth-order polynomial.

1.4 APPLICATIONS

1.4.1 TAYLOR SERIES EXPANSION OF THE SQUARE ROOT

The square-root function, which is used in this section to illustrate the Taylor series expansion, can be expressed as

(1.18)
$$f(x) = \sqrt{x}$$

To evaluate the Taylor series, the derivatives of the function are developed as follows:

(1.19)
$$f^{(1)}(x) = \frac{1}{2} x^{-0.5}$$

(1.20)
$$f^{(2)}(x) = -\frac{1}{4} x^{-1.5}$$

(1.21)
$$f^{(3)}(x) = \frac{3}{8} x^{-2.5}$$

Higher order derivatives can be developed similarly. For a base point $x_0 = 1$ and $h = 0.001$, the four terms of the Taylor series produce the following estimate for the square root of 1.001:

(1.22)
$$f(1.001) = \sqrt{1.001} \approx \sqrt{1} + 0.5(0.001)(1)^{-0.5} - \frac{1}{4(2!)}(0.001)^2(1)^{-1.5}$$
$$+ \frac{3}{8(3!)}(0.001)^3(1)^{-2.5}$$

or

(1.23) $f(1.001) \approx 1 + 0.5 \times 10^{-3} - 0.125 \times 10^{-6} + 0.625 \times 10^{-10} = 1.0004999$

which equals the true value to the number of decimal points shown. Because the interval h is so small, the linear approximation would have been accurate to five decimal points.

1.4.2 EXAMPLE SERIES

In this section, example series are provided. These series might be familiar to the readers from previous courses in mathematics. For example, the exponential evaluation to the base e of x can be expressed by the following series:

(1.24)
$$e^x = 1 + x + \frac{x^2}{2!} + \frac{x^3}{3!} + \cdots + \frac{x^k}{k!} + \cdots = \sum_{k=0}^{\infty} \frac{x^k}{k!}$$

This equation is valid only for x that is finite. Similarly, the natural logarithm of x can be expressed using a Taylor series as

(1.25)
$$\ln(x) = (x-1) - \frac{(x-1)^2}{2!} + \frac{(x-1)^3}{3!} - \cdots = \sum_{k=0}^{\infty} (-1)^{k-1} \frac{(x-k)^k}{k!}$$
$$\text{for } 0 < x \leq 2$$

The sine and cosine functions can also be expressed using the Taylor series as follows, respectively:

(1.26)
$$\sin(x) = x - \frac{x^3}{3!} + \frac{x^5}{5!} - \cdots = \sum_{k=0}^{\infty} (-1)^k \frac{x^{2k+1}}{(2k+1)!}$$

(1.27)
$$\cos(x) = 1 - \frac{x^2}{2!} + \frac{x^4}{4!} - \cdots = \sum_{k=0}^{\infty} (-1)^k \frac{x^{2k}}{(2k)!}$$

Equations 1.26 and 1.27 are valid only for x values that are finite. The reciprocal of $(1 - x)$, where $|x| < 1$, can be expressed by the following Taylor series:

(1.28)
$$\frac{1}{1-x} = 1 + x + x^2 + x^3 + \cdots = \sum_{k=0}^{\infty} x^k$$

These example Taylor series can be used to evaluate their corresponding functions at any point x using a base value of $x_0 = 0$ and increment h.

PROBLEMS

1.1 The concept of repetition is part of the definition of the word *iterate*. Discuss how this applies to the example in Section 1.2.

1.2 Engineering problems can be solved using analytical or numerical methods. Describe the differences between these two classes of methods. What are the limitations, advantages, and disadvantages of the two classes?

1.3 Using the Taylor series expansion for $\cos(x)$ provided in Section 1.4.2, evaluate the series starting at $x_0 = 0$ radian with $h = 0.1$ to $x = 1$ radian. Use one term, two terms, and three terms of the series. Evaluate the error in each value and discuss the effect of incrementally adding the terms to the series. Compare your results with the true solution.

1.4 Using the Taylor series for $\sin(x)$ of Equation 1.26, evaluate the series starting at $x_0 = 0$ radian with $h = 0.1$ to $x = 1$ radian. Use one term, two terms, and three terms of the series. Evaluate the error in each value and discuss the effect of incrementally adding the terms to the series. Compare your results with the true solution.

1.5 Using the Taylor series for e^x provided by Equation 1.24, evaluate the series starting at $x_0 = 0$ with $h = 0.1$ to $x = 1$. Use one term, two terms, and three terms of the series. Evaluate the error in each value and discuss the effect of incrementally adding the terms to the series. Compare your results with the true solution.

1.6 Develop a Taylor series expansion of the following polynomial:

$$f(x) = x^3 - 3x^2 + 5x + 10$$

Use $x = 2$ as the base (or starting) point and h as the increment. Evaluate the series for $h = 0.1, 0.2, 0.3, 0.4, 0.5, 0.6, 0.7, 0.8, 0.9$, and 1.0. Discuss the accuracy of the method as the terms of the Taylor series are added incrementally.

1.7 Develop a Taylor series expansion of the following polynomial:

$$f(x) = x^5 - 5x^4 + x^2 + 6$$

Use $x = 2$ as the base (or starting) point and h as the increment. Evaluate the series for $h = 0.1, 0.2, 0.3, 0.4, 0.5, 0.6, 0.7, 0.8, 0.9$, and 1.0. Discuss the accuracy of the method as the terms of the Taylor series are added incrementally.

1.8 Develop a Taylor series expansion of the following function:

$$f(x) = x^2 - 5x^{0.5} + 6$$

Use $x = 2$ as the base (or starting) point and h as the increment. Evaluate the series for h = 0.1, 0.2, 0.3, 0.4, 0.5, 0.6, 0.7, 0.8, 0.9, and 1.0. Discuss the accuracy of the method as the terms of the Taylor series are added incrementally.

1.9   The Maclaurin series for the function

$$\frac{1}{1-x}$$

is the power series

$$1 + x + x^2 + x^3 + x^4 + \dots$$

For $x = 0.2$, compute the value of the function for each term of the power series; also compute the absolute and relative errors. Complete the analysis for terms to x^6.

1.10   The Maclaurin series for the function

$$\frac{1}{1+x^2}$$

is the power series

$$1 - x^2 + x^4 - x^6 + x^8 - \dots$$

For $x = 0.2$, compute the value of the function for each term of the power series; also compute the absolute and relative errors. Complete the analysis for terms to x^8.

1.11   Write a computer program to evaluate the Taylor series expansion for Problem 1.2.

1.12   Write a computer program to evaluate the Taylor series expansion for Problem 1.4.

1.13   Write a computer program to evaluate the Taylor series expansion for Problem 1.6.

1.14   Write a computer program to evaluate the Taylor series expansion for Problem 1.7.

Matrices

2

2.1 INTRODUCTION

2.1.1 DEFINITION OF A MATRIX

What is a matrix? Simply stated, it is a rectangular array of numbers. Each column has the same number of values r, where r is the number of rows. Similarly, each row has the same number of values c, where c is the number of columns. It is not necessary for the number of rows to equal the number of columns. A rectangular matrix with three rows and two columns is shown in Figure 2.1a. A capital letter will be used to denote the name of the array; the values contained in the matrix are called elements and are denoted as a_{ij}, where the lowercase letter a is used to indicate the element of matrix **A**, i refers to the row, and j refers to the column. Figure 2.1b shows a matrix with r rows and c columns; thus, element b_{57} refers to the value in the fifth row and seventh column of matrix **B**.

2.1.2 FORMATION OF A MATRIX

What produces a matrix? Matrices are useful for representing data sets. Figure 2.2 shows a data matrix of concentrations of the following three water-quality indicators from five wells: hardness (H), alkalinity (A), and pH. The matrix can be represented with the well number either as the row variable (Figure 2.2a) or the column variable (Figure 2.2b). The form of the matrix depends on style or use. It is easier to compare the values of pH by reading down the columns (Figure 2.2a) than across them (Figure 2.2b), so Figure 2.2a is probably the preferred form. The form may also be dictated by computational requirements, which are illustrated later. The variables can be properties that vary with time, space, or individuals, or on the basis of any characteristic. For example, if measurements of average temperature, wind speed, humidity, and evaporation were made daily for a month, then the 30 by 4, denoted as 30×4, matrix would have the time of measurement as the row variable and the climatic variable as the column variable.

Matrices can also be the result of matrix computations. For example, if we were interested in the correlation between the water-quality

$$\text{(a) } 3 \times 2 \text{ matrix} \qquad \mathbf{A} = \begin{bmatrix} a_{11} & a_{12} \\ a_{21} & a_{22} \\ a_{31} & a_{32} \end{bmatrix}$$

$$\text{(b) } r \times c \text{ matrix} \qquad \mathbf{B} = \begin{bmatrix} b_{11} & b_{12} & \cdots & b_{1c} \\ b_{21} & b_{22} & \cdots & b_{2c} \\ \vdots & \vdots & \vdots\vdots\vdots & \vdots \\ b_{r1} & b_{r2} & \cdots & b_{rc} \end{bmatrix}$$

FIGURE 2.1   Example matrices.

$$\text{(a) } 5 \times 3 \text{ matrix} \qquad \text{Well} \begin{array}{c} \\ 1 \\ 2 \\ 3 \\ 4 \\ 5 \end{array} \begin{array}{ccc} H & A & pH \\ \begin{bmatrix} 140 & 35 & 7.7 \\ 195 & 12 & 7.1 \\ 283 & 53 & 6.4 \\ 132 & 188 & 8.3 \\ 60 & 55 & 6.5 \end{bmatrix} \end{array}$$

$$\text{(b) } 3 \times 5 \text{ matrix} \qquad \begin{array}{c} \\ H \\ A \\ pH \end{array} \begin{array}{ccccc} & & \text{Well} & & \\ 1 & 2 & 3 & 4 & 5 \\ \begin{bmatrix} 140 & 195 & 283 & 132 & 60 \\ 35 & 12 & 53 & 188 & 55 \\ 7.7 & 7.1 & 6.4 & 8.3 & 6.5 \end{bmatrix} \end{array}$$

FIGURE 2.2   Data matrix of water-quality concentrations of hardness (H, mg/L), alkalinity (A, mg/L), and pH from five wells.

variables of Figure 2.2, where correlation is an indicator of the strength of linear association between variables (see Section 10.2 in Chapter 10), matrix computations on the data matrix of Figure 2.2a would produce the following correlation matrix:

$$\text{(2.1)} \qquad \begin{array}{c} \\ H \\ A \\ pH \end{array} \begin{array}{ccc} H & A & pH \\ \begin{bmatrix} 1.00 & -0.23 & -0.27 \\ -0.23 & 1.00 & 0.64 \\ -0.27 & 0.64 & 1.00 \end{bmatrix} \end{array}$$

This matrix shows that the measurements of hardness and alkalinity have an inverse relationship (-0.23) that is not strong; pH and alkalinity have a positive moderate relationship (0.64). The value of 1 on the principal diagonal indicates that each variable is perfectly correlated with itself.

Matrices can also result from the physical structure of a problem. Figure 2.3 shows a simple structure that is subject to a distributed load w and a horizontal point load P. The two joints rotate through angles θ_1 and θ_2 and are displaced by distance d, with the sign of the angles assuming counterclockwise movement as positive and the displacement d shown.

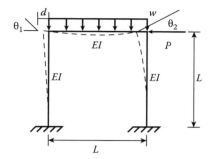

FIGURE 2.3   Development of a matrix for a portal frame.

For given values of w, P, length (L), moment of inertia (I), and modulus of elasticity (E), the rotations and displacement are determined by the solution of the following system of three equations:

$$(2.2) \qquad 4\theta_1 + \theta_2 - \frac{3}{L}d = \frac{wL^3}{24E_1}$$

$$(2.3) \qquad \theta_1 + 4\theta_2 - \frac{3}{L}d = \frac{-wL^3}{24E_1}$$

$$(2.4) \qquad \theta_1 + \theta_2 - \frac{4}{L}d = \frac{PL^2}{6EI}$$

For the three unknowns (θ_1, θ_2, and d), the three equations can be represented by the following 3×4 matrix with the fourth column representing the stipulation:

$$(2.5) \qquad \begin{bmatrix} 4 & 1 & \dfrac{-3}{L} & \dfrac{wL^3}{24EI} \\[2ex] 1 & 4 & \dfrac{-3}{L} & \dfrac{-wL^3}{24EI} \\[2ex] 1 & 1 & \dfrac{-4}{L} & \dfrac{PL^2}{6EI} \end{bmatrix}$$

This matrix is a numerical shorthand for Equations 2.2 through 2.4, and the matrix representation is useful for numerical calculations.

2.1.3 TYPES OF MATRICES

A *square matrix* is a special case in which the number of rows equals the number of columns; the correlation matrix of Equation 2.1 is an example. The set of elements a_{ij} for which $i = j$ is called the *principal diagonal*.

For the correlation matrix of Equation 2.1, all the values on the principal diagonal equal 1.

An *upper triangular* matrix has values on the principal diagonal and above, but contains zero values for the matrix elements below the principal diagonal. All the elements above the principal diagonal of a *lower triangular* matrix are zero. For a *strictly upper triangular* matrix or *strictly lower triangular* matrix, the elements of the principal diagonal would also be zero.

A *diagonal matrix* is a square matrix with all the elements equal to zero except for the elements on the principal diagonal. The following is an $n \times n$ diagonal matrix that contains n eigenvalues (λ_i, $i = 1, 2, ..., n$):

$$(2.6) \qquad \begin{bmatrix} \lambda_1 & 0 & 0 & \cdots & 0 \\ 0 & \lambda_2 & 0 & \cdots & 0 \\ 0 & 0 & \lambda_3 & \cdots & 0 \\ \vdots & \vdots & \vdots & \ddots & \vdots \\ 0 & 0 & 0 & \cdots & \lambda_n \end{bmatrix}$$

A *banded matrix* is a square matrix with elements of zero except for the principal diagonal and values in the positions adjacent to the principal diagonal. For example, a *tridiagonal matrix* is the special case of a banded matrix that has zeros except in the principal diagonal and the two adjacent diagonals:

$$(2.7) \qquad \begin{bmatrix} a_{11} & a_{12} & 0 & 0 & 0 \\ a_{21} & a_{22} & a_{23} & 0 & 0 \\ 0 & a_{32} & a_{33} & a_{34} & 0 \\ 0 & 0 & a_{43} & a_{44} & a_{45} \\ 0 & 0 & 0 & a_{54} & a_{55} \end{bmatrix}$$

A *unit matrix* ($\mathbf{I}$) is a commonly used diagonal matrix; $\mathbf{I}$ has ones on the principal diagonal, with zeros for all other elements.

All the elements of a *null matrix* are zero. It is usually denoted as $\mathbf{0}$. It does not have to be a square matrix.

A *symmetric matrix* is a square matrix in which $a_{ij} = a_{ji}$; the correlation matrix of Equation 2.1 is symmetric.

A *skew-symmetric matrix* is a square matrix with all values on the principal diagonal equal to zero and with off-diagonal values such that $a_{ij} = -a_{ji}$.

The *transpose* of matrix $\mathbf{A}$ is the matrix that results if the rows of $\mathbf{A}$ are written as the columns of the new matrix denoted as $\mathbf{A}^T$. In terms of the elements, $a_{ji}^T = a_{ji}$. If matrix $\mathbf{A}$ has r rows and c columns, then $\mathbf{A}^T$ has c rows and r columns. Note that $(\mathbf{A}^T)^T = \mathbf{A}$. The matrix of Figure 2.2a is the transpose of the matrix of Figure 2.2b.

2.2 MATRIX OPERATIONS

The primary arithmetic operations are addition, subtraction, multiplication, and division. Matrix algebra has operations called matrix addition, subtraction, and multiplication; there is no operation called matrix division. Instead, a matrix operation called matrix inversion is available. Each of these matrix operations is discussed in this section.

2.2.1 MATRIX EQUALITY

Two matrices are said to be equal—that is, $\mathbf{A} = \mathbf{B}$, if every element in matrix $\mathbf{A}$ is equal to the corresponding element in matrix $\mathbf{B}$. That is, $a_{ij} = b_{ij}$ for all i and j. When performing matrix computations on a computer, it is often necessary to store the original contents of a matrix for later use prior to performing computations on the matrix. In such cases, matrix $\mathbf{A}$ can be formed immediately after matrix $\mathbf{B}$ is inputted; this would involve two logic loops, one for the rows and one for the columns. At that point, matrix $\mathbf{A}$ equals matrix $\mathbf{B}$.

2.2.2 MATRIX ADDITION AND SUBTRACTION

To perform matrix addition, that is, $\mathbf{C} = \mathbf{A} + \mathbf{B}$, the two or more matrices must have the same number of rows and the same number of columns. The elements of $\mathbf{C}$ are then the sum of the corresponding elements of the matrices being summed; that is,

(2.8) $$c_{ij} = a_{ij} + b_{ij}$$

To perform matrix subtraction, $\mathbf{C} = \mathbf{A} - \mathbf{B}$, the matrices must have the same number of rows and the same number of columns. The elements of $\mathbf{C}$ are the differences between the corresponding elements of the matrices being subtracted; that is,

(2.9) $$c_{ij} = a_{ij} - b_{ij}$$

Matrix addition is not directional (i.e., commutative), that is,

(2.10) $$\mathbf{A} + \mathbf{B} = \mathbf{B} + \mathbf{A}$$

Matrix subtraction depends on the direction of the operation (i.e., noncommutative):

(2.11) $$\mathbf{A} - \mathbf{B} \neq \mathbf{B} - \mathbf{A}$$

Example 2.1: Matrix Addition and Subtraction

Given the following two matrices, **A** and **B**,

(2.12)
$$\mathbf{A} = \begin{bmatrix} 2 & 4 & 6 \\ 1 & 3 & 5 \\ 9 & 10 & 11 \end{bmatrix}$$

(2.13)
$$\mathbf{B} = \begin{bmatrix} 0 & 2 & 3 \\ 7 & 8 & 4 \\ 2 & 1 & 1 \end{bmatrix}$$

The resulting matrix (**C**) of the addition of **A** and **B** is

(2.14)
$$\mathbf{C} = \mathbf{A} + \mathbf{B} = \begin{bmatrix} 2 & 6 & 9 \\ 8 & 11 & 9 \\ 11 & 11 & 12 \end{bmatrix}$$

The subtraction of **B** from **A** is the following matrix **D**:

(2.15)
$$\mathbf{D} = \mathbf{A} - \mathbf{B} = \begin{bmatrix} 2 & 2 & 3 \\ -6 & -5 & 1 \\ 7 & 9 & 10 \end{bmatrix}$$

It should be noted that when adding or subtracting matrices, the matrices must be of the same size. In this example, the matrices **A**, **B**, **C**, and **D** have the same 3×3 size.

2.2.3 MATRIX MULTIPLICATION: AN INTRODUCTORY EXAMPLE

Matrix multiplication is slightly more complex than matrix addition and subtraction, so it is introduced with an example. A contractor is laying a pipe and has three primary components of cost: excavation, pipe, and inlets. These costs are $5, $2, and $3 per unit, respectively, and can be represented by the one-dimensional cost matrix **C**:

(2.16)
$$\mathbf{C} = \begin{bmatrix} 5 \\ 2 \\ 3 \end{bmatrix}$$

On one project, the contractor uses 4, 6, and 1 units of the three components; this is represented by the one-dimensional matrix **S**:

(2.17)
$$\mathbf{S} = [4 \quad 6 \quad 1]$$

The total project cost **T** can be represented as the product of **S** and **C**:

(2.18)
$$\mathbf{T} = \mathbf{S} \cdot \mathbf{C} = [4 \quad 6 \quad 1] \cdot \begin{bmatrix} 5 \\ 2 \\ 3 \end{bmatrix}$$

(2.19)
$$\mathbf{T} = 4(5) + 6(2) + 1(3) = [\$35]$$

If the contractor has a second project on which 3, 8, and 4 units of the three components are used, this second one-dimensional matrix **S** could be multiplied with **C** to get the total cost for the second project. However, the problem can be simplified by writing **S** as a two-dimensional matrix before multiplying with **C**. The multiplication of **S** and **C** will yield the total cost for each project:

(2.20)
$$\mathbf{T} = \mathbf{S} \cdot \mathbf{C} = \begin{bmatrix} 4 & 6 & 1 \\ 3 & 8 & 4 \end{bmatrix} \cdot \begin{bmatrix} 5 \\ 2 \\ 3 \end{bmatrix}$$

(2.21)
$$\mathbf{T} = \begin{bmatrix} 4(5)+6(2)+1(3) \\ 3(5)+8(2)+4(3) \end{bmatrix} = \begin{bmatrix} \$35 \\ \$43 \end{bmatrix}$$

If the contractor is interested in computing the labor requirements as well as the cost, the **C** matrix can be expanded into a two-dimensional matrix of two columns. The first column contains the cost per unit, and the second column contains the labor requirements per unit. For this example, we can assume that labor levels of 7, 3, and 9 for the three components are required; that is, 7 units of labor are required per unit of excavation, 3 units of labor are required to install 1 unit of pipe, and 9 units of labor are required to install each inlet. Multiplying **S** and **C** yields both the total cost and the total labor requirements for each project:

(2.22)
$$\mathbf{T} = \mathbf{S} \cdot \mathbf{C} = \begin{bmatrix} 4 & 6 & 1 \\ 3 & 8 & 4 \end{bmatrix} \cdot \begin{bmatrix} 5 & 7 \\ 2 & 3 \\ 3 & 9 \end{bmatrix}$$

(2.23)
$$\mathbf{T} = \begin{bmatrix} 4(5)+6(2)+1(3) & 4(7)+6(3)+1(9) \\ 3(5)+8(2)+4(3) & 3(7)+8(3)+4(9) \end{bmatrix} = \begin{bmatrix} \$35 & 55 \\ \$43 & 81 \end{bmatrix}$$

The **T** matrix, which has two rows and two columns, shows that project 1 costs \$35 and requires 55 units of labor and project 2 costs \$43 and requires 81 units of labor. Row i corresponds to project i, while column j corresponds to project criterion j ($j = 1$ for cost and $j = 2$ for labor).

This problem could be expanded to include as many projects as the contractor has and to indicate requirements other than cost and labor. The **S** matrix would have three columns and as many rows as there are projects. The **C** matrix would have three rows and one column for each project criterion. The resulting **T** matrix would have one row for each project and one column for each project criterion.

2.2.4 MATRIX MULTIPLICATION: GENERAL RULES

Using **A**, **B**, and **C** to denote three matrices for the matrix product **C** = **A** · **B**, the following are the rules of matrix multiplication:

1. The number of columns in the first matrix **A** must equal the numbers of rows in the second matrix **B**.
2. The number of rows in the product matrix **C** equals the number of rows in the first matrix **A**.
3. The number of columns in the product matrix **C** equals the number of columns in the second matrix **B**.
4. The element of matrix **C** in row i and column j, c_{ij}, is equal to the sum of the products $a_{ik}b_{kj}$:

(2.24)
$$c_{ij} = \sum_{k=l}^{m} a_{ik} b_{kj}$$

where m is the number of columns in **A** (which is the number of rows in **B**).

5. Matrix multiplication is not commutative; that is,

(2.25)
$$\mathbf{A} \cdot \mathbf{B} \neq \mathbf{B} \cdot \mathbf{A}$$

6. Matrix multiplication is associative; that is,

(2.26)
$$(\mathbf{A} \cdot \mathbf{B}) \cdot \mathbf{C} = \mathbf{A} \cdot (\mathbf{B} \cdot \mathbf{C})$$

The example of Equations 2.22 and 2.23 illustrate the first four rules. In this case, **S** is a 2×3 matrix and **C** is 3×2; therefore, **T** is 2×2, and each summation involves three products, with each element of **T** computed using Equation 2.24.

The terms *premultiplication* and *postmultiplication* are used in expressing matrix multiplication operations. For the two matrices **A** and **B**, the following apply:

1. Premultiplication of **B** by **A** means **A** · **B**.
2. Premultiplication of **A** by **B** means **B** · **A**.
3. Postmultiplication of **A** by **B** means **A** · **B**.
4. Postmultiplication of **B** by **A** means **B** · **A**.

Example 2.2: Matrix Multiplication

The matrix **E** in Example 2.1 can be computed from the following two matrices, **A** and **B**:

$$A = \begin{bmatrix} 2 & 4 & 6 \\ 1 & 3 & 5 \\ 9 & 10 & 11 \end{bmatrix}$$

and

$$B = \begin{bmatrix} 0 & 2 & 3 \\ 7 & 8 & 4 \\ 2 & 1 & 1 \end{bmatrix}$$

The matrix **E** that results from the product of **A** and **B** is

$$(2.27) \qquad E = A \cdot B = \begin{bmatrix} 40 & 42 & 28 \\ 31 & 31 & 20 \\ 92 & 109 & 78 \end{bmatrix}$$

The matrix **F** is the product of **B** and **A**

$$(2.28) \qquad F = B \cdot A = \begin{bmatrix} 29 & 36 & 43 \\ 58 & 92 & 126 \\ 14 & 21 & 28 \end{bmatrix}$$

Therefore, matrix multiplication is not commutative; that is, $A \cdot B \neq B \cdot A$.

2.2.5 MATRIX MULTIPLICATION BY A SCALAR

The multiplication of a matrix **A** by a scalar s has the effect of multiplying each element in the matrix (a_{ij}) by the scalar. The resulting elements of a matrix **B** can be expressed as

$$(2.29) \qquad b_{ij} = sa_{ij}$$

Example 2.3: Matrix Multiplication by a Scalar

For the matrices **A** discussed in Example 2.1,

$$\mathbf{A} = \begin{bmatrix} 2 & 4 & 6 \\ 1 & 3 & 5 \\ 9 & 10 & 11 \end{bmatrix}$$

the resulting matrix **G** of the product of **A** by a scalar s that has a value of 10 is

(2.30) $$\mathbf{G} = s\mathbf{A} = \begin{bmatrix} 20 & 40 & 60 \\ 10 & 30 & 50 \\ 90 & 100 & 110 \end{bmatrix}$$

2.2.6 MATRIX INVERSION

For any real number a, its inverse is defined by the operator $a^{-1}a = 1$. A matrix has a corresponding operation, matrix inversion, which applies to square matrices of n rows and n columns. Matrix inversion is another matrix operation and is defined by

(2.31) $$\mathbf{A}^{-1} \cdot \mathbf{A} = \mathbf{I}$$

in which $\mathbf{A}^{-1}$ is the inverse of **A**, and **I** is the unit matrix. The inverse can be determined by forming n^2 simultaneous equations and solving for the n^2 unknowns. For the case of $n = 2$, the four equations are (letting the elements of $\mathbf{A}^{-1}$ be indicated by c_{ij})

(2.32)
$$
\begin{aligned}
c_{11}a_{11} + c_{12}a_{21} &= 1 \\
c_{11}a_{12} + c_{12}a_{22} &= 0 \\
c_{21}a_{11} + c_{22}a_{21} &= 0 \\
c_{21}a_{12} + c_{22}a_{22} &= 1
\end{aligned}
$$

Solving Equation 2.32 yields the values of c_{ij}, which form the inverse. Equation 2.32 can be solved as two pairs of two simultaneous equations.

Example 2.4: Matrix Inversion

For the matrix $\mathbf{A} = \begin{bmatrix} 2 & 3 \\ 5 & 7 \end{bmatrix}$, Equation 2.31 appears as

(2.33)
$$\begin{bmatrix} c_{11} & c_{12} \\ c_{21} & c_{22} \end{bmatrix} \cdot \begin{bmatrix} 2 & 3 \\ 5 & 7 \end{bmatrix} = \begin{bmatrix} 1 & 0 \\ 0 & 1 \end{bmatrix}$$

which yields the following simultaneous equations:

(2.34)
$$
\begin{aligned}
2c_{11} + 5c_{12} &= 1 \\
3c_{11} + 7c_{12} &= 0 \\
2c_{21} + 5c_{22} &= 0 \\
3c_{21} + 7c_{22} &= 1
\end{aligned}
$$

The solution of Equations 2.33 and 2.34 yields the inverse $\mathbf{A}^{-1} = \begin{bmatrix} -7 & 3 \\ 5 & -2 \end{bmatrix}$. Performing the multiplication of Equation 2.31 with $\mathbf{A}^{-1}$ and $\mathbf{A}$ verifies that the solution of the simultaneous equations yields the inverse matrix.

2.2.7 MATRIX SINGULARITY

A matrix can be referred to as being singular or nonsingular. If the inverse of a matrix exists, then the matrix is said to be nonsingular. Conversely, if the inverse of the matrix does not exist, then the matrix is said to be singular. One implication of matrix singularity in solving a system of simultaneous equations is that a unique solution for the equations does not exist. This is most easily understood by illustrating it using a system of equations. Consider, for example, the following two linear simultaneous equations:

(2.35)
$$2X_1 + 3X_2 = a$$

(2.36)
$$4X_1 + 6X_2 = b$$

If $2a = b$, then an infinite number of solutions exist; for example, three possibilities are (1) $X_1 = 2$, $X_2 = 1$, $a = 7$, and $b = 14$; (2) $X_1 = -1$, $X_2 = 4$, $a = 10$, and $b = 20$; and (3) $X_1 = 0$, $X_2 = -2$, $a = -6$, and $b = -12$. If $2a \neq b$, then there is no feasible solution; for example, if $a = 2$ and $b = 3$, there are no values of X_1 and X_2 that can satisfy the equalities of the two equations. In both cases, the matrix **A** formed by the coefficients on the left side of the equations is said to be a singular matrix. The matrix is given by

(2.37)
$$\mathbf{A} = \begin{bmatrix} 2 & 3 \\ 4 & 6 \end{bmatrix}$$

The inability to find a solution is easily shown using Equation 2.31 as

(2.38)
$$\begin{bmatrix} c_{11} & c_{12} \\ c_{21} & c_{22} \end{bmatrix} \cdot \begin{bmatrix} 2 & 3 \\ 4 & 6 \end{bmatrix} = \begin{bmatrix} 1 & 0 \\ 0 & 1 \end{bmatrix}$$

After forming the four simultaneous equations, it can easily be shown that the equations do not have a solution; therefore, the matrix **A** is singular.

2.2.8 ADDITIONAL TOPICS IN MATRIX ALGEBRA

The *trace* of a square matrix equals the sum of the diagonal elements:

(2.39)
$$tr(\mathbf{A}) = \sum_{i=1}^{n} a_{ii}$$

Matrix augmentation is the addition of a column or columns to the initial matrix. For example, the following shows matrix **A** and an augmented matrix:

(2.40)
$$\mathbf{A} = \begin{bmatrix} 2 & 3 & 1 \\ 1 & 4 & 1 \\ 2 & 3 & 4 \end{bmatrix}$$

(2.41)
$$\mathbf{A}_a = \begin{bmatrix} 2 & 3 & 1 & 1 & 0 & 0 \\ 1 & 4 & 1 & 0 & 1 & 0 \\ 2 & 3 & 4 & 0 & 0 & 1 \end{bmatrix}$$

A matrix can be partitioned by separating it into smaller matrices. For example, matrix **A** can be partitioned into four other matrices:

(2.42)
$$\mathbf{A} = \begin{bmatrix} A_{11} & A_{12} \\ A_{21} & A_{22} \end{bmatrix}$$

where, for example,

(2.43)
$$\mathbf{A}_{11} = \begin{bmatrix} 2 & 3 \\ 1 & 4 \end{bmatrix}$$

(2.44)
$$\mathbf{A}_{12} = \begin{bmatrix} 1 \\ 1 \end{bmatrix}$$

(2.45)
$$\mathbf{A}_{21} = \begin{bmatrix} 2 & 3 \end{bmatrix}$$

(2.46)
$$\mathbf{A}_{22} = \begin{bmatrix} 4 \end{bmatrix}$$

Example 2.5: Matrix Partition

A correlation matrix is often partitioned into a matrix of correlation coefficients between one set of variables called the predictor variables and a vector of predictor–criterion correlation coefficients. For example, a correlation analysis of an engineering problem resulted in the following correlation matrix $\mathbf{R}$ for three variables X_1, X_2, and Y, where Y is the variable to be predicted from measurements of X_1 and X_2:

(2.47)
$$\mathbf{R} = \begin{matrix} & X_1 & X_2 & Y \\ X_1 & \begin{bmatrix} 1 & 0.2 & 0.9 \\ X_2 & 0.2 & 1 & 0.8 \\ Y & 0.9 & 0.8 & 1 \end{bmatrix} \end{matrix}$$

The variables X_1 and X_2 are considered to be the predictor (independent) variables, and Y is the criterion (dependent) variable. The correlation matrix can be partitioned as follows:

(2.48)
$$\mathbf{R} = \begin{matrix} & X_1 & X_2 & Y \\ X_1 & \begin{bmatrix} 1 & 0.2 & | & 0.9 \\ X_2 & 0.2 & 1 & | & 0.8 \\ Y & 0.9 & 0.8 & | & 1 \end{bmatrix} \end{matrix}$$

The matrix

(2.49)
$$\mathbf{R}_{11} = \begin{bmatrix} 1 & 0.2 \\ 0.2 & 1 \end{bmatrix}$$

is called the intercorrelation matrix for the predictor variables. The reason for partitioning the matrix in this case is because the intercorrelation coefficients are assessed differently than the predictor–criterion correlation coefficients. Generally, it is preferable to have low intercorrelation and high predictive-criterion correlations.

2.3 VECTORS

2.3.1 DEFINITIONS

In the example used to introduce matrix multiplication, the **C** and **S** matrices were initially presented as one dimensional, with **C** having one column and **S** having one row. Such matrices are called vectors. A vector of one row is called a *row vector*. A vector of one column is called a *column vector*. An element of a vector can be indicated with a single subscript, so a_3 is the third element of the vector **A**, and b_j is the jth element of the vector **B**.

The name vector indicates that a one-dimensional matrix with n elements can be represented as a vector in n dimensions, with one end of the vector at the origin and the other end at the point defined by the values of the elements of the vector. For example, the vector [2, 3] would be a line from the origin (0, 0) to the point located two units along the first axis and three units along the second axis. Figure 2.4 provides another example; in this case the vectors have two ordinates, measured on the axes X_1 and X_2. Vector **A** is indicated by (2, 1), and vector **B** is indicated by $\left(1, \dfrac{-2}{3}\right)$. In each case, the first element of the vector corresponds to X_1 while the second element corresponds to X_2.

A *null vector* is a vector that has zeros for all its elements. It corresponds to the null matrix.

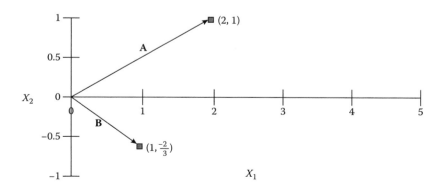

FIGURE 2.4 Vectors.

2.3.2 VECTOR OPERATIONS

The matrix operations of addition, subtraction, and multiplication can be applied to vectors. Two row (column) vectors can be added or subtracted. A row vector with n elements can be postmultiplied by a column vector with n elements to equal a scalar value. In Equations 2.18 and 2.19, row vector **S** is postmultiplied by the column vector **C** to produce a scalar value. The number of rows in the column vector must equal the number of columns in the row vector. This is easily understood if the row vector is viewed as a $1{\times}n$ matrix and the column vector is viewed as an $n{\times}1$ matrix. Postmultiplication of a $(1{\times}n)$ by a $(n{\times}1)$ yields a $(1{\times}1)$ matrix, which is a scalar. It also shows why the number of columns and rows in the vector must be equal. A column vector with n elements can be multiplied by a row vector with m elements to produce an $n{\times}m$ matrix. If the $n{\times}1$ column vector is postmultiplied by the $1{\times}m$ row vector, then the resulting matrix is $n{\times}m$. Thus, in this case, the number of rows and number of columns do not have to be equal. It is important to note that the order of the multiplication is not arbitrary; it depends on the concept that underlies the problem that was the basis for the development of the vectors.

2.3.3 ORTHOGONAL AND NORMALIZED VECTORS

Two vectors are said to be *orthogonal* if their product equals zero. For the vector product $\mathbf{A} \cdot \mathbf{B}$, $\mathbf{A}$ is a row vector and $\mathbf{B}$ is a column vector; the resulting vector product is a scalar value. If two vectors that are orthogonal are plotted in the *n-dimensional* space, the vectors will be perpendicular to each other. For example, the vector product of $\mathbf{A}$ and $\mathbf{B}$ in Figure 2.5 is zero:

(2.50)
$$[2 \quad 3] \cdot \begin{bmatrix} 1 \\ -\dfrac{2}{3} \end{bmatrix} = [0]$$

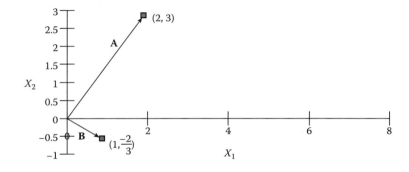

FIGURE 2.5 Orthogonal vectors.

Since the vector product is zero, the vectors are perpendicular to each other.

The *length of a vector* **V** equals the square root of the sum of the squares of its elements; that is,

$$(2.51) \qquad \text{vector length} = \left(\sum_{i=1}^{n} v_i^2 \right)^{0.5}$$

where $v_1, v_2, \ldots, v_n$ are the n elements of the vector **V**. The length of a vector is a scalar quantity. For example, vectors **A** and **B** of Figure 2.5 are 3.606 and 1.444 long, respectively.

A vector can be *normalized* by dividing each element of the vector by its length. Therefore, a normalized vector has a length that is equal to 1. Two vectors that are both normalized and orthogonal to each other are said to be *orthonormal* vectors.

Example 2.6: Vectors

For the two vectors

$$(2.52) \qquad \mathbf{V}_1 = [2 \quad -3 \quad 5]$$

and

$$(2.53) \qquad \mathbf{V}_2 = \begin{bmatrix} -1 \\ 1 \\ 1 \end{bmatrix}$$

the vector lengths are

$$(2.54) \qquad \text{length of } \mathbf{V}_1 = \sqrt{(2)^2 + (-3)^2 + (5)^2} = \sqrt{38} = 6.164$$

$$(2.55) \qquad \text{length of } \mathbf{V}_2 = \sqrt{(-1)^2 + (1)^2 + (1)^2} = \sqrt{3} = 1.732$$

The normalized vectors are

$$(2.56) \qquad \mathbf{V}_{1n} = \begin{bmatrix} \dfrac{2}{\sqrt{38}} & \dfrac{-3}{\sqrt{38}} & \dfrac{5}{\sqrt{38}} \end{bmatrix}$$

and

$$(2.57) \qquad \mathbf{V}_{2n} = \begin{bmatrix} \dfrac{-1}{\sqrt{3}} & \dfrac{1}{\sqrt{3}} & \dfrac{1}{\sqrt{3}} \end{bmatrix}$$

Since the vector product $\mathbf{V}_1 \cdot \mathbf{V}_2$ equals zero, $\mathbf{V}_1$ and $\mathbf{V}_2$ are orthogonal, whereas the vectors $\mathbf{V}_{1n}$ and $\mathbf{V}_{2n}$ are orthonormal.

2.4 DETERMINANTS

The *determinant* of a square matrix is a unique scalar number that represents the value of a matrix. A determinant of a matrix is denoted by $|\mathbf{A}|$. By definition, the determinant of a 2×2 matrix is computed by

$$(2.58) \qquad \begin{vmatrix} a & b \\ c & d \end{vmatrix} = ad - bc$$

The right side of Equation 2.58 is referred to as the *expansion* of the determinant shown on the left side.

For a 3×3 matrix, the value of the determinant is found by the following expansion:

$$(2.59) \quad \begin{vmatrix} a_{11} & a_{12} & a_{13} \\ a_{21} & a_{22} & a_{23} \\ a_{31} & a_{32} & a_{33} \end{vmatrix} = a_{11} \begin{vmatrix} a_{22} & a_{23} \\ a_{32} & a_{33} \end{vmatrix} - a_{12} \begin{vmatrix} a_{21} & a_{23} \\ a_{31} & a_{33} \end{vmatrix} + a_{13} \begin{vmatrix} a_{21} & a_{22} \\ a_{31} & a_{32} \end{vmatrix}$$

The 2×2 determinants shown in the expansion of Equation 2.59 are called *minors*. The minor of an element is formed by eliminating the row and column in which the element appears. For example, the minor of element a_{ij} is the matrix that results when row i and column j are eliminated; the minor for element a_{ij} of matrix $\mathbf{A}$ is denoted $\mathbf{A}_{ij}$. For an $n \times n$ matrix, any minor will be an $(n - 1) \times (n - 1)$ matrix.

The determinant of an $n \times n$ matrix is given by

$$(2.60) \qquad |\mathbf{A}| = a_{11}|\mathbf{A}_{11}| - a_{12}|\mathbf{A}_{12}| + a_{13}|\mathbf{A}_{13}| - \cdots + (-1)^{(n+1)}a_{1n}|\mathbf{A}_{1n}|$$

where the matrix $\mathbf{A}_{ij}$ is the minor of element a_{ij}. The sign of the term is plus if the sum of the row number and column number in which the element lies is even and is minus if it is odd. The product of $(-1)^{i+j}$ and the minor of a_{ij} is called the *cofactor* of a_{ij}. It can be stated that the determinant in Equation 2.60 was performed using the first row of the matrix. In general, any row can be used, but the signs of the different terms need to be revised. The general expression for calculating the determinant of an $n \times n$ matrix using the ith row is

$$|\mathbf{A}| = (-1)^{i+1}a_{i1}|\mathbf{A}| - (-1)^{i+1}a_{i2}|\mathbf{A}_{i2}| + (-1)^{i+1}a_{i3}|\mathbf{A}_{i3}| - \cdots + (-1)^{i+1}(-1)^{n+1}a_{in}|\mathbf{A}_{in}|$$
$$(2.61)$$

The following are properties of a determinant:

1. If the elements of any two rows (columns) are equal, the determinant equals zero. For example, the following matrix has a zero determinant because the first column is equal to the third column:

$$A = \begin{bmatrix} 1 & 2 & 1 \\ 2 & 14 & 2 \\ 3 & 5 & 3 \end{bmatrix}, \quad \text{where } |A| = 0$$

2. If the values in any row (column) are proportional to the corresponding values in another row (column), the determinant equals zero. For example, the following matrix has a zero determinant because the first column is proportional to the third column:

$$A = \begin{bmatrix} 1 & 2 & 2 \\ 2 & 14 & 4 \\ 3 & 5 & 6 \end{bmatrix}, \quad \text{where } |A| = 0$$

3. If all the elements in any row (column) equal zero, the determinant equals zero.
4. If all the elements of any row (column) are multiplied by a constant c, the value of the determinant is multiplied by c. The following matrix illustrates this property:

$$A = \begin{bmatrix} 6 & 4 \\ 4 & 5 \end{bmatrix} = \begin{bmatrix} 2(3) & 2(2) \\ 4 & 5 \end{bmatrix}, \quad \text{where } |A| = 2[3(5) - 2(4)] = 14$$

5. The value of the determinant is not changed by adding any row (column) multiplied by a constant c to another row (column). The following matrix illustrates this property:

$$A = \begin{bmatrix} 3 & 2 \\ 4 & 5 \end{bmatrix}, \quad \text{where } |A| = 3(5) - 2(4) = 7$$

Multiplying the second row by (−1) and adding it to the first row produces the following revised matrix **B**:

$$B = \begin{bmatrix} -1 & -3 \\ 4 & 5 \end{bmatrix}, \quad \text{where } |B| = -1(5) + 3(4) = 7$$

6. If any two rows (columns) are interchanged, the sign of the determinant is changed. In the following example, the determinant is computed twice without and with interchanged columns:

$$\begin{vmatrix} 3 & 2 \\ 4 & 5 \end{vmatrix} = 3(5)-2(4)=7 \quad \text{and} \quad \begin{vmatrix} 2 & 3 \\ 5 & 4 \end{vmatrix} = 2(4)-3(5)=-7$$

7. Multiplying an $n \times n$ matrix by a constant c results in multiplying the value of the determinant by c^n. The following matrix illustrates this property:

$$\mathbf{A} = 2^2 \begin{bmatrix} 3 & 2 \\ 4 & 5 \end{bmatrix}, \quad \text{where } |\mathbf{A}| = 2^2[3(5)-2(4)]=28$$

or

$$\mathbf{A} = \begin{bmatrix} 2(3) & 2(2) \\ 2(4) & 2(5) \end{bmatrix}, \quad \text{where } |\mathbf{A}| = 6(10)-4(8)=28$$

8. The determinant of a matrix equals that of its transpose; that is, $|\mathbf{A}| = |\mathbf{A}^T|$. The following determinants illustrate this property:

$$\begin{vmatrix} 3 & 2 \\ 4 & 5 \end{vmatrix} = 3(5)-2(4)=7 \quad \text{and} \quad \begin{vmatrix} 3 & 4 \\ 2 & 5 \end{vmatrix} = 3(5)-4(2)=7$$

9. If a matrix $\mathbf{A}$ is placed in diagonal form using property 5, then the product of the elements on the diagonal equals the determinant of $\mathbf{A}$. Consider the following matrix:

$$\mathbf{A} = \begin{bmatrix} 3 & 2 \\ 4 & 5 \end{bmatrix}, \quad \text{with } |\mathbf{A}| = 3(5)-2(4)=7$$

Multiplying the first row by $-\dfrac{4}{3}$ and adding it to the second row produces a matrix with a zero element in the second row and first column, as follows:

$$\begin{bmatrix} 3 & 2 \\ 0 & \dfrac{7}{3} \end{bmatrix}$$

Then, multiplying the second row by $-\dfrac{6}{7}$ and adding it to the first row results in the following diagonal matrix:

$$\begin{bmatrix} 3 & 0 \\ 0 & \dfrac{7}{3} \end{bmatrix}$$

Therefore, the determinant of **A** is

$$|\mathbf{A}| = \begin{vmatrix} 3 & 0 \\ 0 & \dfrac{7}{3} \end{vmatrix} = 3\left(\dfrac{3}{7}\right) = 7$$

10. If a matrix **A** has a zero determinant, then **A** is a singular matrix; that is, the inverse of **A** does not exist.

Example 2.7: Matrix Determinant

The matrix **A** discussed in Example 2.1 can be used to illustrate the calculation of the determinant. The matrix **A** is given by

$$\mathbf{A} = \begin{bmatrix} 2 & 4 & 6 \\ 1 & 3 & 5 \\ 9 & 10 & 11 \end{bmatrix}$$

The determinant is computed using the elements of the first row and their minors as follows:

(2.62) $$|\mathbf{A}| = a_{11}|\mathbf{A}_{11}| - a_{12}|\mathbf{A}_{12}| + a_{13}|\mathbf{A}_{13}|$$

or

(2.63) $$|\mathbf{A}| = \begin{vmatrix} 2 & 4 & 6 \\ 1 & 3 & 5 \\ 9 & 10 & 11 \end{vmatrix} = 2\begin{vmatrix} 3 & 5 \\ 10 & 11 \end{vmatrix} - 4\begin{vmatrix} 1 & 5 \\ 9 & 11 \end{vmatrix} + 6\begin{vmatrix} 1 & 3 \\ 9 & 10 \end{vmatrix}$$

(2.64) $|\mathbf{A}| = 2[3(11) - 5(10)] - 4[1(11) - 5(9)] + 6[1(10) - 3(9)] = 0$

The determinant could also have been computed using the elements of the second column and their minors as follows:

(2.65) $|\mathbf{A}| = -a_{12}|\mathbf{A}_{12}| + a_{22}|\mathbf{A}_{22}| - a_{32}|\mathbf{A}_{32}|$

or

(2.66) $|\mathbf{A}| = \begin{vmatrix} 2 & 4 & 6 \\ 1 & 3 & 5 \\ 9 & 10 & 11 \end{vmatrix} = -4\begin{vmatrix} 1 & 5 \\ 9 & 11 \end{vmatrix} + 3\begin{vmatrix} 2 & 6 \\ 9 & 11 \end{vmatrix} - 10\begin{vmatrix} 2 & 6 \\ 1 & 5 \end{vmatrix}$

(2.67) $|\mathbf{A}| = -4[11 - 45] + 3[22 - 54] - 10[10 - 6] = 0$

Therefore, **A** is a singular matrix; that is, the inverse of **A** does not exist.

2.5 RANK OF A MATRIX

A matrix of r rows and c columns is said to be of *order r* by c. If it is a square matrix, r by r, then the matrix is of order r.

The rank of a matrix is another term used to classify matrices. There are several concepts used to explain the rank of a matrix, three of which are presented here.

First, if we examine all possible square submatrices (by partitioning) of an r by c matrix and evaluate those having nonzero determinants, then the rank equals the order of the highest order nonsingular submatrix. For example, a 2×3 matrix has a rank of 2 if one of its 2×2 submatrices has a nonzero determinant.

The second definition uses vector terminology as its basis. Specifically, the rank of a matrix is the minimum number of dimensions of an orthogonal coordinate system that is needed to graph the rows or columns of the matrix as vectors, permitting the axis system to be rotated to minimize the necessary number of dimensions.

The third definition is closely related to the first two definitions. The rank of a matrix is the number of vectors (rows or columns) in the matrix that is linearly independent. The rows (or columns) of a matrix are called linearly independent if none of them can be produced by adding any row (column) multiplied by a constant to another row (column) or other rows (or columns).

Example 2.8: Rank of Matrix

The three definitions of the rank of a matrix can be illustrated with a single example. Consider the following matrix:

$$(2.68) \qquad 2 \times 3 \text{ order matrix, } \mathbf{R} = \begin{bmatrix} 1 & 2 & 4 \\ 2 & 4 & 8 \end{bmatrix}$$

Three square submatrices can be formed from $\mathbf{R}$ as follows:

$$(2.69) \qquad \mathbf{R}_1 = \begin{bmatrix} 1 & 2 \\ 2 & 4 \end{bmatrix}$$

$$(2.70) \qquad \mathbf{R}_2 = \begin{bmatrix} 1 & 4 \\ 2 & 8 \end{bmatrix}$$

$$(2.71) \qquad \mathbf{R}_3 = \begin{bmatrix} 2 & 4 \\ 4 & 8 \end{bmatrix}$$

Each of these has a determinant of zero, so the rank must be less than 2. Thus the first definition indicates that $\mathbf{R}$ has a rank of 1.

If matrix $\mathbf{R}$ is viewed as three vectors in a two-dimensional space (see Figure 2.6), it is evident that the vectors overlap each other; the angle between them is zero. If the axis system is rotated so that it passes through the first vector (1, 2), it is evident that the axis will coincide with the other two vectors. Thus only one axis is needed to represent the three vectors, and so matrix $\mathbf{R}$ is of rank 1.

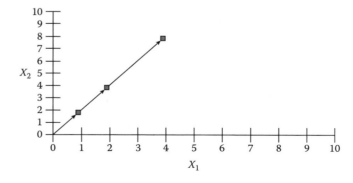

FIGURE 2.6   Graphical portrayal of the rank of a matrix.

The elements of the three vectors are related by the following equations:

(2.72) $v_{i2} = 2v_{i1},$ for $i = 1$ and 2

(2.73) $v_{i3} = 4v_{i1},$ for $i = 1$ and 2

(2.74) $v_{i3} = 2v_{i2},$ for $i = 1$ and 2

where the first subscript refers to the element number of the column vector indicated by the second subscript. Since the three equations are linear, the three vectors are linearly dependent; thus there is only one linearly independent vector, which means that matrix **R** is of rank 1.

Example 2.9: Rank of a Matrix

The matrix of Example 2.7, which is of the third order (that is, a 3×3 square matrix), is used to illustrate the computation of the rank. The matrix is

$$\mathbf{A} = \begin{bmatrix} 2 & 4 & 6 \\ 1 & 3 & 5 \\ 9 & 10 & 11 \end{bmatrix}$$

Since the determinant of **A** is zero, the rank is not 3. The following submatrix has a nonzero determinant:

$$\begin{vmatrix} 2 & 4 \\ 1 & 3 \end{vmatrix} = 2(3) - 4(1) = 2$$

Therefore, the rank is 2. It should be noted that one submatrix of order 2 with a nonzero determinant is needed to obtain a rank of 2.

2.6 APPLICATIONS

2.6.1 REACTIONS OF A BEAM DUE TO LOADS

A beam is supporting a concentrated load (P), a distributed load (q), and a moment (M) as shown in Figure 2.7. The span of the beam is $S = 10$. The beam is subject to three types of loads: dead load (D), live load (L), and

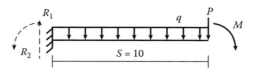

FIGURE 2.7 Loaded beam.

wind load (W). The magnitudes of these loads are given by the following matrix:

$$\begin{array}{cccc} & D & L & W \\ \text{loads} = \begin{array}{c} P \\ q \\ M \end{array} & \left[\begin{array}{ccc} 10 & 12 & -2 \\ 1 & 2 & 1 \\ 2 & 4 & 3 \end{array}\right] \end{array}$$

(2.75)

The loads cause structural reactions where the beam is attached to the support. Two reactions are of interest: R_1 = force reaction at the fixed end, and R_2 = moment reaction at the fixed end. The computations of the two reactions depend on the load type. For an applied force (P) by itself, the resulting reactions are

(2.76)
$$R_1 = P$$

(2.77)
$$R_2 = PS$$

For an applied uniform load (q) by itself, the resulting reactions are

(2.78)
$$R_1 = qS$$

(2.79)
$$R_2 = \frac{qS^2}{2}$$

For an applied moment (M) at the free end by itself, the resulting reactions are

(2.80)
$$R_1 = 0$$

(2.81)
$$R_1 = M$$

The beam reactions can be expressed as

(2.82)
$$\begin{bmatrix} R_1 \\ R_2 \end{bmatrix} = \begin{bmatrix} 1 & S & 0 \\ S & \dfrac{S^2}{2} & 1 \end{bmatrix} \cdot \begin{bmatrix} P \\ q \\ M \end{bmatrix}$$

For $S = 10$, Equation 2.82 can be written as

$$(2.83) \qquad \begin{bmatrix} R_1 \\ R_2 \end{bmatrix} = \begin{bmatrix} 1 & 10 & 0 \\ 10 & 50 & 1 \end{bmatrix} \cdot \begin{bmatrix} P \\ q \\ M \end{bmatrix}$$

Thus, R_1 and R_2 are computed as follows:

$$(2.84) \qquad R_1 = P + 10q + 0M = P + 10q$$

$$(2.85) \qquad R_2 = 10P + 50q + M$$

The values for P, q, and M due to dead load D are given in Equation 2.75. The reactions due to the dead load $\mathbf{R}_D$ can be computed using Equations 2.76 through 2.81 and are given by the following matrix:

$$(2.86) \qquad \mathbf{R}_D = \begin{matrix} & \begin{matrix} R_1 & R_2 \end{matrix} \\ \begin{matrix} P \\ q \\ M \end{matrix} & \begin{bmatrix} 10 & 100 \\ 10 & 50 \\ 0 & 2 \end{bmatrix} \end{matrix}$$

Similarly, the reactions due to the live load $\mathbf{R}_L$ are given by the following matrix:

$$(2.87) \qquad \mathbf{R}_L = \begin{matrix} & \begin{matrix} R_1 & R_2 \end{matrix} \\ \begin{matrix} P \\ q \\ M \end{matrix} & \begin{bmatrix} 12 & 120 \\ 20 & 100 \\ 0 & 4 \end{bmatrix} \end{matrix}$$

Similarly, the reactions due to the wind load $\mathbf{R}_W$ are given by the following matrix:

$$(2.88) \qquad \mathbf{R}_W = \begin{matrix} & \begin{matrix} R_1 & R_2 \end{matrix} \\ \begin{matrix} P \\ q \\ M \end{matrix} & \begin{bmatrix} -2 & -20 \\ 10 & 50 \\ 0 & 3 \end{bmatrix} \end{matrix}$$

In the matrices $\mathbf{R}_D$, $\mathbf{R}_L$, and $\mathbf{R}_W$, the rows represent the types of loads and the columns represent the types of reactions at the support. The total reactions $\mathbf{R}_T$ are given by

(2.89)
$$\mathbf{R}_T = \mathbf{R}_D + \mathbf{R}_L + \mathbf{R}_W = \begin{matrix} P \\ q \\ M \end{matrix} \overset{\begin{matrix} R_1 & R_2 \end{matrix}}{\begin{bmatrix} 20 & 200 \\ 40 & 200 \\ 0 & 9 \end{bmatrix}}$$

Load combinations can be evaluated, such as $D + L$ and $D + W$. The resulting reactions are

(2.90)
$$\mathbf{R}_{D+L} = \mathbf{R}_D + \mathbf{R}_L = \begin{matrix} P \\ q \\ M \end{matrix} \overset{\begin{matrix} R_1 & R_2 \end{matrix}}{\begin{bmatrix} 22 & 220 \\ 30 & 150 \\ 0 & 6 \end{bmatrix}}$$

and

(2.91)
$$\mathbf{R}_{D+W} = \mathbf{R}_D + \mathbf{R}_W = \begin{matrix} P \\ q \\ M \end{matrix} \overset{\begin{matrix} R_1 & R_2 \end{matrix}}{\begin{bmatrix} 8 & 80 \\ 20 & 100 \\ 0 & 5 \end{bmatrix}}$$

The combined reaction due to the total effects of P, q, and M can be computed by determining the column summations in Equations 2.86 to 2.89. The combined reactions $\mathbf{R}_C$ can be expressed using the following vectors:

(2.92)
$$\mathbf{R}_{CD} = \overset{\begin{matrix} R_1 & R_2 \end{matrix}}{[20 \quad 152]} \quad \text{for dead load}$$

(2.93)
$$\mathbf{R}_{CL} = [32 \quad 224] \quad \text{for live load}$$

(2.94)
$$\mathbf{R}_{CW} = [8 \quad 33] \quad \text{for wind load}$$

and

(2.95)
$$\mathbf{R}_{CT} = [60 \quad 409] \quad \text{for total loads}$$

Load combinations can be evaluated, such as $D + L$ and $D + W$. The resulting combined reactions are

$$R_1 \quad R_2$$

(2.96) $\qquad \mathbf{R}_{C(D+L)} = \quad [52 \quad 376] \qquad$ for dead and live loads

and

(2.97) $\qquad \mathbf{R}_{C(D+W)} = [28 \quad 185] \qquad$ for dead and wind loads

2.6.2 CORRELATION ANALYSIS OF WATER-QUALITY DATA

A 5×3 matrix is given in Figure 2.2a, with water-quality concentrations given for three water-quality characteristics; it will be denoted as matrix **X**. The correlations between the three constituents for the five wells are given in the matrix of Equation 2.1. The correlation matrix is computed using several matrix operations, as follows:

1. Compute a row vector that contains the means for each water-quality constituent; it will be denoted as $\mathbf{X}_m$:

(2.98) $\qquad\qquad \mathbf{X}_m = [162 \quad 68.6 \quad 7.2]$

The mean value is computed for each column as

(2.99) $\qquad\qquad x_{mj} = \dfrac{1}{5}\displaystyle\sum_{i=1}^{5} x_{ij}$

2. Compute a row vector, denoted as **S**, that contains the standard deviations of each of the three constituents:

(2.100) $\qquad\qquad \mathbf{S} = [82.94 \quad 68.95 \quad 0.806]$

The standard deviation is computed for each column as

(2.101) $\qquad\qquad s_j = \sqrt{\dfrac{1}{5-1}\displaystyle\sum_{i=1}^{5}(x_{ij} - x_{mj})^2}$

3. A rectangular matrix, denoted as **Z**, is computed by transforming the matrix **X** using both $\mathbf{X}_m$ and **S**. The elements z_{ij} are computed by

(2.102) $\qquad z_{ij} = \dfrac{x_{ij} - x_{mj}}{s_j}, \quad$ for $i = 1,2,\dots 5,$ and $j = 1,2,3$

Thus, for the values in Equations 2.98 and 2.100, $\mathbf{Z}$ is

$$(2.103) \qquad \mathbf{Z} = \begin{bmatrix} -0.265 & -0.487 & 0.620 \\ 0.398 & -0.821 & -0.124 \\ 1.459 & -0.226 & -0.993 \\ -0.362 & 1.732 & 1.365 \\ -1.230 & -0.197 & -0.868 \end{bmatrix}$$

where, for example, element z_{23} is computed by

$$z_{23} = \frac{z_{23} - x_{m3}}{s_3} = \frac{7.1 - 7.2}{0.806} = -0.124$$

4. The correlation matrix, $\mathbf{R}$, is computed by the matrix product divided by a scalar n_r, which is the number of rows, as follows:

$$(2.104) \qquad \mathbf{R} = \frac{1}{n_r - 1} \mathbf{Z}^T \mathbf{Z}$$

The resulting matrix is

$$(2.105) \qquad \mathbf{R} = \frac{1}{5-1} \begin{bmatrix} -0.265 & 0.398 & 1.459 & -0.362 & -1.230 \\ -0.487 & -0.821 & -0.226 & 1.732 & -0.197 \\ 0.620 & -0.124 & -0.993 & 1.365 & -0.868 \end{bmatrix} \cdot \begin{bmatrix} -0.265 & -0.487 & 0.620 \\ 0.398 & -0.821 & -0.124 \\ 1.459 & -0.226 & -0.993 \\ -0.362 & 1.732 & 1.365 \\ -1.230 & -0.197 & -0.868 \end{bmatrix}$$

or

$$(2.106) \qquad \mathbf{R} = \begin{bmatrix} 1.000 & -0.228 & -0.272 \\ -0.228 & 1.000 & 0.640 \\ -0.272 & 0.640 & 1.000 \end{bmatrix}$$

This example illustrates matrix operations that involve vectors, a scalar (n_r in step 4), a transpose, and matrix multiplication.

2.6.3 REGRESSION ANALYSIS FOR PREDICTING BUS RIDERSHIP

The objective of this example is to demonstrate that matrices have a wide variety of applications in engineering, especially in developing prediction models using analytical tools such as regression analysis (Section 10.3 in Chapter 10). Regression analysis is a method used to fit a linear equation to a set of data. This example discusses only the matrix-related part of this analysis.

For this application, assume that a prediction equation is necessary to estimate bus ridership (Y) as a function of the population in an area (X_1) and the cost of a one-way trip (X_2):

(2.107) $$\hat{Y} = b_0 + b_1 X_1 + b_2 X_2$$

where $\hat{Y}$ is an estimator of Y. The coefficients b_0, b_1, and b_2 are unknown regression coefficients. Values for the coefficients are usually computed by transforming Equation 2.107 to the following form:

(2.108) $$\hat{Z} = t_1 Z_1 + t_2 Z_2$$

where $\hat{Z}$, Z_1, and Z_2 are linear transformations of $\hat{Y}$, X_1, and X_2, respectively. The t values, which are called standardized partial regression coefficients, are linear transformations of the b values. A solution to the unknown t values in Equation 2.108 is usually found by using the matrix of measured values of Y, X_1, and X_2 to compute the correlation matrix, which in this case would be a 3×3 matrix because three variables are involved in the analysis: Y, X_1, and X_2.

Assume for the bus ridership problem that the correlation matrix $(\mathbf{R})$ is

(2.109)
$$\mathbf{R} = \begin{array}{c} \\ X_1 \\ X_2 \\ Y \end{array} \begin{array}{ccc} X_1 & X_2 & Y \\ \left[\begin{array}{ccc} 1.00 & -0.30 & 0.64 \\ -0.30 & 1.00 & -0.42 \\ 0.64 & -0.42 & 1.00 \end{array}\right] \end{array}$$

A solution to the problem is found using the system of equations represented by the following matrix expression:

(2.110) $$\mathbf{R}_{11}\mathbf{t} = \mathbf{R}_{12}$$

where $\mathbf{R}_{11}$ is the 2×2 matrix of the intercorrelations between X_1 and X_2, $\mathbf{t}$ is the column vector of t values, and $\mathbf{R}_{12}$ is the column vector of predictor–criterion correlation. To find the solution to Equation 2.110, it is necessary to make the following matrix-algebra formulation:

1. Premultiply both sides of Equation 2.110 by $\mathbf{R}_{11}^{-1}$:

(2.111) $$\mathbf{R}_{11}^{-1}\mathbf{R}_{11}\mathbf{t} = \mathbf{R}_{11}^{-1}\mathbf{R}_{12}$$

2. Based on Equation 2.31, Equation 2.111 becomes

(2.112) $$\mathbf{It} = \mathbf{R}_{11}^{-1}\mathbf{R}_{12}$$

3. The product of a vector and the unit matrix (**I**) equals the vector—that is, **It** = **t**; therefore, Equation 2.112 becomes

(2.113)
$$\mathbf{t} = \mathbf{R}_{11}^{-1} \mathbf{R}_{12}$$

In Equation 2.113, the known quantities are on the right side and the unknowns are on the left side. To find the t values, it is necessary to find the inverse.

For the intercorrelation matrix of Equation 2.109, Equation 2.31 can be used to compute the inverse as

(2.114)
$$\begin{bmatrix} c_{11} & c_{12} \\ c_{21} & c_{22} \end{bmatrix} \cdot \begin{bmatrix} 1.00 & -0.30 \\ -0.30 & 1.00 \end{bmatrix} = \begin{bmatrix} 1 & 0 \\ 0 & 1 \end{bmatrix}$$

Solving Equation 2.114 for the c_{ij} values yields the following inverse matrix:

(2.115)
$$\mathbf{R}_{11}^{-1} = \begin{bmatrix} 1.10 & 0.33 \\ 0.33 & 1.10 \end{bmatrix}$$

Substituting into Equation 2.113 and using the vector of predictor–criterion correlation yields

(2.116)
$$\begin{bmatrix} t_1 \\ t_2 \end{bmatrix} = \begin{bmatrix} 1.10 & 0.33 \\ 0.33 & 1.10 \end{bmatrix} \cdot \begin{bmatrix} 0.64 \\ -0.42 \end{bmatrix} = \begin{bmatrix} 0.57 \\ -0.25 \end{bmatrix}$$

Thus Equation 2.108 becomes

(2.117)
$$\hat{Z} = 0.57 Z_1 - 0.25 Z_2$$

The coefficients of Equation 2.107 can be found by making the necessary linear transformations used in regression analysis.

PROBLEMS

2.1   Using weather summaries from the daily newspapers for the last week, create a 7×2 matrix of the daily maximum and minimum temperatures. Compute the mean for each day and present the daily means as a column vector. Compute the means of the daily maximums and minimums and present the values as a row vector with two elements.

2.2   Using weather summaries, for the last 2 weeks, create a 14×3 matrix of the mean daily temperature (column 1), the mean daily humidity (column 2), and the mean wind speed (column 3). Compute the mean value for each column and place the value in a row vector (1×3).

2.3   Create a matrix using scores for all tests that you had in a course as well as the m scores for two or more other students; make sure that all scores are on a 0 to 100 scale. Create a column vector that has the same number of elements m as there are tests, with the elements of the vector being the weights assigned by the instructor, with the weights standardized so that they sum to 1.0. For a matrix of n students and m tests, postmultiply the test score matrix by the column vector of weights. Discuss the resulting vector.

2.4   During one semester, the grades on four of five quizzes in a thermodynamics class were 6, 4, 9, and 8 and the grades on quizzes in a dynamics class were 6, 8, 7, and 5. If the vector mean quiz grades is [7.6, 6.4] create a (5×2) matrix that shows the 10 quiz grades.

2.5   Obtain data on the calorie content and saturated fat content for five foodstuffs. Create one row vector of the means and a second row vector of the maximum values.

2.6   A mechanical engineer is studying the efficiency of a two-step process for producing a one-piece machine part. First, the part is cut from the raw material and then the rough-cut part is precision machined. The engineer observes that the time required for machining is longer if the machinist takes very little time to rough-cut the material. Similarly, if the machinist spends more time cutting the material, less time is required to machine the part. Create a rectangular matrix in which the two rows represent the cutting time (short, long) and three columns represent the machining time (short, moderate, long). Values in the matrix represent the probability p_{ij} that it takes j time units to machine a part, given that i time units were taken to cut the part; thus the matrix elements in the rows must sum to 1.

2.7   An agricultural engineer is establishing a policy on irrigation. The policy statement will indicate the number of units of irrigated water that will be applied to crops depending on recent rainfall and rainfall that is expected in the future. The probabilities of rainfall rates expected in the future (low, average, high) are correlated with past rainfall rates (low, average, high). Square matrices can be created using the three rows to represent the rainfall in the most recent time period and the three columns to represent the rainfall in the next time period. The matrix elements are the probabilities p_{ij} that rainfall for the next time period is at level j, given that rainfall for the most recent time period is at level i. The matrix elements in the rows must sum to 1. Create two matrices: one in which low rainfalls tend to follow high rainfalls and a second in which low rainfalls follow low rainfalls.

2.8   Show the transpose of each of the following matrices:

(a) $\mathbf{A} = \begin{bmatrix} 2 & 6 \\ -3 & 4 \end{bmatrix}$

(b) $\mathbf{B} = \begin{bmatrix} -1 & 9 & 6 \\ 3 & 0 & -7 \end{bmatrix}$

(c) $\mathbf{C} = \begin{bmatrix} 0.2 & -0.5 \\ -0.8 & 0.1 \\ -0.3 & 0.7 \end{bmatrix}$

2.9   Show the transpose of each of the following matrices:

(a) $\begin{bmatrix} 2 & -1 & 4 \\ 3 & -6 & 0 \\ -2 & 5 & 3 \end{bmatrix}$

(b) $\begin{bmatrix} 2 & 5 \\ -8 & 1 \\ 4 & -3 \end{bmatrix}$

(c) $\begin{bmatrix} 3 & 6 & -2 & 1 \\ -1 & 4 & 5 & -3 \\ 0 & -6 & 4 & -5 \end{bmatrix}$

2.10   Using the following matrices, show that matrix addition is not directional, but that matrix subtraction is directional:

$$\mathbf{A} = \begin{bmatrix} 3 & -2 \\ 0 & 4 \end{bmatrix}, \quad \mathbf{B} = \begin{bmatrix} -2 & 3 \\ 1 & 2 \end{bmatrix}$$

2.11   Using the following matrices, show that matrix addition is not directional, but that matrix subtraction is directional:

$$\mathbf{A} = \begin{bmatrix} 5 & -1 \\ -2 & 6 \\ 3 & -4 \end{bmatrix} \quad \mathbf{B} = \begin{bmatrix} 2 & 1 \\ -2 & -3 \\ 5 & -2 \end{bmatrix}$$

2.12   Given matrices $\mathbf{C}$ and $\mathbf{E}$, find matrix $\mathbf{D}$ such that $\mathbf{C} + \mathbf{D} = \mathbf{E}$. The matrices are given by

$$\mathbf{C} = \begin{bmatrix} 0.4 & -0.3 & 0.5 \\ 0.0 & 0.7 & -0.2 \end{bmatrix} \quad \text{and} \quad \mathbf{E} = \begin{bmatrix} 0.1 & -0.8 & 0.9 \\ -0.6 & 0.3 & -0.4 \end{bmatrix}$$

2.13 Given matrices **X** and **Y**, find matrix **Z** such that **Y** − **Z** = **X**:

$$\mathbf{X} = \begin{bmatrix} 3.1 & -0.7 \\ 2.6 & 4.1 \\ -1.4 & -3.2 \end{bmatrix} \quad \mathbf{Y} = \begin{bmatrix} 1.7 & 2.9 \\ 0.0 & -1.6 \\ 5.3 & -3.7 \end{bmatrix}$$

2.14 Using the matrices in Problem 2.10, show that matrix multiplication is not commutative.

2.15 Using the matrices in Problem 2.11, show that matrix multiplication is not commutative.

2.16 Using the following matrices, show that matrix multiplication is not commutative:

$$\mathbf{F} = \begin{bmatrix} 0.1 & 0.7 \\ -0.3 & 0.4 \end{bmatrix} \quad \text{and} \quad \mathbf{G} = \begin{bmatrix} -0.2 & 0.8 \\ 0.5 & -0.6 \end{bmatrix}$$

2.17 Using the following matrices, show that matrix multiplication is not commutative.

$$\mathbf{H} = \begin{bmatrix} 2 & 4 & 0 \\ 3 & -5 & -1 \\ -1 & 2 & 4 \end{bmatrix} \quad \mathbf{K} = \begin{bmatrix} 6 & -3 & 5 \\ -1 & 4 & -4 \\ -2 & 2 & 0 \end{bmatrix}$$

2.18 Explain why the following two matrices cannot be used to show that matrix multiplication is not commutative:

$$\mathbf{M} = \begin{bmatrix} 2 & 4 \\ 1 & -3 \end{bmatrix} \quad \mathbf{N} = \begin{bmatrix} 1 & 0 & -4 \\ -2 & 2 & 5 \end{bmatrix}$$

2.19 For each of the following matrix pairs, compute the matrix product **C** = **A**·**B**:

(a) $\mathbf{A} = \begin{bmatrix} 0.1 & -0.3 & 0.2 \\ 0.6 & -0.1 & 0.4 \end{bmatrix}, \quad \mathbf{B} = \begin{bmatrix} 1 \\ -4 \\ 2 \end{bmatrix}$

(b) $\mathbf{A} = \begin{bmatrix} 0.3 \\ 1.2 \\ -0.5 \end{bmatrix}, \quad \mathbf{B} = [-0.2 \quad 0.7 \quad 0.5]$

(c) $\mathbf{A} = [-0.2 \quad 0.7 \quad 0.5], \quad \mathbf{B} = \begin{bmatrix} 0.3 \\ 1.2 \\ -0.5 \end{bmatrix}$

(d) $\mathbf{A} = \begin{bmatrix} 0.4 & 0.6 \\ 1.3 & -0.2 \\ -0.8 & 1.5 \end{bmatrix}, \quad \mathbf{B} = \begin{bmatrix} 2 & 3 & -1 & 4 \\ 6 & -4 & 0 & 5 \end{bmatrix}$

2.20 For each pair of matrices, compute the product $\mathbf{C} = \mathbf{A} \cdot \mathbf{B}$:

(a) $\mathbf{A} = \begin{bmatrix} 1 & -6 & 0 \\ 2 & 3 & -1 \end{bmatrix} \quad \mathbf{B} = \begin{bmatrix} 4 & -3 \\ -2 & 5 \\ 1 & 0 \end{bmatrix}$

(b) $\mathbf{A} = [1.4 \quad -0.7 \quad -1.1] \quad \mathbf{B} = \begin{bmatrix} 1.6 & -4.3 \\ -2.1 & -3.7 \\ 0.8 & 0.2 \end{bmatrix}$

2.21 Compute the inverse of matrix $\mathbf{B}$ in Problem 2.10.
2.22 Compute the inverse of matrix $\mathbf{M}$ in Problem 2.18.
2.23 Compute the inverse of matrix $\mathbf{H}$ in Problem 2.17.
2.24 Compute the inverse of the following matrix:

$$\begin{bmatrix} 3 & -1 & 2 \\ 1 & 4 & -3 \\ 2 & 5 & 3 \end{bmatrix}$$

2.25 Show that the inverse of the inverse of matrix $\mathbf{X}$ is $\mathbf{X}$.
2.26 Show which of the following matrices is singular:

(a) $\mathbf{A} = \begin{bmatrix} 0.1 & 0.3 \\ 0.4 & 1.2 \end{bmatrix}$

(b) $\mathbf{B} = \begin{bmatrix} -2 & 6 & 2 \\ 1 & -3 & 2 \\ 2 & -6 & -2 \end{bmatrix}$

2.27 Determine if the following matrix is singular:

$$\mathbf{R} = \begin{bmatrix} 2 & 1 & -4 \\ 3 & -1 & -6 \\ -4 & -3 & 8 \end{bmatrix}$$

2.28 Determine if the following matrix is singular:

$$\begin{bmatrix} -1 & 3 \\ -3 & 1 \end{bmatrix}$$

2.29 Determine the element a_{22} needed for matrix **A** to be singular:

$$A = \begin{bmatrix} 1 & 3 \\ -5 & a_{22} \end{bmatrix}$$

2.30 Determine the element b_{22} needed for matrix **B** to be singular:

$$B = \begin{bmatrix} 6 & 0.3 & -2 \\ -4 & b_{22} & \dfrac{4}{3} \\ 2 & 0.1 & -1.5 \end{bmatrix}$$

2.31 What does the trace of a correlation matrix equal?

2.32 Compute the trace of matrix **B** in Problem 2.26 part (b).

2.33 Compute the trace of the matrix in Problem 2.27.

2.34 Compute the trace of the matrices **H** and **K** in Problem 2.17.

2.35 Add the following vectors:
   (a) $V_1 = [2 \quad 6 \quad -3]$ and $V_2 = [-3 \quad 1 \quad 2]$

   (b) $V_1 = \begin{bmatrix} 0.6 \\ 0.2 \\ 0.0 \\ 0.3 \end{bmatrix}$ and $V_2 = \begin{bmatrix} -0.4 \\ 0.1 \\ -0.5 \\ 0.2 \end{bmatrix}$

2.36 Compute the sum of the two vectors
   (a) $V_1 = \begin{bmatrix} 3.4 \\ 1.6 \\ -5.2 \end{bmatrix}$ $V_2 = \begin{bmatrix} 0.8 \\ -2.5 \\ -1.6 \end{bmatrix}$

   (b) $V_3 = [1.3 \quad 0.8 \quad -2.2 \quad -4.0] \quad V_4 = [-1.1 \quad 0.3 \quad -1.6 \quad 2.7]$

2.37 Subtract vector V_1 from vector V_2 in Problem 2.35 parts (a) and (b).

2.38 Subtract (a) vector V_1 from V_2 in Problem 2.36 part (a) and (b) vector V_4 from V_3 in Problem 2.36 part (b).

2.39 Show the transpose of the following matrix:

$$X = \begin{bmatrix} 2.1 & -1.2 & 3.5 & -0.8 \\ -1.7 & 0.0 & 4.1 & 2.6 \end{bmatrix}$$

2.40 Find the vector product $V_1^T V_2$ using the vectors in Problem 2.35 parts (a) and (b).

2.41 Find the vector product $V_1 V_2^T$ using the vectors in Problem 2.35 parts (a) and (b).

2.42 Find the vector product $V_1^T V_2$ using the vectors in Problem 2.36 part (a).

2.43 Find the vector product $V_4^T V_3$ using the vectors in Problem 2.36 part (b).

2.44 Determine the element a_2 for the vector $\mathbf{V}_5$ to be a normal vector:

$$\mathbf{A} = [-0.162 \quad a_2 \quad 0.365 \quad 0.189]$$

2.45 Determine the elements a_{13} and a_{21} for the column vectors to be normal vectors.

$$\mathbf{A} = \begin{bmatrix} 0.222 & a_{21} \\ 0.333 & -0.234 \\ -a_{13} & -0.345 \\ 0.111 & -0.456 \end{bmatrix}$$

2.46 If the following matrix is viewed as three column vectors, which of the vectors are normalized? Which pair, if any, of the column vectors is orthonormal?

$$\begin{bmatrix} 0.5883 & -0.5921 & -0.1830 \\ 0.7845 & -0.1502 & 0.3649 \\ -0.1961 & 0.7917 & 0.9129 \end{bmatrix}$$

2.47 If the following matrix is viewed as three row vectors, which of the vectors are normalized? Which pair, if any, of the row vectors is orthonormal?

$$\mathbf{P} = \begin{bmatrix} 0.521 & -0.128 & 0.844 \\ -0.602 & 0.672 & 0.474 \\ 0.605 & 0.779 & 0.165 \end{bmatrix}$$

2.48 If the matrix in Problem 2.47 is viewed as three column vectors, which vectors are normalized? Which pair, if any, of the column vectors is orthonormal.

2.49 Find a column vector that is orthogonal to both the column vectors of matrix $\mathbf{A}$:

$$\mathbf{A} = \begin{bmatrix} 1 & 2 \\ 1 & 2 \\ -2 & 2 \end{bmatrix}$$

2.50 Normalize matrix $\mathbf{B}$—first by rows and then by columns. Compare the two results.

$$\mathbf{B} = \begin{bmatrix} 2 & -2 & 4 \\ 3 & 2 & -1 \\ -1 & 3 & 2 \end{bmatrix}$$

2.51 The coordinates of a point in a three-dimensional axis system are the elements of a row vector

$$\mathbf{X} = [X_1 \quad X_2 \quad X_3]$$

For two points B and A in the axis system, the elements of a vector of the direction cosines of the line connecting the two points are (i.e., a vector from B to A)

$$\cos\theta_i = \frac{a_i - b_i}{L}$$

$$L = \sqrt{\sum_{i=1}^{3}(a_i - b_i)^2}$$

Compute the θ_i for the following two points (i.e., a vector from $\mathbf{S}$ to $\mathbf{T}$):

$$\mathbf{S} = [3 \quad 4 \quad -2]$$

$$\mathbf{T} = [-1 \quad 3 \quad 3]$$

2.52 Compute the value of x that is necessary for the following vector $\mathbf{A}$ to be normal:

$$\mathbf{A} = [0.7 \quad 0.4 \quad -0.2 \quad -0.3 \quad x]$$

2.53 Compute the value of b_{22} that yields a normalized column vector for the following matrix. Is the matrix orthonormal?

$$\mathbf{B} = \begin{bmatrix} 0.23 & 0.66 \\ -0.58 & b_{22} \\ 0.78 & 0.26 \end{bmatrix}$$

2.54 Compute the values of a_{12} and a_{32} that are necessary for matrix $\mathbf{A}$ to be orthonormal in the columns:

$$\mathbf{A} = \begin{bmatrix} 0.5 & a_{12} \\ 0.2 & -0.6 \\ -0.843 & a_{32} \end{bmatrix}$$

2.55 Compute the determinants of matrices $\mathbf{A}$ and $\mathbf{B}$ in Problem 2.10.

2.56 Compute the determinant of the matrix in Problem 2.24.

2.57 Compute the determinant of matrix $\mathbf{B}$ in Problem 2.26.

2.58 Compute the determinant of matrix $\mathbf{P}$ in Problem 2.47.

2.59 Compute the determinant of matrix **R** in Problem 2.27.

2.60 Compute the determinant of matrices **H** and **K** in Problem 2.17.

2.61 Determine the rank of the matrices in Problem 2.19.

2.62 Determine the rank of the following matrices:

(a) $\begin{bmatrix} 1 & 2 & 4 \\ 1.5 & 4 & 6 \end{bmatrix}$

(b) $\begin{bmatrix} 3 & 2 & 1 \\ 1 & 3 & 2 \\ 2 & 1 & 3 \end{bmatrix}$

(c) $\begin{bmatrix} 2.0 & 8.0 & 8.0 \\ 1.5 & 4.5 & 6.0 \\ 3.5 & 24.5 & 14.0 \\ 1.5 & 4.5 & 6.0 \end{bmatrix}$

2.63 A soil scientist collects the erosion rate Y from four test plots. The plots differ in percent slope (X_1) and soil type; the mean diameter of the soil particles (X_2) is used to represent the soil type. Compute the correlation matrix method of Section 2.6.3. The collected data are

$$\begin{array}{ccc} X_1 & X_2 & Y \end{array}$$
$$\begin{bmatrix} 2 & 0.10 & 11 \\ 5 & 0.09 & 14 \\ 6 & 0.04 & 12 \end{bmatrix}$$

Introduction to Numerical Methods

3.1 INTRODUCTION

Many operations in mathematical analysis can be identified as belonging to one of two types: analytical or numerical. For example, when we need to solve a second-order polynomial $aX^2 + bX + c = 0$, for the values of X that satisfy the equality, it is common to use the closed-form solution $X = \left(-b \pm \sqrt{b^2 - 4ac}\right)/2a$. This is an analytical solution. Alternatively, we could assume some value for the solution (X), and through a systematic method of trial and error, we could obtain essentially the same solution as the analytical approach. However, this numerical approach is much more tedious, so when an analytical solution exists, it is usually the preferred method. For many types of problems, such as solving a fifth-order polynomial, a closed-form or analytical solution does not exist. Then the iterative, or numerical, approach must be used.

In spite of the simplicity of most numerical methods, it is somewhat difficult to give a specific definition for the term *numerical method*. Instead of trying to define the term, we will discuss the characteristics, advantages, and disadvantages of numerical methods and then provide an example to illustrate these characteristics.

The solution to many engineering problems involves the use of complex functional forms. Numerical methods can be used to solve a variety of quantitative problems faced by engineers. For many engineering problems, an analytical solution is either impossible or the cost or effort of performing an analytical solution would be prohibitive. Furthermore, in an effort to achieve greater accuracy in engineering solutions, more complex analyses are being used. Very often, the complexity of the problem becomes such that an analytical solution is not possible. In cases where analytical solutions are not practical or the problem is very complex, numerical methods can often be used to find a solution.

3.1.1 CHARACTERISTICS OF NUMERICAL METHODS

Numerical methods have most of the following characteristics:

1. The solution procedure is iterative, with the accuracy of the estimated solution improving with each iteration.

2. The solution procedure provides only an approximation to the true, but unknown, solution.
3. An initial estimate of the solution may be required.
4. The solution procedure is conceptually simple, with algorithms representing the solution procedure that can be easily programmed on a digital computer.
5. The solution procedure may occasionally diverge from rather than converge to the true solution.

Each of these characteristics will be evident in the numerical methods introduced in many of the remaining chapters of this book. Example 3.1 is used to illustrate the concept of a numerical method.

Example 3.1: Square Root

Finding the square root of a number is a frequent task. On a hand calculator, we simply enter the number and then press the $\sqrt{x}$ key. On a digital computer, the function SQRT is used. For the purposes of illustrating numerical methods, let us assume that SQRT is not available and that we need to develop a method for estimating $\sqrt{x}$, where x is any positive real value. We can start by assuming that we have an initial estimate of $\sqrt{x}$, which we will denote as x_0, and x_0 is in error by an unknown amount Δx. If we know Δx, then we would have the following equality:

$$(3.1) \qquad x_0 + \Delta x = \sqrt{x}$$

If both sides of Equation 3.1 are squared and we assume that $(\Delta x)^2$ is much smaller than Δx and can be neglected, we can solve for Δx:

$$(3.2) \qquad \Delta x = \frac{x - x_0^2}{2x_0}$$

The value of Δx computed with Equation 3.2 can be added to x_0 to get a revised estimate of x. Thus the new estimate x_1 of the true solution is

$$(3.3) \qquad x_1 = x_0 + \Delta x$$

Generalizing the notation, the concept behind Equations 3.2 and 3.3 yields

$$(3.4) \qquad x_{i+1} = x_i + \Delta x_i$$

where x_i and x_{j+1} are the estimates of x on trials i and $(i + 1)$, respectively, and

(3.5)
$$\Delta x = \frac{x - x_i^2}{2x_i}$$

To illustrate the use of Equations 3.3 to 3.5, assume that $x = 150$. We know that $12^2 = 144$, so $x_0 = 12$ is a reasonable initial estimate. Therefore, the first iteration of Equation 3.5 yields

(3.6)
$$\Delta x = \frac{x - x_0^2}{2x_0} = \frac{150 - 12^2}{2(12)} = 0.25$$

Thus, from Equation 3.4, x_1 equals 12.25. A second iteration of Equation 3.5 can now be applied using the revised estimate x_1:

(3.7)
$$\Delta x = \frac{x - x_1^2}{2x_1} = \frac{150 - (12.25)^2}{2(12.25)} = -0.00255$$

Applying Equation 3.4 again yields $x_2 = 12.24745$. A third iteration of Equation 3.5 yields

(3.8)
$$\Delta x = \frac{x - x_2^2}{2x_2} = \frac{150 - (12.24745)^2}{2(12.24745)} = -0.12861 \times 10^{-5}$$

The revised value x_3 equals 12.2474487, which is identical to x_2 to seven significant digits. While the true value is not usually known, the true solution of $\sqrt{150}$ equals 12.24744871. Thus x_2 was accurate to seven significant digits, and x_3 was accurate to nine digits.

The purpose of the preceding example is not to provide a numerical method for estimating $\sqrt{x}$; rather, the objective is to illustrate the characteristics of numerical methods. With respect to the first characteristic identified at the beginning of this section, the solution procedure is iterative in that Equations 3.3 to 3.5 were applied three times, with the accuracy improving with each iteration. Second, the solution procedure provided an approximation of the true value; the initial estimate of 12 was a first approximation, while x_1, x_2, and x_3 were subsequent approximations. Third, the solution procedure required an initial estimate; for our example problem, an initial estimate of 12 was used. Fourth, the solution procedure is conceptually simple and can be easily programmed. Figure 3.1 provides an outline of the solution procedure. Fifth, the procedure would have converged even if a less accurate initial estimate, such as 10 or 20, were used. Other numerical methods may not converge if the initial estimate is not a reasonably good estimate.

```
Inputs:
  x₀  =  initial value of the solution
  x   =  value needed
  ΔT  =  tolerance (a convergence criterion)
Computation steps:
      Function SQRTN(x, x₀, TOL)
  c   x = value for which square root is needed
  c   x₀ = an input, initial estimate of square root of x
  c   x₀ = final estimate of square root of x
  c   TOL = maximum allowable (tolerable) error in
  c   square root of x
  1   delx = (x − x₀**2)/(2.0*x₀)
      x₀ = x₀ + delx
      if (abs(delx).gt.TOL)go to 1
      SQRTN = x₀
      end
Decision check:
Compare absolute value of Δx and ΔT:
      If (|Δx| > ΔT), re-do computation
      If (|Δx| < ΔT), assume convergence
Solution:
The last value of x₀ is the solution
```

FIGURE 3.1   Outline of the solution procedure for the numerical solution of $\sqrt{x}$.

3.2 ACCURACY, PRECISION, AND BIAS

When dealing with solutions from numerical analyses, we need to address the issues of *accuracy, precision,* and *bias.* The term *accuracy* is commonly used, for example, as in reference to the accuracy of the media's report on a particular issue or to the accuracy of an opinion poll taken just before a presidential election. Of interest here is the meaning of the term as it applies to numerical solutions.

Consider the four targets of Figure 3.2. Assuming that those shooting at the targets were aiming at the center, the person shooting at target *A* was successful. While not every bullet went through the exact center, the distances between the holes in the target and the center are small. The holes in target *B* are similarly clustered as in target *A*, but they show large deviation from the center. The large deviations or errors between the holes and the center of the target suggest a lack of exactness or correctness. While all the holes deviate significantly from the center, there is, however, a measure of consistency in the holes. In summary, the holes in target *B* show two important characteristics: They tend to agree with each other, but they deviate considerably from where the shooter was aiming.

The holes in target *C* are very different in character from the holes in either target *A* or *B*. First, most are not near the center and, second, they are not near each other. Thus, they lack both correctness and consistency. Because most of them are not near the center of the target, the shooter is not exact. Because of the wide scatter, there is a lack of consistency. In comparing the holes on targets *B* and *C*, they both lack exactness, but there is a measure of consistency for the holes in target *B* that is missing in target *C*.

A fourth distribution of holes is shown in target *D*. Like target *B*, all of the holes are to the left of the center; however, unlike target *B*, the holes lack consistency. Both targets *C* and *D* show considerable scatter, but with target *D* the scatter is concentrated to one side of the target.

A comparison of the four targets indicates that there are three important characteristics of data. The holes in targets *A* and *B* show a measure of consistency. In data analyses, this consistency is more formally referred to as precision. *Precision* is defined as the ability to give multiple estimates that are near to each other. In terms of the targets *A* and *B*, the shooters were precise. The shooters of targets *C* and *D* were imprecise since the holes show a lot of scatter.

Bias is a second important characteristic of data. The holes in targets *B* and *D* are consistently to the left of center; that is, there is a systematic distortion of the holes with respect to the center of the target. If we found a point that could represent the center of the holes in any target, then the difference between the centers of the holes and the center of the target would be a measure of the shooter's bias. More formally, *bias* is a systematic deviation of values from the true value. The holes in targets *B* and *D* show a systematic deviation to the left, while targets *A* and *C* are considered to be unbiased because there is no systematic deviation.

Precision and bias are elements of accuracy. Specifically, bias is a measure of systematic deviation, and precision is a measure of random deviations. Inaccuracy can result from either a bias or a lack of precision. In referring to the targets, the terms correctness and exactness were used. More formally, *accuracy* is the degree to which the measurements deviate from the true value. If the deviations are small, then the method or process that led to the measurement is accurate.

With reference to the numerical method of Example 3.1, it provides accurate estimates because it closely approaches the true value regardless of the initial estimate. If the method consistently led to estimates that exceeded the true value by some amount δ, then the numerical method would have a positive bias of δ. If the estimates were consistently less than the true value by an amount δ, then the bias would be negative. If the deviations were often greater than the true value and just as often less than the true value, then the method would be imprecise and inaccurate.

Table 3.1 provides a summary of the concepts of accuracy, precision, and bias as they characterize the holes in the targets of Figure 3.2. Of course, terms like high, moderate, and low are somewhat subjective. Statistical measures of systematic and nonsystematic variations are provided in usage for particular applications.

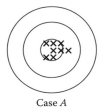
Case *A*

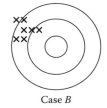
Case *B*

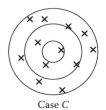
Case *C*

Case *D*

FIGURE 3.2   Accuracy, precision, and bias.

TABLE 3.1   Summary of Bias, Precision, and Accuracy for Figure 3.2

Target	Bias	Precision	Accuracy
A	None (unbiased)	High	High
B	High	High	Low
C	None (unbiased)	Low	Low
D	Moderate	Low	Low

3.3 SIGNIFICANT FIGURES

Consider the problem of measuring the distance between two points using a ruler that has a scale with 1 mm between the finest divisions. If we record our measurements in centimeters and if we estimate fractions of a millimeter, then a distance recorded as 3.76 cm gives two precise digits (i.e., the 3 and the 7) and one estimated digit (i.e., the 6). If we define a significant digit to be any number that is relatively precise, then the measurement of 3.76 cm has three significant digits. Even though the last digit could be a 5 or a 7, it still provides some information about the length, so it is assumed to be significant. If we recorded the number as 3.762, we would still have only three significant digits since the 2 is not precise. Only one imprecise digit can be considered as a significant digit.

The number of significant digits is used to reflect the accuracy of a number. For example, if the number 46.23 is exact to the four digits shown, it is said to have four significant digits. The error is then no more than 0.005.

The number of significant digits is set either by the scale measurement or the physical significance of the numbers. For example, a digital bathroom scale that shows weight to the nearest pound uses up to three significant digits. If the scale shows, for example, 159 pounds, then the individual assumes that his or her weight is within 0.5 pound of the observed value. In this case, the scale has set the number of significant digits. The number of significant digits in someone's height is set by the physical significance of the numbers. The ruler may provide for measurements to $\frac{1}{32}$ in., but the number may be recorded to the nearest inch because more precise values are not important. In that case we would have two significant digits, since a measurement of 5 ft, $11\frac{3}{32}$ in. would be recorded as 71 in. In this case, the meaningfulness of the information content of the number set the number of significant digits.

The rule for identifying the number of significant digits is as follows: The digits from 1 to 9 are always significant, with 0 being significant when it is not being used to set the position of the decimal point. For example, each of the following have three significant digits: 2410; 2.41; and 0.00241. In the first example, the 0 is only used to set the decimal place. Confusion can be avoided by using scientific notation: 2.41×10^3. This means that 2.41×10^3 has three significant digits, where 2.410×10^3 has four significant digits. The numbers 13 and 13.00 differ in that the former is recorded at two significant digits, and the latter has four significant digits. In this case, the two zeros are significant.

The location of the decimal point does not influence the number of significant digits. The measurements of 3.76 cm could be written as 37.6 mm or 0.0376 m; each of these three values has three significant digits.

When performing computations, the following is a general rule on setting the number of significant digits in a computed value: Any mathematical operation using an imprecise digit is imprecise. When combined with the rule that only one imprecise digit can be considered significant, then the number of significant digits is set. Consider the following multiplication of two numbers (4.2**6** and 8.**39**), each having three significant digits, with the last digit of each being imprecise:

4.2**6**	Starting number
8.**39**	Starting number
0.**3834**	0.0**9** times 4.2**6**
1.27**8**	0.3 times 4.2**6**
34.0**8**	8 times 4.2**6**
35.**7414**	Total (product result)

The digits that depend on imprecise digits are underlined. In the final answer, only the first two digits (35) are not based on imprecise digits. Since one and only one imprecise digit can be considered as significant, the result is recorded as 35.7.

A second example shows the potential error:

$$1.4(2.4) = 3.36$$

Rounding 3.36 to 3.4 and multiplying by 3.4 yields:

$$3.4(3.4) = 11.56$$

Rounding 11.56 to 11.6 and multiplying by 0.6 yields:

$$11.6(0.6) = 6.96$$

Multiplying 1.4 by 2.4 and then by 0.6 yields 6.8544. Thus, 6.96 is in error by 1.5%.

Just because a set of numbers is only accurate to a specified number of digits does not mean that computations based on those numbers should be rounded and reported to that number of digits. Computed values should only be rounded at the end of the computation process. For intermediate steps, the computed numbers should be determined and reported to more digits than the number of significant digits in the original set of numbers. The reporting of coefficients of equations provides an illustration of the problem. Consider the equation with values of the coefficients reported to five significant digits:

(3.9) $$\hat{Y} = 11.587 + 1.9860x$$

TABLE 3.2   Rounding Numerical Calculations

x	$\hat{Y}_3$	$\hat{Y}_4$	$\hat{Y}_5$
1	13.59	13.576	13.573
20	51.40	51.310	51.307
40	91.20	91.030	91.027
100	210.60	210.19	210.187

where $\hat{Y}$ is the predicted value of the variable y in the regression model of Equation 3.9. For three significant digits the equation would be

$$\hat{Y} = 11.6 + 1.99x \tag{3.10}$$

Using three, four, and five digits for the regression coefficients yields the predicted values ($\hat{Y}_3$, $\hat{Y}_4$, and $\hat{Y}_5$) for given values of x as shown in Table 3.2. Rounding the regression coefficients to three digits does not cause an error for $x = 1$, but errors of 0.1, 0.2, and 1 occur for $x = 20$, $x = 40$, and $x = 100$, respectively, assuming that the final values are rounded to three significant digits. Using the regression coefficients with four digits produces estimated values for all of the values of x that are without rounding error when evaluated at three significant digits. In this case, estimated values of Y can be rounded to three significant digits, but the regression coefficients should not be rounded to three digits; the rounding should be made at the end of the computation, rather than at intermediate calculations.

Example 3.2: Arithmetic Operations and Significant Digits

When performing mathematical operations as a part of the solution of a numerical analysis, it is important to consider the number of significant digits in the solution. If we are measuring the lengths of the base and height of a triangle with an instrument that provides values to the nearest tenth of an inch, then the measured lengths of $b = 12.3$ in. and $h = 17.2$ in. are expressed to three significant digits. If we use these values to compute the area of a triangle, then we should express the solution as a number having the smallest number of significant digits; thus the area would be

$$A = 0.5bh = 0.5(12.3)(17.2) = 106 \text{ in.}^2 \tag{3.11}$$

Note that if we ignore the concept of significant digits, the area would be 105.78 in.^{2}; however, using this value would imply that there are five significant digits. We only know that the true value is expected to lie between

$$0.5(12.25)(17.15) = 105.04375 \text{ cm}^2 \tag{3.12}$$

and

$$0.5(12.35)(17.25) = 106.51875 \text{ cm}^2 \tag{3.13}$$

It is also of interest to note that, whereas the 0.5 in Equation 3.11 appears to only be expressed with one significant digit, it is considered to be an exact value.

It is common in engineering to deal with quantities that cannot be expressed in a finite number of significant digits—for example, e (for exponential), π, $\sqrt{3}$, and $\sqrt{11}$. Although these quantities have specific numerical meaning, their absolutely correct values can only be expressed in an infinite number of significant digits. Therefore, for a specified engineering application, an engineer needs to specify the number of significant digits that should be used to express these quantities. For example, π = 3.14159265358979323 ... can be expressed to three significant digits as 3.14. As a result, a round-off error is introduced, which is the true value minus the approximate value. The round-off error in this case is −0.00159265358979323.... This example illustrates the importance of significant digits and their role in introducing errors or accuracy in numerical solutions.

3.4 ANALYSIS OF NUMERICAL ERRORS

3.4.1 ERROR TYPES

An error in estimating or determining a quantity of interest can be defined as a deviation from its unknown true value. In general, errors can be classified based on their sources as non-numerical and numerical errors.

Non-numerical errors include (1) modeling errors, (2) blunders and mistakes, and (3) uncertainty in information and data. Prediction models used for assessing the parameters of an engineering problem are commonly based on a set of assumptions and limitations. Therefore, the models can give predicted values that might deviate from the corresponding true values of the problem. Unfortunately, the true values of the problem are unknown. Assumptions and limitations that are made for formulating the model reduce the analytical complexity to a reasonable level. The deviations of the predicted values from the true values are the modeling errors.

Blunders and gross mistakes can generally be attributed to human errors. They can take several forms—for example, arithmetic mistakes, omissions, incorrect use of equations or computational procedures, incorrect development of computational algorithms including computer programs, incorrect input, incorrect understanding of output, incorrect interpretation of computational algorithms and computer programs, and the use of prediction models beyond the range of their validity. This source of error is usually not considered in error analysis. However, it can be controlled using, for example, checking procedures, simplified algorithms, and simplified human–machine interfaces, such as control panels for equipment or input formats for computer programs.

Uncertainty in information can be attributed to two main sources: objective and subjective. The objective source includes inherent variability in some engineering parameters and the use of limited information; for example, a sample is used to draw conclusions about a problem. This source of uncertainty can be dealt with using probability and statistics, and it is not numerically based. The subjective uncertainty in information is encountered in cases where an engineer has to make a subjective assessment—for example, the condition of a corroded steel beam. In this case, the assessment might vary from one engineer to another, resulting in subjective uncertainty in information.

Numerical errors include (1) round-off errors, (2) truncation errors, (3) propagation errors, and (4) mathematical-approximation errors. Round-off errors were discussed in Section 3.3. They are due to expressing some quantities in a limited number of significant digits, whereas the true values have a large number of significant digits. Round-off errors can be computed as the difference between a true value and its rounded value.

In numerical analysis, it is common to deal with mathematical expressions that consist of the addition of many terms, such as the first term of a Taylor series or the sum of the first five terms of an infinite series. The numerical contribution of these terms is often of diminishing value as more terms are added to the expression. To keep a computational effort to a reasonable level, it is common to truncate such mathematical expressions to a certain number of terms. The truncated terms contribute to the total error. However, terms that are truncated are considered to contribute insignificantly to error in the final estimate of the model. The difference between the evaluation of the complete (nontruncated) expression and the truncated expression is called the truncation error. An example of these mathematical expressions is a mathematical infinite series (e.g., the Taylor series). The Taylor series, which was discussed in Chapter 1, contains an infinite number of terms. The first term is called the zero-order term. The second term is the first-order term because it is based on the first derivative. The third term uses the second derivative and is called the second-order term. The additional terms are referred to similarly. It is common to truncate the Taylor series to the first-order terms or, sometimes, to the second-order terms. Therefore, truncation errors are introduced. The effects of these errors can be significant unless the numerical algorithm is properly formulated.

In Equations 3.11 to 3.13, it was noted that an arithmetic operation such as multiplication of two numbers that are expressed to three significant digits results in a product that can only be expressed to three significant digits. The product is uncertain to any larger number of significant digits. The product expressed to three significant digits in Equations 3.11 to 3.13 is 106, but the true value is expected to lie between 105.04375 and 106.51875. Therefore, arithmetic operations result in a propagation of errors. Sometimes, by establishing a computational order for the arithmetic operations in an algorithm, the propagation errors can be reduced. For example, in evaluating the summation of several numerical values using a specified number of significant digits, it is desirable to rank order the numerical values in ascending order before performing the summation. The result of this summation can therefore have a reduced propagation error as compared to an arbitrary order for performing the summation.

Mathematical-approximation errors result from using approximate mathematical expressions for a relationship or solution. For example, numerical methods sometimes call for the use of a linear model for representing a nonlinear mathematical expression. The difference in prediction between the nonlinear and linear models is an error of this type.

In this chapter, only numerical errors are of interest and need to be evaluated because their assessment provides a means of determining when a numerical iterative solution should be terminated. Also, the understanding of the errors in numerical methods can assist an engineer in selecting appropriate numerical methods for solving a problem.

3.4.2 MEASUREMENT AND TRUNCATION ERRORS

When dealing with engineering measurements, the accuracy and precision of an estimate may be just as important as the estimate itself. We can define the error (e) as the difference between the computed (x_c) and true (x_t) values of a number x:

(3.14)
$$e = x_c - x_t$$

A relative true error (e_r) is defined as the error e relative to the true value:

(3.15)
$$e_r = \frac{x_c - x_t}{x_t} = \frac{e}{x_t}$$

The relative true error could be expressed as a percentage by multiplying e_r of Equation 3.15 by 100.

To illustrate the concept of error analysis, consider the atomic weight of oxygen, which is 15.9994 and is believed to be accurate to ±0.0001. The relative error would then be

(3.16)
$$e_r = \frac{0.0001}{15.9994} = \pm\,0.6 \times 10^{-5}$$

Even though the atomic weight is expressed to six significant digits, the relative error is only expressed to one significant digit because the error is only given to one significant digit.

Example 3.3: Truncation Error in Atomic Weight

In addition to measurement errors, such as the measurement of atomic weights, engineers may also be concerned with errors due to truncation. If we round the atomic weight of oxygen to 16, which is common, we introduce a truncation error, which can be computed using Equation 3.14 as

(3.17)
$$e = 16 - 15.9994 = 0.0006$$

Relative to the true value, this results in a relative error, which is computed with Equation 3.15 as

(3.18)
$$e_r = \frac{0.0006}{15.9994} = 0.4 \times 10^{-4}$$

Thus the relative error due to truncation (Equation 3.18) is seven times greater than the relative error due to measurement (Equation 3.16).

3.4.3 ERROR ANALYSIS IN NUMERICAL SOLUTIONS

As indicated in Section 3.1, numerical methods provide an approximation to a problem's solution, rather than the true value. Furthermore, numerical methods are iterative. Based on these two characteristics of numerical methods, it would be of interest to have an estimate of the error with each iteration so that the computations could be discontinued when sufficient accuracy has been achieved. The purpose of a tolerance in Figure 3.1 was to provide a means of exiting from the loop where the error is within some user-specified tolerance.

In practice, the true value is not known, so Equations 3.14 and 3.15 cannot be used to compute the error. In such cases, the best estimate of the number x should be used. Unfortunately, the best estimate is the computed estimate, so Equations 3.14 and 3.15 lose their meaning when directly applied to the accuracy of solutions from numerical methods. Instead, we will apply Equation 3.14 iteratively and thus it is rewritten as

(3.19)
$$e_i = x_i - x_t$$

where e_i is the error in x at iteration i, and x_i is the computed value of x from iteration i; similarly, the error for iteration $(i + 1)$ is

(3.20)
$$e_{i+1} = x_{i+1} - x_i$$

Therefore, the change in the error (called simply the relative error) can be computed from Equations 3.19 and 3.20 as

(3.21)
$$\Delta e_i = e_{i+1} - e_i = (x_{i+1} - x_t) - (x_i - x_t) = x_{i+1} - x_i$$

It can be shown that e_{i+1} is expected to be smaller than Δe_i, so if we continue to iterate until Δe is smaller than our tolerable error, we can be reasonably sure that x_{i+1} is sufficiently close to x_t.

To illustrate the calculation of the error using Equation 3.21, assume that we wish to find the value of x that makes the following third-order polynomial an identity:

(3.22)
$$x^3 + ax^2 + bx + c = 0$$

If we divide Equation 3.22 by x and solve for x using the x^2 term, we get

(3.23)
$$x = \sqrt{-ax - b - \frac{c}{x}}$$

We can find the positive value of x for Equation 3.22 by selecting an initial estimate of x and solving the right side of Equation 3.23 to provide a new estimate of x (i.e., the left side of Equation 3.23). This new estimate can be inserted into the right side of Equation 3.23 to find a better estimate. The iterations are continued until the change, Δe, which is given by Equation 3.21, is less than some stated tolerance.

Example 3.4: Numerical Errors Analysis

The following polynomial is used to illustrate the numerical solution according to Equation 3.23:

(3.24)
$$x^3 - 3x^2 - 6x + 8 = 0$$

Equation 3.24 can be expressed in the following form:

(3.25)
$$x = \sqrt{3x + 6 - \frac{8}{x}}$$

If we choose $x_0 = 2$ as the initial estimate of the positive value of x, Equation 3.25 becomes

(3.26)
$$x_1 = \sqrt{3x_0 + 6 - \frac{8}{x_0}} = 2.828427$$

Thus Equation 3.21 yields an error of

(3.27)
$$e_1 = x_1 - x_0 = 0.828427$$

Using the value of x_1 from Equation 3.26 as input to Equation 3.25 yields a revised estimate of x_2 of 3.414214, which has an error from Equation 3.21 of $e_2 = 0.585786$. The results of 10 iterations are given in Table 3.3. It is evident that the solution approaches $x = 4$, and substitution of $x = 4$ into Equation 3.24 shows that it is a valid solution. On trial 1, the change Δe is actually smaller than the true error, which is also shown in Table 3.3, but after trial 2, Δe_i is always larger than or equal to the true error. If the tolerable error were set at 0.025, the iterative solution would be stopped after trial 7; for trial 6, the change $\Delta e = 0.030$ was larger than the tolerance and therefore one more iteration was necessary. However, the estimate on trial 6 is actually within the stated tolerance with respect to the true value, but since the true value was not known, trial 7 was necessary. Ten trials were necessary to achieve a level of accuracy within a tolerance of 0.002.

TABLE 3.3   Error Analysis of the Solution of the Polynomial of Equation 3.24 with $x_t = 4$

Trial i	x_i	Δe_i	$x_i - x_i$
0	2.000000	–	2.000000
1	2.828427	0.828427	1.171573
2	3.414214	0.585786	0.585786
3	3.728203	0.313989	0.271797
4	3.877989	0.149787	0.122011
5	3.946016	0.068027	0.053984
6	3.976265	0.030249	0.023735
7	3.989594	0.013328	0.010406
8	3.995443	0.005849	0.004557
9	3.998005	0.002563	0.001995
10	3.999127	0.001122	0.000873

In numerical analysis, the number of significant digits that should be used in computation is a common question. Even though the numbers are given in Equation 3.24 to only one significant digit, the values are reported to seven digits in Table 3.3; this occurs only because these were the values stored in the hand calculator used in developing Table 3.3. If fewer digits were used, the solution to this problem would have converged in the same number of trials. In practice, truncating the values after each iteration is not necessary if the values can be stored to a large number of digits. It is only necessary to carry the number of digits that would be needed to check the tolerable error for the accuracy considered desirable.

3.5 ADVANTAGES AND DISADVANTAGES OF NUMERICAL METHODS

As indicated in the introduction to this chapter, numerical methods are an alternative to analytical solutions. When it is possible, analytical solutions are usually preferred. With analytical problem solving, the solution is found directly, rather than iteratively as with numerical methods; but it is often not possible to obtain an analytical solution, and a numerical method of solution is necessary. Analytical methods of problem solving yield an exact solution, if one exists; a numerical solution only yields an approximation, although if a sufficient number of iterations are performed, there may be no difference between the numerical solution and the true value. Since analytical methods yield the solution directly, the computation time required is usually less than the time required for a numerical solution.

In addition to advantages, analytical problem solving has disadvantages. First, the solution procedures are only valid for simple problems and cannot be used for many types of problems that engineers need to solve. Second, with many engineering problems, the solution is subject to a number of constraints. In general, it is usually difficult to solve a problem with an analytical method when the problem involves constraints; in such cases, numerical methods have the advantage because constraints are relatively easy to handle.

In summary, numerical methods have the following advantages: (1) They are conceptually simple, (2) they are easily programmed on a digital computer, (3) constraints are easily handled, and (4) solutions are possible for complex problems. However, numerical methods also have disadvantages, including that (1) they are solved iteratively and thus more computation effort may be required, (2) an exact solution may not be found, and (3) initial estimates of the solution are often required.

3.6 APPLICATIONS

3.6.1 PIPE DESIGN

Consider the case shown in Figure 3.3, where a long pipe connects two reservoirs, with a pump located in the system to pump the water from the low-elevation reservoir to the high-elevation reservoir. The reservoirs are open to the atmosphere, and the water surfaces are denoted as points 1 and 3. The location of the pump is denoted as point 2. The elevation of any point i in the system above a datum is indicated as Z_i; thus Z_1 and Z_3 are the elevations of the water surfaces in the reservoirs, and Z_2 is the elevation of the pump.

The following equation, which is called the energy equation, governs the flow of water through the system:

$$(3.28) \qquad \frac{P_1}{\gamma} + \frac{V_1^2}{2g} + Z_1 - h_f + E_p = \frac{P_3}{\gamma} + \frac{V_3^2}{2g} + Z_3$$

in which p_i = pressure at point i, γ = specific weight of the fluid, V_i = velocity of the fluid at point i, h_f = friction losses, and E_p = energy head supplied by the pump. Each term in Equation 3.28 has dimensions of length (i.e.,

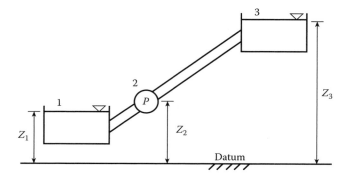

FIGURE 3.3 Schematic of a pipe system.

feet or meters). For the system of Figure 3.3, several assumptions can be made to simplify the energy equation of Equation 3.28. First, since both reservoirs are open to the atmosphere, $P_1 = P_3$, so the pressure head terms can be dropped. If both reservoirs have large surface areas, the surface levels will fall and rise very slowly, so V_1 and V_3 are essentially zero. If we let $h = Z_3 - Z_1$, then Equation 3.28 reduces to

(3.29)
$$E_p = h + h_f$$

Equation 3.29 can be interpreted as follows: the pump must supply sufficient energy E_p to overcome the energy lost due to frictional forces h_f and the energy needed to increase the potential energy by h. We will assume that most of the physical characteristics of the system are known; this includes h, the length of the pipe L, the roughness of the pipe e_f, the viscosity of the fluid v, and the specific weight of the fluid γ. Additionally, the flow rate Q through the pipe and the horsepower of the pump h_p are known. The horsepower input h_p is related to the energy head of the pump and the flow rate by

(3.30)
$$h_p = \frac{\gamma Q E_p}{550e}$$

in which e is the efficiency of the pump and motor. Friction losses in the system are related to the physical characteristics of the system and the velocity of the fluid in the pipe as

(3.31)
$$h_f = f \frac{L}{D}\left(\frac{V^2}{2g}\right)$$

in which D is the diameter of the pipe, f is the friction coefficient, V is the velocity of the flow in the pipe, and g is the gravitational acceleration. Based on the continuity equation $Q = AV$, where A is the cross-sectional area of the circular pipe, Equation 3.31 can be rewritten as

(3.32)
$$h_f = f \frac{L}{D}\left(\frac{Q^2}{2gA^2}\right) = \frac{8 fLQ^2}{\pi^2 gD^5}$$

Substituting Equations 3.32 and 3.30 into Equation 3.29 yields

(3.33)
$$\frac{550eh_p}{\gamma Q} = h + \frac{8 fLQ^2}{\pi^2 gD^5}$$

The goal of the problem is to find the diameter of the pipe that is necessary to pass flow rate Q for a specified pump. Equation 3.33 could be solved directly for D if the friction factor f were not a function of D. The relationship between f and D is nonlinearly related to the pipe roughness e_f and the Reynolds number, R, as

$$(3.34) \qquad\qquad R = \frac{VD}{v}$$

where v = fluid viscosity. To simplify the solution, R is assumed to be large so that f is independent of R but is a function of the ratio e_f/D (see Table 3.4). Thus the diameter D must be known to find f and f must be known to find D. Therefore an iterative solution is required. An initial estimate of D is assumed and used to find f from Table 3.4. Then the following equation, which is a rearrangement of Equation 3.33, is used to compute D:

$$(3.35) \qquad D = \left[\frac{8 f L Q^2}{\pi^2 g \left(\dfrac{550 h_p e}{\gamma Q} - h \right)} \right]^{0.2} = \left[\frac{8 f L \gamma Q^3}{\pi^2 g (550 h_p e - \gamma Q h)} \right]^{0.2}$$

To illustrate the design process, consider a flow Q of 12 cfs of water through a 3000 ft length (L) of cast iron pipe (with e_f = 0.00085 ft). A 100 hp pump (h_p) with an efficiency, e, of 65% is used to pump the water from one reservoir to another that has a surface elevation 22 ft higher (h). Using g = 32.2 ft/sec² and g = 62.4 lb/ft³, Equation 3.35 becomes

$$(3.36) \qquad\qquad D = 3.3508\, f^{0.2}$$

To begin the iterative process, a diameter of 1 ft will be assumed; thus e_f/D = 0.00085. From Table 3.4, the friction factor is 0.019; thus, from Equation 3.36, the diameter is 1.52 ft. This gives a relative roughness e_f/D of 0.00056, which yields f = 0.017. With Equation 3.36, we get a diameter of 1.48 ft. Another iteration yields e_f/D = 0.00057, f = 0.017, and D = 1.5 ft. Thus the engineer should specify an 18 in. pipe in the design plans.

TABLE 3.4   Relationship between Relative Roughness (e_f/D) and the Friction Factor (f) of a Pipe

e_f/D	f	e_f/D	f	e_f/D	f
0.05	0.072	0.01	0.038	0.0004	0.016
0.04	0.065	0.006	0.032	0.0002	0.0137
0.03	0.057	0.002	0.0235	0.0001	0.012
0.02	0.0485	0.001	0.0196	0.00005	0.0106
0.015	0.044	0.0008	0.0185	0.00001	0.00815

3.6.2 BENDING MOMENT FOR A BEAM

A simply supported beam AB has an overall span of 9 m. The beam is loaded with a variable load intensity over a portion of the span of 6 m in length, as shown in Figure 3.4a. The entire beam is considered to be a free body, and the following conditions of statics are used:

(3.37) $$\Sigma M_A = 0$$

(3.38) $$\Sigma F_Y = 0$$

The first condition states that the summation of moments about point A is zero, and the second condition states that the summation of forces along the y axis is zero. These two conditions can be used to obtain the reactions at the supports at A and B. The reaction R_B can be determined using Equation 3.37 as

$$-R_B(9) + 0.5(6)(20)\left(\frac{2}{3}\right)(6) = 0$$

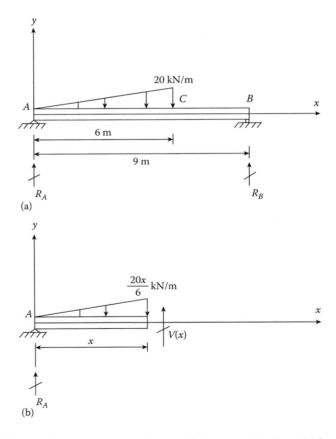

FIGURE 3.4   Simply supported beam: (a) beam and loading; (b) free-body diagram.

Therefore, R_B = 26.6667⁺ kN. The reaction R_A can be determined using Equation 3.38 as

$$R_A + R_B - (0.5)(6)(20) = 0$$

Therefore, R_A = 33.3333⁺ kN. In structural engineering, the location of the maximum bending moment is commonly of interest. The maximum-moment location corresponds to the point of zero shear force. Numerical methods can be used to determine this point along the span of the beam. Although an analytical expression can be evaluated in this simple case, a numerical method is used to illustrate its use. For a more complex loading condition, the numerical method can still be used, whereas an analytical solution might not be possible.

The numerical solution starts by assuming a value for the location of the maximum bending moment—that is, zero shear force (V) location, say x_0. Using the condition of statics of Equation 3.38 for the free-body diagram shown in Figure 3.4b, the shear force $V(x_0)$ is

$$(3.39) \qquad V(x_0) = R_A - \frac{1}{2}x_0\frac{x_0}{6}(20)$$

The shear force $V(x_0)$ based on this initial estimate of x_0 might not be zero. Then a new value for the location (distance), x_{i+1}, can be computed based on x_i (with a starting value of x_0) by adjusting the original distance using the following adjustment (Δx):

$$(3.40) \qquad \Delta x = \frac{R_A - \frac{1}{2}x_i\frac{x_i}{6}(20)}{R_A}$$

Therefore, the new distance is

$$(3.41) \qquad x_{i+1} = x_i + \Delta x = x_i + \frac{R_A - \frac{1}{2}x_i\frac{x_i}{6}(20)}{R_A}$$

The resulting new value from Equation 3.41 (i.e., x_{i+1}), should be used again in Equation 3.41 as an x_i value to obtain a better estimate of the distance. This process should be continued until some tolerance level (error level) is achieved. Starting with, for example, x_0 = 4.5 m, the resulting values for x_{i+1} are shown in Table 3.5.

The exact solution for the maximum-moment location can be computed analytically based on the condition that $V(x_0)$ from Equation 3.39 equals zero at the location of maximum moment. Therefore, the exact or true distance (x_t) can be determined by solving for the distance from $V(x_0) = 0$, which is

$$(3.42) \qquad x_t = \sqrt{2R_A\frac{6}{20}} = \sqrt{20} = 4.4721359 \text{ m}$$

TABLE 3.5   Iterations with an Initial Value of 4.5 m

Iteration i	x_i	Error (%)
0	4.5	–
1	4.4875	0.34355
2	4.480617	0.189646
3	4.476821	0.104753
4	4.474725	0.057882
5	4.473567	0.031989
6	4.472927	0.017681
7	4.472573	0.009773
8	4.472378	0.005402
9	4.472269	0.002986
10	4.47221	0.001651
11	4.472177	0.000912
12	4.472159	0.000504
13	4.472148	0.000279
14	4.472143	0.000154
15	4.47214	8.52E-05

where $R_A = 33.3333^+$ kN. The percent of error is also shown in Table 3.5. The effect of the initial value x_0 on the convergence of the estimation process is shown in Figure 3.5. The effect of the initial value on the percent of error is also shown in Figure 3.6. In this case, it is evident from these two figures that the process converges to the true value (x_i) regardless of the initial value (x_0). However, the efficiency of the process, which is defined as the rate of convergence, increases as the initial value approaches the true value.

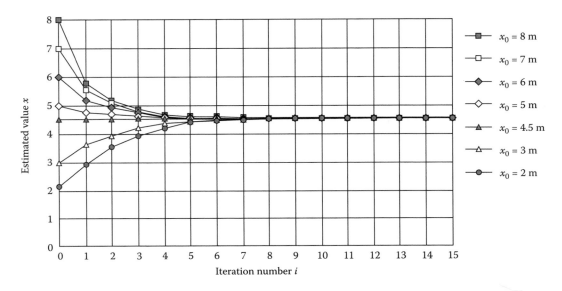

FIGURE 3.5   Effect of initial value x_0 on estimated value and convergence.

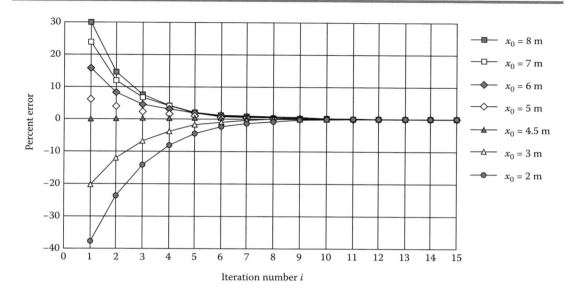

FIGURE 3.6   Effect of initial value x_0 on error.

PROBLEMS

3.1 Formulate a numerical solution to find the cube root of a number x. Using an initial value of $x_0 = 3$, estimate the cube root of 31 with an accuracy of 0.000005.

3.2 Formulate a numerical solution to find the fourth root of a number X. Using an initial value of $X_0 = 4$, estimate the fourth root of 52 with an accuracy of 0.0001.

3.3 Example 3.1 provides the numerical adjustment factor Δx for the square root (see Equation 3.5). Problems 3.1 and 3.2 require adjustment factors for the cube and fourth root of a real number. Based on these adjustment factors and without doing the development, extend the results to the case of finding the fifth root. Show that the algorithm works by finding the fifth root of 94 using $X_0 = 3$ as an initial estimate.

3.4 A carnival employee has a booth where she estimates the weights of attendees. They get a prize if she is incorrect by more than 10 pounds. The following are 10 of the employee's guesses:

Guess	127	183	98	154	113	222	67	142	122	165
True value	134	196	101	147	121	219	75	154	127	176

Provide a definition that the carnival owner could use to assess whether or not the employee provides accurate estimates of the weight of the attendees. Does the employee provide biased estimates? Assess the employee's prediction accuracy.

3.5 A chemical engineer has a model that predicts the temperature change in an experiment. The engineer conducts five experiments, measures the actual change, and compares the model-predicted values to the actual. The following data are provided:

Actual	47.2	53.3	64.4	58.7	43.8
Predicted	45.8	51.2	61.3	58.3	42.8

Assess the accuracy and bias of the prediction model.

3.6 An environmental engineer creates a mixture that has a concentration of exactly 50 mg/L of a pollutant over a period of several months. Eight samples are sent to each of three laboratories for assessment, with the following results:

Lab 1	48.6	52.1	55.6	46.2	53.1	46.8	49.7	50.6
Lab 2	51.3	50.9	52.1	51.6	50.6	51.0	51.3	50.2
Lab 3	43.7	49.2	44.6	45.8	50.2	48.5	42.3	45.0

Compare the performance of the laboratories in terms of bias, precision, and accuracy.

3.7 The bias is the average error; predicted values can be adjusted for systematic variation by algebraically adding the mean bias to each predicted value. Adjust the predicted values in Problem 3.5 to remove the bias and then recompute the average error.

3.8 Compute the error for each of estimated weights in Problem 3.4. Then compute the bias. Adjust each of the estimated weights by the bias when the bias is rounded to the nearest integer pound. What is the bias in the adjusted weights?

3.9 If g is given to three significant digits and the height to two digits, find the accuracy of the terminal velocity V, where V is given by $V = \sqrt{2gh}$. If $g = 32.2$ ft/sec^2 and $h = 9.7$ ft, indicate the value of V and its expected accuracy.

3.10 The volume (V) of a block is given by $V = Lwh$. If L, w, and h are given to four significant digits, what is the accuracy of V? Find V and its accuracy if $L = 3.217$ m, $w = 0.7924$ m, and $h = 1.302$ m.

3.11 The surface area A of a right cylinder is ($2\pi r^2 + 2\pi rh$), where r and h are the radius and height of the cylinder, respectively. If $r = 3.172$ in. and $h = 7.41$ in., find A and its accuracy.

3.12 The area of an ellipse is given by $A = \pi ab$, where a and b are the half widths of the major and minor axes, respectively. If a and b are recorded to three significant digits, what is the accuracy of A? Find A and its accuracy if $a = 26$ cm and $b = 132$ mm. How many digits should be used with π?

3.13 The volume of a cone is $V = \pi r^2 h/3$ where r is the radius of the top circular portion of the cone and h is the height of the cone. If r and h are recorded to four significant digits, what is the accuracy of V? Find the volume of an ice cream cone V and its accuracy if $r = 7$ cm and the height is 15 cm. How many digits should be used with π?

3.14   An engineer is studying the effect of using fertilizer on crop yield. An experiment on eight fields, with the amount of fertilizer (X) varied, results in the following crop yields (Y) of the fields:

Field	1	2	3	4	5	6	7	8
X	0	0.5	1.4	2.9	4.1	6.2	7.8	9.3
Y	6.1	7.9	12.3	12.5	16.2	17.5	16.8	17.7

The engineer plots Y versus X, approximates a straight line through the points, and computes the following equation from the line: $\hat{Y} = 6 + 2X$. A colleague uses a statistical method for fitting a linear equation that results in the following model: $\hat{Y} = 8.7 + 1.16X$. Make two plots of the data, graph one equation on each plot, and discuss the representativeness of the two equations in terms of bias, precision, and accuracy. Use these assessments to discuss modeling error.

3.15   The data shown were generated from the power relation $y = 2.1X^{0.75}$, which was unknown to the two scientists. One scientist assumed the data could be represented by the linear function $y = 1 + 1.2X$. A second scientist used the data to fit a power function $y = 1.9X^{0.82}$. Discuss the bias and accuracy of the two models.

X	1	2	3	4	5
Y	2.10	3.53	4.79	5.94	7.02

3.16   The data shown were generated from the exponential relation $y = 2.3\,e^{-0.6X}$, which is unknown to two scientists. One scientist assumed that the data could be represented by the linear model $y = 2 - 0.5X$. A second scientist used the data to fit the exponential model $y = 2.55\,e^{-0.54X}$. Discuss the bias and accuracy of the two models:

X	0.5	1.0	1.5	2.0	2.5	3.0
Y	1.70	1.26	0.935	0.693	0.513	0.380

3.17   Compute the errors and relative errors for the two models and data in Problem 3.16. Also compute the average error, which is a mathematical estimate of the bias, and the standard deviation of the errors (use Equation 2.101 in Chapter 2), which is a measure of the accuracy. Compare the bias and accuracy of the two models.

3.18   For Example 3.4, find a valid solution to Equation 3.24 using the numerical procedure of Equation 3.25 and an initial estimate of $x_0 = -4$. Assume that convergence occurs when the change in the solution from one trial to the next is less than 0.05. What is the true error?

Roots of Equations

4.1 INTRODUCTION

The solution of many scientific and engineering problems requires finding the roots of equations that can be nonlinear. Any nonlinear equation can, after suitable algebraic manipulation, be expressed as a function of the following form:

(4.1)
$$f(x) = 0$$

The values of x that satisfy the condition of Equation 4.1 are termed the roots of the function; these roots are the solution to the equation $f(x) = 0$. Figure 4.1 shows graphically the roots for selected functions within an interval of interest $[A, B]$. Figure 4.1a shows a nonlinear function without roots since the function does not intersect the x axis. Figures 4.1b through d show functions with one, two, and three roots, respectively.

For very simple functions, the roots can often be found analytically. For example, for the general second-order polynomial $ax^2 + bx + c = 0$, the roots x_1 and x_2 can be found analytically using the following equation:

(4.2)
$$x_1 = \frac{-b + \sqrt{b^2 - 4ac}}{2a}$$

(4.3)
$$x_2 = \frac{-b - \sqrt{b^2 - 4ac}}{2a}$$

The roots for a third-order polynomial can also be found analytically. However, a general solution does not exist for other higher order polynomials.

It is more difficult to obtain analytical solutions for the roots of non-polynomial, nonlinear functions. For example, the function $(x/2)^2 - \sin(x) = 0$ cannot be solved analytically. To obtain the roots of this function, some type of numerical method must be employed. This, in general, requires a large number of calculations, particularly if the roots are to be determined to a high degree of precision. Thus, the problem is well suited to numerical analysis.

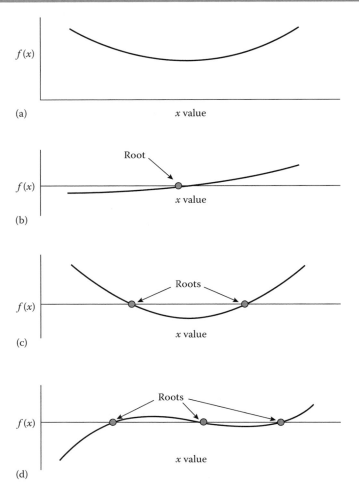

FIGURE 4.1   Roots for selected functions: (a) no roots, (b) one root, (c) two roots, and (d) three roots.

4.2 EIGENVALUE ANALYSIS

A large number of engineering problems require the determination of a set of values called *eigenvalues* or *characteristic values*. These are values, usually denoted as λ, for which the following matrix system has a nonzero (i.e., nontrivial) solution vector **X**:

(4.4) $$[\mathbf{A} - \lambda\mathbf{I}]\,\mathbf{X} = 0$$

The matrices **A** and **I** are of the order $n \times n$, with **I** being an identity diagonal matrix. The parameter λ is called the eigenvalue. The electrical engineer, for example, uses eigenvalue analysis in the solution of two-terminal-pair networks and in the optimization of adjustments of a control system. The chemical engineer uses eigenvalue analysis in the design of reactor systems. The engineering geologist must find eigenvalues in the analysis of sediment deposits in river beds. And the aeronautical engineer applies eigenvalue analyses in analyzing the flutter of an airplane wing. In each of these examples, the problem solution sets up the matrix **A** and seeks to find the solution of the following determinant equation:

(4.5) $$[\mathbf{A} - \lambda\mathbf{I}] = 0$$

where the vertical bars in the expression $|\mathbf{A} - \lambda\mathbf{I}|$ indicate the determinant of the matrix $[\mathbf{A} - \lambda\mathbf{I}]$. The determinant of Equation 4.5 can be expanded into a polynomial called the characteristic equation, with the roots of the *characteristic equation* being the eigenvalues.

Consider the case where **X** is a third-order vector and **A** is a third-order matrix. Equation 4.5 then becomes

(4.6) $$|\mathbf{A} - \lambda\mathbf{I}| = \begin{vmatrix} a_{11} - \lambda & a_{12} & a_{13} \\ a_{21} & a_{22} - \lambda & a_{23} \\ a_{31} & a_{32} & a_{33} - \lambda \end{vmatrix} = 0$$

Expanding this determinant would provide a third-order polynomial (characteristic equation) having the form

(4.7) $$(a_{11} - \lambda)[(a_{22} - \lambda)(a_{33} - \lambda) - a_{23}a_{32}] - a_{12}[a_{21}(a_{33} - \lambda) - a_{23}a_{31}] + a_{13}[a_{21}a_{32} - (a_{22} - \lambda)a_{31}] = 0$$

or

(4.8) $$\lambda^3 + b_2\lambda^2 + b_1\lambda + b_0 = 0$$

where b_0, b_1, and b_2 are a function of the elements of **A**, which are denoted as a_{ij}. The solution (i.e., roots) of the characteristic equation would provide three eigenvalues.

Example 4.1: Soil Properties

If the solution of eigenvalue problems were the only reason, it would still be worthwhile knowing a numerical solution procedure for nonlinear equations; however, other engineering problems also require such a solution. Consider the case of an engineer interested in identifying independent factors that affect sediment yield from agricultural lands. The eigenvalues of a matrix of correlation coefficients can be used to evaluate the degree of dependency between measured variables. Consider the simple case where the engineer measures a soil particle size index (x_1), the slope (x_2), and the mean annual sediment loss (y) from 50 agricultural fields, and the 50 sets of data yield the following correlation matrix:

(4.9)

	x_1	x_2	y
x_1	1.00	−0.42	−0.61
x_2	−0.42	1.00	0.37
y	−0.61	0.37	1.00

Denoting the correlation matrix as $\mathbf{A}$, the determinant $|\mathbf{A} - \lambda\mathbf{I}|$ becomes

$$(4.10) \qquad \begin{vmatrix} 1-\lambda & -0.42 & -0.61 \\ -0.42 & 1-\lambda & 0.37 \\ -0.61 & 0.37 & 1-\lambda \end{vmatrix}$$

Expanding the determinant yields the polynomial

$$(4.11) \qquad (1 - \lambda)[(1 - \lambda)^2 - 0.37^2] + 0.42[-0.42(1-\lambda) - 0.37(-0.61)] \\ - 0.61[-0.42(0.37) + 0.61(1 - \lambda)]$$

or

$$(4.12) \qquad \lambda^3 - 3\lambda^2 + 2.3146\lambda - 0.504188$$

The eigenvalues are found by setting the polynomial to zero and solving for the three roots. The roots (or λ's) provide useful information about the dependency of soil particle size index, the slope, and the sediment yield.

The correlation matrix of Equation 4.9 is a (3 × 3) matrix, which produced a third-order polynomial. Correlation matrices for engineering data are often (10 × 10) and larger, which produce high-order polynomials. Unlike second-order polynomials that can be solved using the quadratic formula of Equations 4.2 and 4.3, such high-order polynomials cannot be solved analytically. It can be concluded from this discussion that an alternative to analytical solutions is required to find the roots of high-order polynomials that represent important engineering systems.

4.3 DIRECT-SEARCH METHOD

One obvious method for finding the roots of a general nonlinear function is a trial-and-error approach in which the function $f(x)$ is evaluated at many points over some range of x until $f(x)$ equals zero; that is, Equation 4.1 is satisfied. Of course, we would have to be very lucky to guess the exact root of the function using this method. Even if the trial-and-error search is confined to a relatively small range for x, there are an infinite number of trial values for x to be considered. Fortunately, most problems do not require an *exact* value for the root; instead, an *estimate* of the root (to within some specified degree of *precision*) is often sufficient.

The direct-search method is perhaps the clearest, but it also can be the least efficient technique for estimating the roots of functions. The steps in this method can be outlined as follows:

1. Specify a range or *interval* for x within which the root is assumed to occur. Some knowledge of the behavior of the function is required to specify this interval. Smaller initial intervals will require fewer calculations to obtain the root to the desired accuracy.
2. Divide the interval into smaller, uniformly spaced subintervals. The size of these subintervals will be dictated by the required precision for the estimate of the root. For example, if the root must be estimated to within ±0.001, then the subintervals for the root will be 0.001 units long.
3. Search through all subintervals until the subinterval containing the root is located; this occurs when the equality of Equation 4.1 exists within the interval.

The last step is the key to the method. A simple test for determining if the root occurs within a given subinterval is illustrated in Figure 4.2. The function $f(x)$ is evaluated at the beginning and end points, A and B, respectively, of the subinterval. If a root does not lie within the subinterval, then the values of $f(A)$ and $f(B)$ will have the same sign—that is, either both negative or both positive (Figures 4.2a and b). However, if a root does lie within the subinterval, then the values of $f(A)$ and $f(B)$ will have different signs (Figures 4.2c and d). Alternatively stated, if the product of $f(A)$ and $f(B)$ is positive, then the function does not likely cross the x axis and a root does not lie within the subinterval; if the product is negative, then the function crosses the x axis and a root lies within the subinterval.

A FORTRAN program listing for the direct-search method for finding roots of functions is given in Figure 4.3. This program will find all roots of $f(x)$ within the interval $A \leq x \leq B$ as long as the input precision value (that is, tolerance) TOL is sufficiently small that multiple roots will not occur within a single subinterval. Note that $f(x)$ is evaluated in a function subprogram; thus the program can be used to find the roots of any function simply be substituting the appropriate function subprogram for $f(x)$. In Figure 4.3, the function $f(x) = (x/2)^2 - \sin(x)$ is used in the FUNCTION F(X) subprogram for the purpose of illustration.

The direct-search method will find the roots of any function as long as all the roots are real and within the specified interval. However, it should be obvious that this method is very cumbersome. For a high degree of precision, the subinterval size must be very small and a large number of calculations must be performed. To minimize the number of calculations, the subinterval size must be increased, but this obviously reduces the precision of the estimate roots. In addition, some roots might be missed entirely if the subinterval size is so large that more than one root

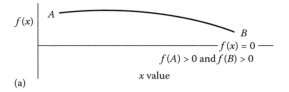

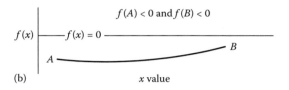

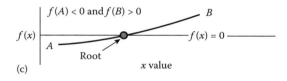

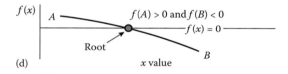

FIGURE 4.2   Direct-search method: (a) no roots in the interval [A, B], (b) no roots in the interval [A, B], (c) one root in the interval [A, B], and (d) one root in the interval [A, B].

occurs in a single subinterval. Nevertheless, the direct-search method is the most straightforward technique for determining all roots within a given interval.

The direct-search method assumes that there is one and only one root within each subinterval. If there is an even number of roots within the search interval, then $f(A)$ and $f(B)$ will have the same sign, and the search process will miss the roots within the interval. For example, given the function

(4.13) $f(x) = (x - 1.1)(x - 1.2)(x + 0.8) = x^3 - 1.52x - 0.52x + 1.056$

the two positive roots will not be found if the search interval is 0.25, with $A = 1.0$ and $B = 1.25$. In this case, $f(A) = 0.036$ and $f(B) = 0.0154$,

```
        PROGRAM ROOTS
c       Computes roots of a function using direct-search method
c       Read input data
        PRINT *, 'ENTER STARTING POINT FOR SEARCH INTERVAL:'
        READ *, A
        PRINT *, 'ENTER ENDING POINT FOR SEARCH INTERVAL:'
        READ *, B
        PRINT *, 'ENTER REQUIRED PRECISION:'
        READ *, PREC
c       Compute number and size of subintervals
        NINTVL = INT((B-A)/PREC) + 1
        DELTAX = (B-A)/REAL(NINTVL)
c       Initialize subinterval end points
        XB = A
        XE = XB + DELTAX
c       Loop over all subintervals
        DO 10 N = 1,NINTVL
            FXB = F(XB)
            FXE = F(XE)
            IF (FXB*FXE.LE.0.0) THEN
                PRINT 100, XB, XE
100             FORMAT('A ROOT LIES BETWEEN X=', E12.4,'AND', E12.4)
            END IF
            XB = XE
            XE = XB + DELTAX
10      CONTINUE
        END
            FUNCTION F(X)
c       Evaluates function at specified point
            F = (X/2.0)**2 - SIN(X)
        RETURN
        END
```

FIGURE 4.3   Program listing for finding the roots of a function using the direct-search method.

and so the search will proceed to the next search interval and miss the roots $X = 1.1$ and $X = 1.2$. This is especially critical where the function has two equal roots (called *multiple roots*), because the direct-search method will not identify any roots regardless of the size of the search subinterval. The multiple-root case is illustrated graphically in Figure 4.4 as the tangent point to the x axis. This root cannot be determined by the direct-search method. An example function that has multiple roots (double roots) is

$$(4.14) \qquad f(x) = (x - 1)^2 = x^2 - 2x + 1$$

Another case that might result in the failure of the direct-search method in finding the roots is shown in Figure 4.5. In this case, the function has a discontinuity at a point within the interval $[A, B]$ of interest.

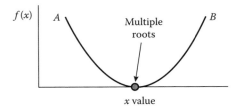

FIGURE 4.4   Multiple roots.

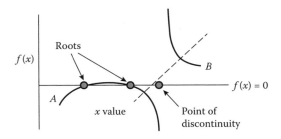

FIGURE 4.5   Point of discontinuity.

Example 4.2: Direct Search Solution of Eigenvalues

Consider the polynomial for Equation 4.11, which was the expansion of the determinant given in Section 4.1 as

$$\lambda^3 - 3\lambda^2 + 2.3146\lambda - 0.504188 = 0$$

The sum of the eigenvalues for a matrix with values of 1.0 on the principal diagonal equals the rank of the matrix, which equals the order of the characteristic polynomial; thus the three roots for this polynomial must be between 0.0 and 3.0, since all roots must be located in the range $0 \leq \lambda \leq 3$. We will assume for the purposes of illustration that an accuracy of 0.1 is sufficient. Thus the direct-search procedure can begin with a value of $\lambda = 0$ and continue in increments of 0.1 until estimates of all three roots are found. Table 4.1 gives the values of λ and $f(\lambda)$ using the direct-search method.

The smallest root occurs in the interval $0.3 < \lambda < 0.4$ since $f(0.3)$ is negative and $f(0.4)$ is positive. The second root occurs in the interval $0.6 < \lambda < 0.7$ since $f(0.6)$ is positive and $f(0.7)$ is negative. The largest root has a value between 1.9 and 2.0 since $f(1.9) < 0$ and $f(2.0) > 0$. We could assume roots of 0.4, 0.7, and 1.9, since these represent the values of λ where $f(\lambda)$ is closer to zero. Linear interpolation could be used to get better estimates, but we assumed that an accuracy of 0.1 was sufficient. If greater precision is required, the direct search method could be applied to each of the three relevant intervals.

TABLE 4.1   Results of Direct-Search Method

λ	f(λ)	λ	f(λ)	λ	f(λ)
0.0	−0.504	0.7	−0.011	1.4	−0.400
0.1	−0.302	0.8	−0.061	1.5	−0.407
0.2	−0.153	0.9	−0.122	1.6	−0.385
0.3	−0.053	1.0	−0.190	1.7	−0.326
0.4	0.006	1.1	−0.257	1.8	−0.226
0.5	0.199	1.2	−0.319	1.9	−0.077
0.6	0.021	1.3	−0.368	2.0	0.125

4.4 BISECTION METHOD

The bisection method is an extension of the direct-search method for cases when it is known that only one root occurs within a given interval of x. For the same level of precision, the bisection method will, in general, require fewer calculations than the direct-search method. Using Figure 4.6 as a reference, the steps in the bisection method can be outlined as follows:

1. For the interval of x from the starting point x_s to the end point x_e, locate the midpoint x_m at the center of the interval. These points correspond to the starting and ending points of the half-intervals.
2. At the starting point x_s, midpoint x_m, and the ending point x_e of the interval, evaluate the function resulting into $f(x_s)$, $f(x_m)$, and $f(x_e)$, respectively.
3. Compute the products of the functions evaluated at the ends of the two half-intervals, that is, $f(x_s)\,f(x_m)$ and $f(x_m)\,f(x_e)$. The root lies in the interval for which the product is negative, and the midpoint x_m is used as the estimate of the root for this iteration.
4. Check for convergence as follows:
   a. If the convergence criterion (that is, tolerance) is satisfied, then use x_m as the final estimate of the root.
   b. If the tolerance has not been met, specify the ends of the half-interval in which the root is located as the starting and ending points for a new interval and return to step 1.

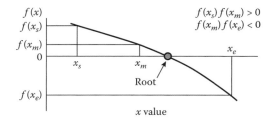

FIGURE 4.6   Schematic of bisection method.

The bisection method will always converge on the root, provided that only one root lies within the starting interval for x.

4.4.1 ERROR ANALYSIS AND CONVERGENCE CRITERION

Since numerical methods for finding the roots of functions are iterative, it is important to include a convergence criterion into the process of finding a root; otherwise, a computerized solution could continue to iterate indefinitely. Since the value of a root has significance to the engineering problem under consideration, the allowable error would depend on the tolerance permitted by the problem. Thus the absolute value between an estimate and the true value would be the ideal criterion to indicate convergence. But since the true value is not known, we need a surrogate value to represent the true value. In practice, the change in the estimate of the root from one trial to the next is used as the convergence criterion; this can be expressed as either the absolute value or as a percentage of the estimate. If the change between estimates on successive iterations is used as a convergence criterion, it is important to recognize that this criterion may not reflect the true accuracy of the estimate—that is, the absolute difference between the estimated value and the true value of the root.

To ensure closure of the iteration loop, a convergence criterion is needed to terminate the iterative procedure of a numerical method for finding the roots of a function. The convergence criterion used in step 4 of the bisection method can be expressed in terms of either the absolute value of the difference (ϵ_d) or the percent relative error (ϵ_r) for each iteration. The convergence criterion when the error is expressed as an absolute value is

$$(4.15) \qquad\qquad \epsilon_d = |x_{m,i+1} - x_{m,i}|$$

For the case when the convergence criterion is expressed as a relative percent error, the criterion is

$$(4.16) \qquad\qquad \epsilon_r = \left| \frac{x_{m,i+1} - x_{m,i}}{x_{m,i+1}} \right| 100$$

where $x_{m,i}$ = the midpoint in the previous root-search iteration (ith iteration), and $x_{m,i+1}$ = the midpoint in a new root-search iteration ($i + 1$ iteration). The tolerance used with the convergence criterion of Equation 4.15 can be specified depending on the accuracy that is needed for the specific problem. For some engineering problems, a tolerance of 5% according to Equation 4.16 is adequate; other problems may require a tolerance of 0.01%. The true accuracy of the solution at any iteration can be computed if the true solution (root x_t) is known. In such a case, the true error (ϵ_t) in the ith iteration is

$$(4.17) \qquad\qquad \epsilon_t = \left| \frac{x_t - x_{m,i}}{x_t} \right| 100$$

Equation 4.17 requires knowledge of the true solution x_t. In practical use of numerical methods, the true solution is not known. Therefore, an approximate solution of the error can be evaluated using the relative error x_r as given in Equation 4.16.

Example 4.3: Roots of a Polynomial Using the Bisection Method

The following polynomial is used to illustrate the use of the bisection method for finding its roots:

$$(4.18) \qquad f(x) = x^3 - x^2 - 10x - 8 = 0$$

We are interested in finding the roots within the interval $3.75 \leq x \leq 5.00$ to a relative accuracy as an absolute value between successive iterations of 0.01. Using the four steps of the bisection methods, the values of x_s, x_m, and x_e were computed for each iteration i. The results are summarized in Table 4.2. Note that the midpoint is used as the estimate of the root for each iteration. The absolute value of the error is given in the last column of Table 4.2. The error decreases with each iteration, and on the seventh iteration the error equals the tolerance, so convergence is assumed. The final estimate of the root is 3.944.

It is evident that the root is near 4. Substituting $x = 4$ into Equation 4.18 shows that the true value of the root is 4. Therefore, the true accuracy of the estimated root is 0.006 or, in relative terms according to Equation 4.17, 0.15%.

Example 4.4: Location of Maximum Bending Moment

In Section 3.6.2, the location of the maximum bending moment for a simply supported beam was determined based on the condition that the shear force at this location is zero. The exact location was also expressed based on the conditions of statics as the solution of the following equation.

$$(4.19) \qquad f(x) = 20 - x^2 = 0$$

TABLE 4.2   Polynomial Solved Using the Bisection Method for Example 4.3

Iteration i	x_s	x_m	x_e	$f(x_s)$	$f(x_m)$	$f(x_e)$	$f(x_s)f(x_m)$	$f(x_m)f(x_e)$	Error ϵ_d
							Subinterval Containing Root		
1	3.750	4.375	5.000	−6.830	12.850	42.000	−	+	−
2	3.750	4.062	4.375	−6.830	1.903	12.850	−	+	0.313
3	3.750	3.906	4.062	−6.830	−2.724	1.903	+	−	0.156
4	3.906	3.984	4.062	−2.724	−0.477	1.903	+	−	0.078
5	3.984	4.023	4.062	−0.477	0.696	1.903	−	+	0.039
6	3.984	4.004	4.023	−0.477	0.120	0.696	−	+	0.019
7	3.984	3.994	4.004	−0.477	−0.180	0.120	+	−	0.010

TABLE 4.3   Location of Maximum Bending Moment Using Bisection Method for Example 4.4

Iteration i	x_s	x_m	x_e	$f(x_s)$	$f(x_m)$	$f(x_e)$	Absolute Error (%)	Relative Error (%)
1	0	3	6	20	11	−16	–	32.917960
2	3	4.5	6	11	−0.25	−16	33.33333	0.623060
3	3	3.75	4.5	11	5.9375	−0.25	20.00000	16.147450
4	3.75	4.125	4.5	5.9375	2.984375	−0.25	9.09091	7.762196
5	4.125	4.3125	4.5	2.984375	1.402344	−0.25	4.34783	3.569568
6	4.3125	4.40625	4.5	1.402344	0.584961	−0.25	2.12766	1.473255
7	4.40625	4.453125	4.5	0.584961	0.169678	−0.25	1.05263	0.425098
8	4.453125	4.476563	4.5	0.169678	−0.03961	−0.25	0.52356	0.098980
9	4.453125	4.464844	4.476563	0.169678	0.06517	−0.03961	0.26247	0.163059
10	4.464844	4.470703	4.476563	0.06517	0.012814	−0.03961	0.13106	0.032039
11	4.470703	4.473633	4.476563	0.012814	−0.01339	−0.03961	0.06549	0.033470
12	4.470703	4.472168	4.473633	0.012814	−0.00029	−0.01339	0.03275	0.000720
13	4.470703	4.471436	4.472168	0.012814	0.006264	−0.00029	0.01638	0.015662

The location of the maximum bending moment must lie within the span of the uniform loan (i.e., $0 \leq x \leq 6$ m). Therefore, the starting interval for the bisection method is [0, 6]. The computations of the bisection method are shown in Table 4.3. The estimated root (x_m) and errors in percent are shown in the table, as well as in Figures 4.7 and 4.8, respectively. The two types of errors—the absolute error

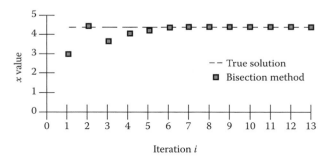

FIGURE 4.7   Location of maximum bending moment.

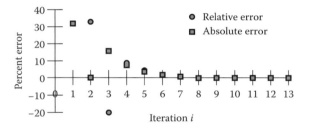

FIGURE 4.8   Errors in the location of maximum bending moment.

and relative error—are shown in Figure 4.8. The relative error in an iteration was computed using Equation 4.16. The actual error for an iteration was computed according to Equation 4.17 using the true root ($x_t = 4.4721359$).

4.5 NEWTON–RAPHSON ITERATION

Although the bisection method will always converge on the root, the rate of convergence is very slow. A faster method for converging on a single root of a function is the Newton–Raphson iteration method.

While the concept of a *Taylor series expansion* is formally introduced in Chapter 1, at this point it is sufficient to state that the equation that is used to develop the Newton–Raphson iteration method is the linear portion of a Taylor series:

$$(4.20) \qquad f(x_1) = f(x_0) + \frac{df}{dx} \Delta x$$

where $f(x_1)$ is the value of the function at a specific value (x_1) of the independent variable x, $f(x_0)$ is the value at x_0, Δx equals the difference ($x_1 - x_0$), and df/dx is the derivative of the relationship. Since the root of the function relating $f(x)$ and x is the value of x where $f(x_1)$ equals zero, Equation 4.20 can be expressed as

$$(4.21) \qquad 0 = f(x_0) + \frac{df}{dx}(x_1 - x_0)$$

where x_1 will now be the root of the function, since $f(x_1)$ was assumed to be zero. Solving Equation 4.21 for the unknown x_1 yields

$$(4.22) \qquad x_1 = x_0 - \frac{f(x_0)}{df/dx}$$

Equation 4.22 can be interpreted as follows: The new estimate of the root, x_1, is equal to the previous estimate x_0 minus the ratio of the function at x_0 to the value of the derivative at x_0. If the derivative df/dx is evaluated at x_0 and we denote this derivative as $f'(x_0)$, and then Equation 4.22 becomes

$$(4.23) \qquad x_1 = x_0 - \frac{f(x_0)}{f'(x_0)}$$

Thus Equation 4.23 provides a revised approximation of the root (x_1) using some initial estimate of the root (x_0) and the values of both the function $f(x_0)$ and the derivative of the function $f'(x_0)$, both evaluated at x_0. Equation 4.23 can be iteratively evaluated as

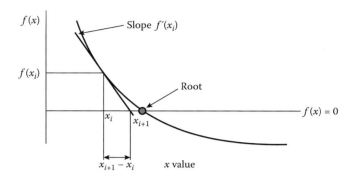

FIGURE 4.9   Newton–Raphson method.

$$(4.24) \qquad x_{i+1} = x_i - \frac{f(x_i)}{f'(x_i)}$$

If the function $f(x)$ is linear, then Equation 4.24 will provide an exact solution on the first trial; however, if $f(x)$ is nonlinear, then the linear Taylor series of Equation 4.20 is only valid over small ranges of Δx. The importance of the nonlinear terms that were truncated for Equation 4.20 will determine the accuracy of the solution of Equations 4.23 and 4.24. In practice, Equations 4.23 and 4.24 are applied iteratively, with the first trial based on an estimate of x_0 and each subsequent trial based on the most recent estimate of the root. The solution procedure is illustrated in Figure 4.9, where the function and independent variable are denoted as $f(x)$ and x, respectively. The subscript on x refers to the trial (or iteration) number, with x_0 being the initial estimate. It can be seen from Figure 4.9 that Equations 4.23 and 4.24 can be obtained by expressing the slope $f'(x_i)$ over the interval $[x_i, x_{i+1}]$ as

$$(4.25) \qquad f'(x_i) = \frac{f(x_i) - 0}{x_{i+1} - x_i} = \frac{f(x_i)}{x_{i+1} - x_i}$$

Therefore, Equations 4.23 and 4.24 can be obtained by rearranging the terms of Equation 4.25.

Example 4.5: Solution of a Third-Order Polynomial

As an example of Newton–Raphson iteration, consider the following third-order polynomial:

$$(4.26) \qquad f(x) = x^3 - x^2 - 10x - 8$$

The derivative of this function is

$$(4.27) \qquad f'(x) = 3x^2 - 2x - 10$$

If an initial estimate of the root of 6.0 is assumed, then the first trial estimate is

$$(4.28) \quad x_1 = 6 - \frac{(6)^3 - (6)^2 - 10(6) - 8}{3(6)^2 - 2(6) - 10} = 6 - 1.3023 = 4.6977$$

Using this value as the revised estimate of x_1, the second estimate would be obtained by

$$(4.29) \quad x^2 = 4.67977 - \frac{(4.6977)^3 - (4.6977)^2 - 10(4.6977) - 8}{3(4.6977)^2 - 2(4.6977) - 10}$$
$$= 4.6977 - 0.5688 = 4.1289$$

A third trial would yield an estimate of the root of 4.0057, and the fourth trial would yield a value of 4.0000. For precision to five significant digits, the value does not change with additional iterations.

Example 4.6: Error Analysis for the Newton–Raphson Method

For the example of Equation 4.26, the difference in absolute value between both the estimated value of the x_i and the true root (x_t = 4.000), and the estimated value for a trial (x_i) and the previous trial (x_{i-1}) are shown in Table 4.4.

For this case, the convergence criterion $x_i - x_{i-1}$ yields a conservative estimate of the true error. Thus it would be a useful convergence criterion for this case. For other problems, such as when the function $f(x)$ does not show large changes even for large changes in the estimate of the root, another criterion may be more appropriate to test for convergence. Regardless of the convergence criterion, the solution should be checked by substituting the calculated root into the function to ensure that the result is rational.

4.5.1 NONCONVERGENCE

The Newton–Raphson iteration usually converges to a root faster than does the bisection method. However, this increase in speed does not

TABLE 4.4   Error Analysis for Example 4.5 Using Newton–Raphson Method

Iteration i	x_i	Relative Error $x_i - x_{i-1}$	Absolute Error $x_i - x_t$
0	6.0000	–	2.0
1	4.6977	1.3023	0.6977
2	4.1289	0.5688	0.1289
3	4.0057	0.1232	0.0057
4	4.0000	0.0057	0.0000

come without liabilities; under certain circumstances, Newton–Raphson iteration may fail to converge. Thus, when programming the method, the program should be written with an iteration counter and the option for discontinuing the search procedure when some preselected number of iterations has been exceeded. This procedure ensures that excessive iteration does not happen.

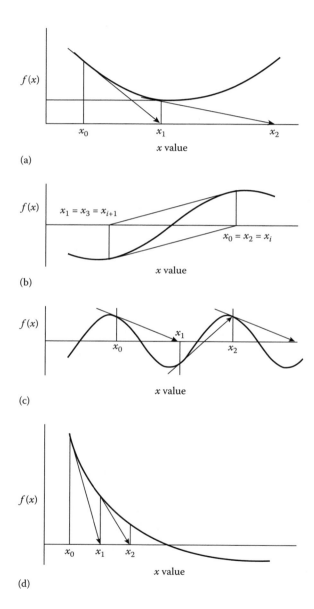

FIGURE 4.10 Nonconvergence of the Newton–Raphson method: (a) $f'(x_i)$ is approaching zero, (b) similarity of slopes, (c) excessive iteration because of the periodic function, and (d) excessive iteration because of the poor initial estimate.

While nonconvergence is not common, it can occur if the initial estimate is selected such that the derivative of the function equals zero. In such cases, $f'(x_i)$ of Equation 4.24 would be zero and $f(x_i)/f'(x_i)$ would equal infinity. Thus convergence would not be achieved and some error message indicating division by zero would be given. For example, a division by zero will occur if an initial estimate of $(2 \pm 124)/6$ were selected for the function for Equation 4.27. In this case, an iteration counter would not be necessary because program execution would be terminated by a division-by-zero error after the first trial. An example case of $f'(x_i) = 0$ is shown in Figure 4.10a.

A second instance of nonconvergence would occur if $f(x_i)/f'(x_i)$ equals $-f(x_{i+1})/f'(x_{i+1})$. In this case, x_i would equal x_{i+2} (where i is the iteration number) and x_{i+1} would equal x_{i+3}. Thus the solution would involve iterating between values of x_i and x_{i+1}. Graphically, this situation is shown in Figures 4.10b and c. The problem of excessive iterations can be prevented by including an iteration counter with a specified maximum number of iterations in the program as a criterion.

A large number of iterations will be required if the value of $f'(x_i)$ is much larger than $f(x_i)$. In such cases, $f(x_i)/f'(x_i)$ is small, which leads to a small adjustment at each iteration. This situation can occur, for example, when the root of a polynomial is near zero. In such cases, the numerator in the second term of Equations 4.23 and 4.24 will be much smaller than the denominator, and thus the adjustment $f(x_i)/f'(x_i)$ will be small. For such cases, it is important to obtain a good initial estimate of the root or allow for a large number of iterations before discontinuing the search. An example case of this excessive iteration is shown in Figure 4.10d.

Example 4.7: Solution for the Eigenvalue Problem

The eigenvalue problem presented with Equation 4.12 can be solved using the Newton–Raphson method. The eigenvalues are the roots of the polynomial. It is known from matrix theory that the sum of the eigenvalues equals the sum of the elements of the principal diagonal. For the correlation matrix, the sum would be 3; thus the sum of the eigenvalues will equal 3.0. Since all of the eigenvalues must be positive, it is reasonable to use a value of 2 as a reasonable initial estimate of λ. The derivative polynomial is

(4.30) $$f'(\lambda) = 3\lambda^2 - 6\lambda + 2.3146$$

Using Equations 4.23 and 4.24 with $\lambda_0 = 2$, the revised estimate of λ is given by

(4.31)
$$\lambda_1 = \lambda_0 - \frac{f(\lambda_0)}{f'(\lambda_0)} = 2 - \frac{(2)^3 - 3(2)^2 + 2.3146(2) - 0.504188}{3(2)^2 - 6(2) + 2.3146}$$
$$= 2 - \frac{0.125}{2.3146} = 1.94599$$

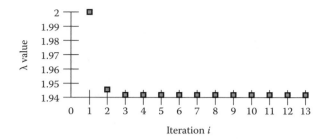

FIGURE 4.11   Eigenvalue solution for Example 4.7.

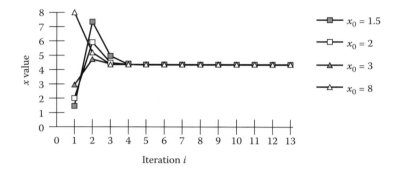

FIGURE 4.12   Location of maximum bending moment (Example 4.8).

The result of the first iteration can be expressed to four significant digits as 1.946. A second iteration yields a value of 1.942, and continued iteration does not result in any change. If $\lambda = 1.942$ is substituted into the third-order polynomial, it is evident that $f(\lambda) = 0$. This is shown in Figure 4.11.

Example 4.8: Location of Maximum Bending Moment

Using the function of Example 4.4, $f(x) = 20 - x^2$, the Newton–Raphson method is used to solve for the unknown roots. This example illustrates the effect of the initial value on the solution for this method. Using $x_0 = 3$, the method converges as shown in Figure 4.12 to the value 4.472136. Using $x_0 = 2$, the method also converges to the same answer. The figure shows other cases using other initial values. In Figure 4.12, the results are connected by line segments to facilitate tracking the progress of the method for different initial values.

4.6 SECANT METHOD

The secant method is similar to the Newton–Raphson method, with the difference that the derivative $f'(x)$ is numerically evaluated, rather than computed analytically. The algorithm for the secant method can be developed using the geometric similarity of two triangles as shown in

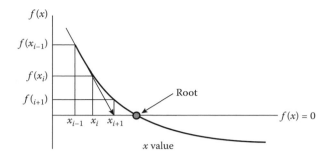

FIGURE 4.13 Secant method.

Figure 4.13. A new estimate of x_{i+1} of the root can be obtained using values of the function $f(x_i)$ and $f(x_{i-1})$ at two other estimates x_i and x_{i-1} of the root, as shown in Figure 4.13. A linear line segment between $[x_{i-1}, f(x_{i-1})]$ and $[x_i, f(x_i)]$ intersects the x axis at x_{i+1}. The following equality results from the triangle for the base $(x_{i+1} - x_{i-1})$ and the triangle for the base $(x_{i+1}, - x_i)$:

(4.32)
$$\frac{f(x_{i-1})}{x_{i+1} - x_{i-1}} = \frac{f(x_i)}{x_{i+1} - x_i}$$

Solving Equation 4.32 for x_{i+1} yields

(4.33)
$$x_{i+1} = \frac{x_{i-1} f(x_i) - x_i f(x_{i-1})}{f(x_i) - f(x_{i-1})}$$

or

(4.34)
$$x_{i+1} = x_i \frac{f(x_i)[x_{i-1} - x_i]}{f(x_{i-1}) - f(x_i)}$$

If the quantity in the denominator of the second term on the right side of Equation 4.34 is divided by $[x_{i-1} - x_i]$, it would be a numerical approximation of the derivative $f'(x)$. Thus Equation 4.34 is an approximation of Equation 4.20. The secant method is especially useful when it is difficult to compute an analytical expression for the derivative. It should be noted that the secant method requires two initial estimates of a root.

Example 4.9: Secant Method

The secant method can be used to determine the root of the following equation:

(4.35)
$$f(x) = \exp(-x) - x^2$$

TABLE 4.5   Secant Method for Example 4.9

Iteration i	$f(x_{i-1})$	x_i	Comments
0	1	0	Initial condition
1	−0.632120	1	Initial condition
2	0.166485	0.612700	
3	0.018994	0.693440	
4	−0.000710	0.703838	
5	2.81E−06	0.703466	
6	4.11E−10	0.703467	
7	−2.20E−16	0.703467	
8	0	0.703467	
9	0	0.703467	

Using initial estimates of $x_0 = 0$ and $x_1 = 1$, the next estimate of the root (x_2) can be determined using Equation 4.34 as

$$x^2 = 1 - \frac{f(1)[0-1]}{f(0)-f(1)} = 1 - \frac{[\exp(-1)-1^2][0-1]}{[\exp(0)-0^2]-[\exp(-1)-1^2]}$$

$$= 0.6127$$

Table 4.5 shows the results for several additional iterations. The method converges to a root of 0.703467.

4.7 POLYNOMIAL REDUCTION

After one root of a polynomial has been found, the process can be repeated using a new initial estimate. However, if proper consideration is not given to selecting the initial estimate of the second root, then application of, for example, the Newton–Raphson iteration may result in the same root being found. Thus there is a need to eliminate the possibility of finding the same root on the second application.

Polynomial reduction can be used to eliminate a root from a polynomial, which then eliminates the possibility of converging to a root that was found on previous applications. Polynomial reduction is essentially a process of dividing the polynomial by the root, which reduces the original polynomial to a new polynomial that does not have the original root as a root (unless, of course, the polynomial has a pair of equal roots—that is, multiple roots).

Formally, polynomial reduction states that if the polynomial $f(x)$ equals zero and root x_1 is a root of $f(x)$, then there is a reduced polynomial $f^*(x)$ such that $(x - x_1)f^*(x) = 0$, where $f^*(x) = f(x)/(x - x_1)$. Thus, if $f(x)$ is a polynomial of order n, then the reduced polynomial is a polynomial of order $n - 1$.

Example 4.10: Polynomial Reduction

For the example used to illustrate the Newton–Raphson method (see Example 4.5), one root of the polynomial of Equation 4.26 was found to be $x = 4$. This polynomial can be reduced as

(4.36)

$$
\begin{array}{r}
x^2 + 3x + 2 \\
x-4\overline{\smash{\big)}\,x^3 -\ \ x^2 -10x-8} \\
\underline{x^3 - 4x^2} \\
3x^2 - 10x \\
\underline{3x^2 - 12x} \\
2x - 8 \\
\underline{2x - 8} \\
0 = \text{error}
\end{array}
$$

Thus the reduced polynomial $x^2 + 3x + 2$ can be used to find additional roots of Equation 4.26. The reduced polynomial will equal zero for roots of the parent polynomial of Equation 4.26. The Newton–Raphson method can then be used to find a root of the reduced equation, and the polynomial can be reduced using polynomial reduction until all of the roots are found. It is important to note that any error in the value of x_1 will lead to error in the coefficients of the reduced equation and, thus, all roots found subsequently.

Example 4.11: Solution of the Eigenvalue Problem

Polynomial reduction can be used to reduce the eigenvalue polynomial of Equation 4.12:

(4.37)

$$
\begin{array}{r}
\lambda^2 - 1.058\lambda + 0.2600 \\
\lambda-1.942\overline{\smash{\big)}\,\lambda^3 -\ \ 3\lambda^2 + 2.3146\lambda - 0.504188} \\
\underline{\lambda^3 - 1.942\lambda^2} \\
-1.058\lambda^2 + 2.3146\lambda \\
\underline{-1.058\lambda^2 + 2.0546\lambda} \\
0.2600\lambda - 0.504188 \\
0.2600\lambda - 0.50492 \\
\hline
0.000732 = \text{error}
\end{array}
$$

Since the error is very small, $\lambda^2 - 1.058\lambda + 0.2600$ is the reduced polynomial. The roots of the reduced polynomial $\lambda^2 - 1.058\lambda + 0.2600$ are also the roots of the original third-order polynomial.

4.8 SYNTHETIC DIVISION

Polynomial reduction assumes that an estimate of the root is reasonably exact and that the aim is to reduce the polynomial. However, the concept underlying polynomial reduction can be used to find the value of a root. This method is called *synthetic division*.

Given the following nth-order polynomial $f_n(x)$, the polynomial can be reduced by dividing it by an initial estimate of the root x_0:

$$(4.38) \qquad f_n(x) = b_n x^n + b_{n-1} x^{n-1} + \dots + b_1 x^1 + b_0$$

This produces a polynomial of order $n - 1$, which will be denoted as $h_{n-1}(x)$:

$$(4.39) \qquad \frac{f_n(x)}{x - x_0} = h_{n-1}(x) + \frac{R_0}{x - x_0}$$

where R_0 is the remainder. Equation 4.39 would be rewritten as

$$(4.40) \qquad f_n(x) = (x - x_0) h_{n-1}(x) + R_0$$

The reduced polynomial $h_{n-1}(x)$ is given by

$$(4.41) \qquad h_{n-1}(x) = c_{n-1} x^{n-1} + c_{n-2} x^{n-2} + \dots + c_1 x + c_0$$

If $h_{n-1}(x)$ is also reduced using the estimate of the root, we get the following $(n - 2)$th-order polynomial, which will be denoted as $g_{n-2}(x)$:

$$(4.42) \qquad \frac{h_{n-1}(x)}{x - x_0} = g_{n-2}(x) + \frac{R_1}{x - x_0}$$

where R_1 is the remainder. The reduced polynomial $g_{n-2}(x)$ can be written as

$$(4.43) \qquad g_{n-2}(x) = d_{n-2} x^{n-2} + d_{n-3} x^{n-3} + \dots + d_1 x + d_0$$

Combining Equations 4.40 and 4.42 yields

$$(4.44) \qquad f_n(x) = (x - x_0)^2 g_{n-2}(x) + (x - x_0) R_1 + R_0$$

The derivative of Equation 4.44 with respect to x [i.e., $f'_n(x)$] is

$$(4.45) \qquad f'_n(x) = 2(x - x_0) g_{n-2}(x) + (x - x_0)^2 g'_{n-2}(x) + R_1$$

where $g'_{n-2}(x)$ is the derivative of $g_{n-2}(x)$ with respect to x. At $x = x_0$, Equations 4.44 and 4.45 reduce to $f_n(x) = R_0$ and $f'_n(x) = R_1$. Thus, based on the Newton–Raphson method, we get

$$(4.46) \qquad x_1 = x_0 - \frac{f_n(x)}{f'_n} = x_0 - \frac{R_0(x_i)}{R_1(x_i)}$$

where $R_0(x_i)$ is the remainder term based on an initial polynomial reduction with a root of x_i, and $R_1(x_i)$ is the remainder term based on a second polynomial reduction using again x_i. Equation 4.46 indicates that, for an initial estimate of the root x_0, an improved estimate can be obtained with Equation 4.46 after computing R_0 and R_1 by polynomial reduction. Therefore, the iterative root estimation equation for this method is

$$(4.47) \qquad x_{i+1} = x_i - \frac{R_0(x_i)}{R_1(x_i)}$$

4.8.1 PROGRAMMING CONSIDERATIONS

The solution procedure for synthetic division is easily programmed using the general form of Equations 4.38 to 4.47. By dividing $f_n(x)$, Equation 4.38, by $x - x_0$, we get

$$(4.48) \qquad x - x_0 \overline{\smash{\big)}\, \begin{array}{l} c_{n-1}x^{n-1} + c_{n-2}x^{n-2} + c_{n-3}x^{n-3} + \cdots + c_0 \\ b_n x^n \quad + b_{n-1}x^{n-1} + b_{n-2}x^{n-2} + \cdots + b_0 \end{array}}$$

It can be shown that the coefficients of the reduced polynomial c_i are related to the b_i values by

$$(4.49) \qquad c_n = b_n$$

$$(4.50) \qquad c_j = b_j + x_i c_{i+1} \quad \text{for } j = (n-1), (n-2), \dots, 1$$

where x_i is an estimate of the root in the ith iteration. Similarly, the coefficients of Equation 4.43 are related to those of Equation 4.41 by

$$(4.51) \qquad d_n = c_n$$

$$(4.52) \qquad d_j = c_j + x_i d_{i+1} \quad \text{for } j = (n-1), (n-2), \dots, 1$$

The remainders R_0 and R_1 in the ith iteration are given by

$$(4.53) \qquad R_0 = b_0 + x_i c_1$$

$$(4.54) \qquad R_1 = c_1 + x_i d_2$$

Thus the procedure is as follows:
1. Input n, x_0, b_j for $j = 0, 1, \dots, n$.
2. Compute the n values of c_j using Equations 4.49 and 4.50.
3. Compute R_0.
4. Compute the $(n-1)$ values of d_j.

5. Compute R_1.
6. Use Equation 4.47 to compute a revised estimate of x_{i+1} of the root x_i.
7. Check for convergence as follows:
   a. If $|x_{i+1} - x_i| \leq$ tolerance, discontinue the iteration and use x_{i+1} as the best estimate of the root; or
   b. If $|x_{i+1} - x_i| >$ tolerance, set $x_i = x_{i+1}$ and return to step 2 and continue the iteration process

Example 4.12: Synthetic Division

Using the example of Equation 4.26 and an initial estimate x_0 of 6, the reductions indicated by Equations 4.39 and 4.42 are as follows:

$$
(4.55) \quad
\begin{array}{r}
x^2 + 5x + 20 \\
x-6 \overline{)\, x^3 - x^2 - 10x - 8} \\
\underline{x^3 - 6x^2} \\
5x^2 - 10x \\
\underline{5x^2 - 30x} \\
20x - 8 \\
\underline{20x - 120} \\
112 = R_0
\end{array}
$$

$$
(4.56) \quad
\begin{array}{r}
x + 11 \\
x-6 \overline{)\, x^2 + 5x + 20} \\
\underline{x^2 - 16x} \\
11x + 20 \\
\underline{11x - 66} \\
86 = R_1
\end{array}
$$

Thus the revised estimate is

$$
(4.57) \qquad x_1 = 6 - \frac{112}{86} = 4.6977
$$

This is the same value obtained on the first iteration of the Newton–Raphson method. While both methods give the same answer, the Newton–Raphson iteration requires evaluation of the first derivative of the polynomial, whereas the synthetic division method requires two polynomial reductions. A second iteration could be performed using 4.6977 as the estimate of the root. The results for the first root are shown in Table 4.6. After six iterations, the root to seven significant digits is 4.

TABLE 4.6   Synthetic Division for Example 4.12

Parameter	Iteration 0	Iteration 1	Iteration 2	Iteration 3	Iteration 4	Iteration 5
x_i	6	4.69767	4.12889	4.00569	4.00001	4
R_0	112	26.62407	4.05173	0.17098	0.00036	1.54E–09
R_1	86	46.80909	32.88552	30.12522	30.00026	30
b_3	1	1	1	1	1	1
b_2	–1	–1	–1	–1	–1	–1
b_1	–10	–10	–10	–10	–10	–10
b_0	–8	–8	–8	–8	–8	–8
c_3	1	1	1	1	1	1
c_2	5	3.69767	3.12889	3.00569	3.00001	3
c_1	20	7.37047	2.91888	2.03984	2.00008	2
d_3	1	1	1	1	1	1
d_2	11	8.39535	7.25779	7.01138	7.00002	7
d_1	86	46.80909	32.88552	30.1252	30.0003	30

Example 4.13: Eigenvalues

The polynomial developed by expanding the determinant $|\mathbf{A} - \lambda\mathbf{I}|$ can be solved using synthetic division. The following is an example of a polynomial expansion of a determinant:

$$(4.58) \qquad \lambda^3 - 3\lambda^2 + 2.3146\lambda - 0.504188$$

The values for b_j are $b_3 = 1$, $b_2 = -3$, $b_1 = 2.3146$, and $b_0 = -0.504188$. Using an initial estimate of 2 for the root, the values of c_j and d_j are computed as shown in Table 4.7. After two iterations the root converged to $\lambda = 1.942$, which is the value that was found using the Newton–Raphson method. The set of c_j values for the last iteration are $c_3 = 1$, $c_2 = -1.058$, and $c_1 = 0.2597$, which can be used to form the second-order polynomial:

$$(4.59) \qquad \lambda^2 - 1.058\lambda + 0.2597 = 0$$

which is the same equation produced by polynomial reduction. This second-order polynomial can then be evaluated for a second root. Using these coefficients as the new b_i values and an initial estimate of 1.0, synthetic division yields the value of a second root in four iterations, $\lambda = 0.670$ (see Table 4.8). The c_j values are $c_2 = 1$ and $c_1 = -0.3860$, which give the polynomial

$$(4.60) \qquad \lambda - 0.3860 = 0$$

This yields the last root of the original third-order polynomial, $\lambda = 0.3860$. The sum of the three roots equals 3, as expected.

TABLE 4.7   Example 4.13 Synthetic Division for a Third-Order Polynomial

Parameter	Iteration 0	Iteration 1	Iteration 2	Iteration 3	Iteration 4	Iteration 5	Iteration 6
λ_i	2	1.94599	1.941691	1.941665	1.941665	1.941665	1.941665
R_0	0.12501	0.00859	5.24E–05	1.99E–09	5.55E–16	–3.30E–16	
R_1	2.31460	1.99929	1.974948	1.974798	1.974798	1.974798	
b_3	1	1	1	1	1	1	
b_2	–3	–3	–3	–3	–3	–3	
b_1	2.31460	2.31460	2.31460	2.31460	2.31460	2.31460	
b_0	–0.50419	–0.50419	–0.50419	–0.50419	–0.50419	–0.50419	
c_3	1	1	1	1	1	1	
c_2	–1	–1.05401	–1.05831	–1.05834	–1.05834	–1.05834	
c_1	0.31460	0.26351	0.25969	0.25967	0.25967	0.25967	
d_3	1	1	1	1	1	1	
d_2	1	0.89198	0.88338	0.88333	0.88333	0.88333	
d_1	2.31460	1.99929	1.97495	1.97480	1.97480	1.97480	

TABLE 4.8   Example 4.13 Synthetic Division for a Second-Order Polynomial

Parameter	Iteration 0	Iteration 1	Iteration 2	Iteration 3	Iteration 4	Iteration 5	Iteration 6
λ_i	2	1.271502	0.914042	0.748042	0.685094	0.672387	0.671824
R_0	2.142998	0.530710	0.127778	0.027556	0.003963	0.000161	
R_1	2.941665	1.484668	0.769748	0.437750	0.311852	0.286439	
b_3	1	1	1	1	1	1	
b_2	–1.058340	–1.058340	–1.058340	–1.058340	–1.058340	–1.058340	
b_1	0.259668	0.259668	0.259668	0.259668	0.259668	0.259668	
c_2	1	1	1	1	1	1	
c_1	0.941665	0.213167	–0.144290	–0.310290	–0.373240	–0.385950	
d_2	1	1	1	1	1	1	
d_1	2.941665	1.484668	0.769748	0.437750	0.311852	0.286439	

4.9 MULTIPLE ROOTS

Figure 4.14 shows a case of multiple roots, where the function $f(x)$ is tangent to the x axis. This case corresponds to having two roots of the same magnitude and sign. In general, there can be triple, quadruple, …, or multiple roots for a function $f(x)$. It can be shown that an even number of multiple roots results in a tangent $f(x)$ to the x axis, whereas an odd number of multiple roots results in a function $f(x)$ that crosses the x axis with an inflection point at the root—that is, a point where the function changes curvature. For example, the following function has three roots at $x = 1$:

$$(4.61) \qquad f(x) = (x-1)(x-1)(x-1)(x-5)$$

This function can be expressed as

$$(4.62) \qquad f(x) = x^4 - 8x^3 + 18x^2 - 16x + 5$$

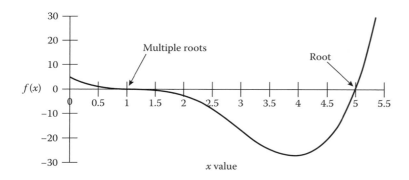

FIGURE 4.14   Multiple roots at $x = 1$, and one root at $x = 5$.

The function in Equations 4.61 and 4.62 has triple roots ($x = 1$) and one root ($x = 5$) as shown in Figure 4.14. The figure shows the inflection point at $x = 1$ that results in $f(x) = 0$.

Multiple roots pose difficulties for the previous methods. The bisection method has difficulties with multiple roots because the function does not change sign at even multiple roots. The Newton–Raphson and secant methods have difficulties because the derivative $f'(x)$ at a multiple root is zero. Since $f(x)$ reaches zero at a faster rate than $f'(x)$ as x approaches the multiple root, it is possible to check for the condition $f(x) = 0$ and terminate the computations before reaching $f'(x) = 0$.

4.10 SYSTEMS OF NONLINEAR EQUATIONS

The methods introduced so far for finding the roots of a function deal with single-variable equations of the type $f(x) = 0$. Some engineering problems have two or more variables for which roots are needed. For the two-variable problem, this would have the form $f_i(x, y) = 0$, where the subscript i denotes the equation number, and both x and y are independent variables. The number of equations should be equal to the number of variables. The following is an example of a two-variable system of equations:

(4.63) $$x^3 - 3x^2 + xy = 0$$

(4.64) $$4x^2 - 4xy^2 + 3y^2 = 0$$

The methods used for single-variable functions cannot be applied directly to find the values of x and y for multiple-variable functions. Instead, the *Jacobi method* is adapted to solve such equations. In this case, each of the i equations is solved for one of the i variables. Using initial estimates, the transformed equations are solved iteratively until the solution converges. The solution procedure is best illustrated by an example. Equations 4.63 and 4.64 are used for this purpose. Although Equation 4.63 can be reduced to $x^2 - 3x + y = 0$ by dividing it by x, the original equation can be used without any effect on the final answer. These equations are solved for x and y, respectively:

$$(4.65) \qquad\qquad x = (3x^2 - xy)^{1/3}$$

$$(4.66) \qquad\qquad y = \left(\frac{4x^2 + 3y^2}{4x} \right)^{0.5}$$

Using initial estimates that $x = y = 3$, Equation 4.65 yields

$$x = [3(3)^2 - (3)(3)]^{1/3} = 2.621$$

Using $x = 2.621$ and $y = 3$ with Equation 4.66 gives

$$y = \left[\frac{4(2.621)^2 + 3(3)^2}{4(2.621)} \right]^{0.5} = 2.2796$$

For the second iteration, $x = 2.621$ and $y = 2.2796$ with Equation 4.65 yields

$$x = [3(2.621)^2 - (2.621)(2.2796)]^{1/3} = 2.4458$$

Using $x = 2.4458$ and $y = 2.2796$ with Equation 4.66 yields

$$y = \left[\frac{4(2.4458)^2 + 3(2.2796)^2}{4(2.4458)} \right]^{0.5} = 2.0098$$

If the iterations were continued, the solution would converge to the following roots: $x = 2.16$ and $y = 1.82$. The iterations are shown in Table 4.9 and Figure 4.15.

Example 4.14: Roots for a System of Equations

The solution procedure for a system of equations is illustrated using a pair of linear equations. While an analytical solution can easily be performed for these equations, the numerical solution is used for the purpose of demonstration. In Example 8.9 (Chapter 8), the following equations are developed to demonstrate the Galerkin method:

$$(4.67) \qquad\qquad \frac{1}{4} b_1 + \frac{7}{15} b_2 = \frac{1}{3}$$

$$(4.68) \qquad\qquad \frac{2}{15} b_1 + \frac{1}{3} b_2 = \frac{1}{4}$$

Solving for b_1 and b_2 yields

$$b_1 = \frac{4}{3} - \frac{28}{15} b_2$$

TABLE 4.9   Results for Example 4.13

Iteration i	x	y
1	2.620741	2.279550
2	2.445807	2.009789
3	2.353163	1.908024
4	2.297176	1.867022
5	2.259935	1.848446
6	2.233679	1.838727
7	2.214562	1.832913
8	2.200396	1.829070
9	2.189798	1.826369
10	2.181825	1.824402
11	2.175806	1.822942
12	2.171253	1.821849
13	2.167804	1.821025
14	2.165189	1.820402
15	2.163204	1.819931
16	2.161698	1.819573
17	2.160554	1.819302
18	2.159684	1.819096
19	2.159024	1.818939
20	2.158522	1.818820

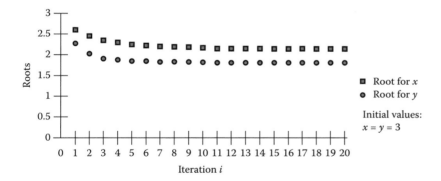

FIGURE 4.15   Results for Example 4.13.

$$b_2 = \frac{3}{4} - \frac{6}{15}b_1$$

Using initial estimates of $b_1 = b_2 = 1$, the first iteration is

$$b_1 = \frac{4}{3} - \frac{28}{15}(1) = -0.53333$$

$$b_2 = \frac{3}{4} - \frac{6}{15}(-0.53333) = 0.96333$$

The second iteration is

$$b_1 = \frac{4}{3} - \frac{28}{15}(0.96333) = -0.46489$$

$$b_2 = \frac{3}{4} - \frac{6}{15}(-0.46489) = 0.93596$$

Table 4.10 and Figure 4.16 show summaries of the iterations. Comparing the values computed for the 20th iteration with the analytically derived solution of Equation 8.76 in Chapter 8

TABLE 4.10   Results for Example 4.14

Iteration i	b_1	b_2
1	−0.53333	0.963333
2	−0.46489	0.935956
3	−0.41378	0.915513
4	−0.37563	0.900250
5	−0.34713	0.888853
6	−0.32586	0.880344
7	−0.30998	0.873990
8	−0.29811	0.869246
9	−0.28926	0.865704
10	−0.28265	0.863059
11	−0.27771	0.861084
12	−0.27402	0.859609
13	−0.27127	0.858508
14	−0.26922	0.857686
15	−0.26768	0.857072
16	−0.26654	0.856614
17	−0.26568	0.856272
18	−0.26504	0.856016
19	−0.26456	0.855825
20	−0.26421	0.855683

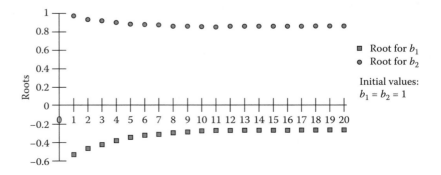

FIGURE 4.16   Results for Example 4.14.

(i.e., $b_1 = -0.26316$ and $b_2 = 0.85526$) indicates that the method does not converge rapidly; however, it does converge. The problem of convergence for the Jacobi method is discussed in Section 5.5 (Chapter 5).

4.11 APPLICATIONS

4.11.1 PIPE FLOW EVALUATION

In Section 3.6.1 (Chapter 3), a method for designing a pipe system was introduced. The same concepts can be used to evaluate the flow rate for a given system. Equation 3.33 (Chapter 3) can be rearranged to the following third-order polynomial:

(4.69)
$$\frac{8\,fL}{\pi^2 gD^5}Q^3 + hQ - \frac{550eh_p}{\gamma} = 0$$

where
  f = friction coefficient of pipe
  L = length of pipe
  g = gravitational acceleration
  D = diameter of pipe
  Q = VA in which V = flow velocity and A = cross-sectional area
  h = potential head
  e = efficiency of pump
  h_p = horsepower of pump
  γ = specific weight of fluid

Two of the three roots of Equation 4.69 are imaginary; the third root of the polynomial is the flow rate for the characteristics of the reservoir system, fluid, and pump.

Consider the case of an 800 ft pipe section (with pipe roughness coefficient $e_f = 0.0004$) connecting two reservoirs containing oil (specific gravity = 0.82). The 6 in. pipe system includes a 6 hp pump (with $e = 0.6$). The difference in the surface elevations of the two reservoirs is $h = 5$ ft. For the 6 in. pipe, $e_f/D = 0.0008$ ft, which gives a friction coefficient f of 0.0185 (see Table 3.4 in Chapter 3). Thus Equation 4.69 is

(4.70)
$$\frac{8(0.0185)(800)}{\pi^2(32.2)(0.5)^5}Q^3 + 5Q - \frac{550(0.6)(6)}{0.82(62.4)} = 0$$

(4.71)
$$11.922Q^3 + 5Q - 38.696 = 0$$

It is evident that $Q = 1$ would produce a negative value for the left side of Equations 4.70 and 4.71, but $Q = 2$ would produce a positive value. Thus a Q of 1.25 cfs will be used as the initial estimate. The Newton–Raphson method can be used. The derivative of Equations 4.70 and 4.71 is

(4.72) $$f'(Q) = 35.766Q^2 + 5$$

Thus Newton's correction formula is

(4.73) $$Q_{i+1} = Q_i - \frac{11.922Q_i^3 + 5Q_i - 38.696}{35.766Q_i^2 + 5}$$

The corrected estimate of Q is

(4.74) $$1.25 - \frac{11.922(1.25)^3 + 5(1.25) - 38.696}{35.766(1.25)^2 + 5} = 1.40 \text{ cfs}$$

A second iteration yields

(4.75) $$1.40 - \frac{11.922(1.40)^3 + 5(1.40) - 38.696}{35.766(1.40)^2 + 5} = 1.386 \text{ cfs}$$

A third iteration yields 1.386 cfs, which is of sufficient accuracy for all practical purposes.

4.11.2 GAS LAW

The ideal gas law $PV = nRT$ is a basic concept covered in introductory chemistry classes. In practice, more accurate $P–V–T$ relationships of gases are available. The following Beattie–Bridgeman equation is one example:

(4.76) $$P = \frac{RT}{V} + \frac{a_1}{V^2} + \frac{a_2}{V^3} + \frac{a_3}{V^4}$$

This can be rearranged to the form of a fourth-order polynomial:

(4.77) $$PV^4 - RTV^3 - a_1V^2 - a_2V - a_3 = 0$$

For values of R, a_1, a_2, and a_3, the volume V for any temperature T and pressure P can be determined from the real roots of the polynomial.

For a particular gas, $a_1 = -1.06$, $a_2 = 0.057$, and $a_3 = -0.00011$. If the pressure P equals 25 atm, then the temperature T equals 293K, and $R = 0.082$ liter-atm/K-g mole. Then Equation 4.77 becomes

(4.78) $$f(V) = 25V^4 - 24.03V^3 + 1.06V^2 - 0.057V + 0.00011 = 0$$

An initial approximation of V can be made with the following ideal gas law:

(4.79) $$V_0 = \frac{RT}{P} = \frac{0.082(293)}{25} = 0.961$$

TABLE 4.11   Results of Gas Law Application

Iteration	V_i	$f(V)$	$f'(V)$
0	0.961	0.9200	18.75
1	0.912	−0.1050	17.76
2	0.918	0.0039	18.49
3	0.918		

Substituting the initial estimate V_0 into Equation 4.78 yields 0.92, not the zero value that would be expected if V_0 were the correct volume. The derivative of Equation 4.78 is

(4.80) $$f'(V) = 100V^3 - 72.09V^2 + 2.12V - 0.057$$

Newton's formula is

(4.81) $$V_{i+1} = V_i - \frac{f(V)}{f'(V)}$$

Using the initial estimate of $V_0 = 0.961$ and Equations 4.78 and 4.80 with Equation 4.81 yields the results in Table 4.11. Thus three rounds are necessary for the solution to converge to a volume of 0.918.

PROBLEMS

The polynomial

$$x^3 + ax^2 + bx + c = 0$$

can be rearranged and solved for x as

$$x^2 + ax + b + \frac{c}{x} = 0$$

(4.82) $$x^2 = -ax - b - \frac{c}{x}$$

$$x = \sqrt{-ax - b - \frac{c}{x}}$$

A root can be found using an initial estimate of x and iterating until the expression on the right of the equal sign equals the estimate of the root. Solve Problems 4.1 through 4.4 for all roots using this iteration technique.

4.1 $x^3 + 0.2x^2 - 0.67x - 0.364 = 0$
4.2 $x^3 - 17.2x^2 - 81.97x - 83.79 = 0$
4.3 $x^3 + 7.2x^2 + 13.55x + 3.444 = 0$
4.4 $-1.456x^3 - 3.672x^2 + 11.208x + 17.68 = 0$

4.5   Develop an equation that corresponds to Equation 4.82 but can be used for solving a fourth-order polynomial.

4.6   Use the result of Problem 4.5 to solve for the root of

$$24x^4 + 112x^3 - 202x^2 - 54x + 36 = 0$$

4.7   Consider the four correlation matrices

$$\mathbf{R}_1 = \begin{bmatrix} 1 & 0 \\ 0 & 1 \end{bmatrix}, \qquad \mathbf{R}_2 = \begin{bmatrix} 1 & 0.2 \\ 0.2 & 1 \end{bmatrix},$$

$$\mathbf{R}_3 = \begin{bmatrix} 1 & 0.7 \\ 0.7 & 1 \end{bmatrix}, \qquad \mathbf{R}_4 = \begin{bmatrix} 1 & 1 \\ 1 & 1 \end{bmatrix}$$

The off-diagonal elements are the intercorrelations between the two predictor variables. Evaluate the eigenvalues of the four matrices and develop general guidelines for interpreting the eigenvalues of a correlation matrix.

4.8   Develop the polynomial for the determinant of $R - \lambda I$ for the following correlation matrix:

$$\begin{bmatrix} 1.0 & 0.6 & 0.8 \\ 0.6 & 1.0 & 0.5 \\ 0.8 & 0.5 & 1.0 \end{bmatrix}$$

4.9   Determine the eigenvalues of the following correlation matrices and discuss the results:

$$\underline{A} = \begin{bmatrix} 1.0 & 0.6 & 0.8 \\ 0.6 & 1.0 & 0.5 \\ 0.8 & 0.5 & 1.0 \end{bmatrix} \qquad \underline{B} = \begin{bmatrix} 1.0 & 0.2 & 0.1 \\ 0.2 & 1.0 & 0.3 \\ 0.1 & 0.3 & 1.0 \end{bmatrix}$$

4.10   The following function contains three real roots between 0 and 2.5:

$$f(x) = x^3 - 3.5x^2 + 3.28x - 0.924 = 0$$

a. Using an increment of 0.25, determine the approximate location of the roots using the direct-search method. Discuss the problem identified from this analysis.

b. Using the increment $\Delta x = 0.25$, plot $f(x)$ versus x and discuss how the graph can be used to suggest the most feasible intervals to search for the roots.

4.11   The following function contains one real root in the interval from 0 to 1:

$$f(x) = x^4 - 1.74x^3 + 0.9419x^2 - 0.4252x + 0.0027 = 0$$

a. Using an increment of 0.10, determine the approximate location of the roots using the direct-search method. Discuss the problem identified from this analysis.

b. Using the interval $\Delta x = 0.10$, plot $f(x)$ versus x and discuss how the graph can be used to suggest the most feasible intervals to search for the roots.

4.12 The following function contains one real root in the interval from 0 to 1:

$$f(x) = x^4 + 4.667x^3 - 8.4167x^2 - 2.25x + 1.5 = 0$$

a. Using an increment of 0.05, determine the approximate location of the roots using the direct-search method. Discuss the problem identified from this analysis.

b. Using the interval $\Delta x = 0.10$, plot $f(x)$ versus x and discuss how the graph can be used to suggest the most feasible intervals to search for the roots.

4.13 Estimate the root of the polynomial in Problem 4.10 that lies in the interval 2 to 2.5 using the bisection method to a precision of (a) 0.02 and (b) 1%.

4.14 Estimate the root of the polynomial in Problem 4.12 that lies in the interval 0 to 0.02 using the bisection method to a precision of (a) 0.002 and (b) 0.1%.

4.15 Estimate the root of the polynomial in Problem 4.4 that lies in the interval −5 to −2 using the bisection method to a precision of (a) 0.1 and (b) 0.5%.

4.16 Estimate the root of the polynomial in Problem 4.6 that lies in the interval 0 to 1.0 using the bisection method to a precision of (a) 0.05 and (b) 1%.

The functions given in Problems 4.17 through 4.26 possess real roots. Using a precision of 2% of a root, find the estimates of all roots using the Newton–Raphson method. After finding a root, reduce the polynomial using polynomial reduction before finding the next root with the Newton–Raphson method.

4.17 $x^3 - 0.39x + 0.07 = 0$

4.18 $x^3 - 1.7x^2 + 0.84x - 0.108 = 0$

4.19 $x^3 - 2x^2 - 5x + 6 = 0$

4.20 $x^3 + x^2 - 17x + 15 = 0$

4.21 $x^3 - 6.1x^2 - 11.26x - 6.336 = 0$

4.22 $x^3 + 3x^2 - 4 = 0$

4.23 $x^4 - 15x^2 - 6x + 24 = 0$

4.24 $x^4 - 10x^2 + 9 = 0$

4.25 $x^4 - 2x^3 - 15x^2 - 4x + 20 = 0$

4.26 $x^4 - 5.6x^3 - 39.97x^2 - 12.33x + 24.288 = 0$

4.27 Using the polynomial in Problem 4.4 determine all of the roots with the Newton–Raphson method.

4.28 Use the Newton–Raphson method to determine the roots of the following polynomial:

$$0.025x^3 - 0.049x^2 - 0.04728x + 0.050112 = 0$$

4.29 For the following function, determine the three roots using the Newton–Raphson method using different initial values for each root. Then redo the problem but after finding a root reduce the function using polynomial reduction before finding the next root using the Newton–Raphson method:

$$x^3 - x^2 + 0.1692x + 0.036432 = 0$$

For the functions given in Problems 4.30 through 4.33, determine the interval for each root in which any initial estímate in the interval will converge to the root.

4.30 $x^3 - 2x^2 - 5x + 6 = 0$
4.31 $x^3 - 3.5x^2 + 3.28x - 0.924 = 0$
4.32 $x^4 - 10x^2 + 9 = 0$
4.33 $x^4 - 15x^2 - 6x + 24 = 0$

The functions given in Problems 4.34 through 4.40 possess real roots. Using a precision of 1% of a root, find the estimates of all roots using the secant method. After finding a root, reduce the polynomial using polynomial reduction before finding the next root with the secant method.

4.34 Use the function in Problem 4.18.
4.35 Use the function in Problem 4.20.
4.36 Use the function in Problem 4.17.
4.37 Use the function in Problem 4.26.
4.38 Use the function in Problem 4.11.
4.39 Use the function in Problem 4.32.
4.40 Use the function in Problem 4.33.

The functions given in Problems 4.41 through 4.52 possess real roots. Using a precision of 2% of a root, find estimates of all roots by synthetic division.

4.41 $x^3 - 0.6x^2 + 0.11x - 0.006 = 0$
4.42 $x^3 + 0.9x^2 - 10.56x - 16.94 = 0$
4.43 $x^3 + 0.6x^2 - 4.77x + 2.43 = 0$
4.44 $x^3 - 13.1x^2 + 48.48x - 46.62 = 0$
4.45 $x^4 - 8.37x^2 - 10.336x - 1.734 = 0$
4.46 $x^4 - 7.8x^3 + 22.08x - 26.738x + 11.6127 = 0$
4.47 $x^4 + 4.5x^3 + 2.64x^2 - 4.528x + 1.0752 = 0$
4.48 $x^4 + 7.4x^3 + 17.99x^2 + 15.31x + 2.4 = 0$
4.49 $24x^4 + 112x^3 - 202x^3 - 54x + 36 = 0$
4.50 $-1.456x^3 - 3.672x^2 + 11.208x + 17.68 = 0$
4.51 $0.025x^3 - 0.049x^2 - 0.04728 + 0.05011 = 0$
4.52 $x^3 - x^2 + 0.1692x + 0.036432 = 0$

Using any of the methods introduced in this chapter, determine the eigenvalues of the matrices in Problems 4.53 through 4.58.

4.53
$$\begin{bmatrix} 1.00 & 0.73 & -0.24 \\ 0.73 & 1.00 & 0.74 \\ -0.24 & 0.74 & 1.00 \end{bmatrix}$$

4.54
$$\begin{bmatrix} 1.00 & -0.28 & 0.84 \\ -0.28 & 1.00 & -0.62 \\ 0.84 & -0.62 & 1.00 \end{bmatrix}$$

4.55
$$\begin{bmatrix} 1.00 & 0.54 & -0.67 \\ 0.38 & 1.00 & 0.55 \\ 0.92 & -0.24 & 1.00 \end{bmatrix}$$

4.56
$$\begin{bmatrix} 2.0 & 1.2 & 0.8 \\ 0.6 & 3.0 & 1.5 \\ 1.1 & 1.4 & 1.0 \end{bmatrix}$$

4.57
$$\begin{bmatrix} 1.0 & 0.2 & 0.6 & 0.3 \\ 0.2 & 1.0 & 0.5 & 0.8 \\ 0.6 & 0.5 & 1.0 & 0.1 \\ 0.3 & 0.8 & 0.1 & 1.0 \end{bmatrix}$$

4.58
$$\begin{bmatrix} 1.0 & -0.1 & 0.4 & -0.6 \\ -0.1 & 1.0 & 0.3 & -0.5 \\ 0.4 & 0.3 & 1.0 & -0.3 \\ -0.6 & -0.5 & -0.3 & 1.0 \end{bmatrix}$$

4.59 The cube root of a positive number c is equivalent to the solution of the equation $x^3 - c = 0$. Develop an algorithm based on Newton–Raphson's method for finding the cube root. Test the algorithm using $c = 15$ with an initial estimate of the cube root of 5.

4.60 The fourth root of a positive number d is equivalent to the solution of the equation $x^4 - d = 0$. Develop an algorithm based on the secant method for finding the fourth root. Test the algorithm using $d = 53$ with an initial estimate of the fourth root of 4.2.

4.61 Write a computer program for the synthetic division method as outlined in Section 4.8. Test the program using Examples 4.12 and 4.13.

4.62 The function given in Equations 4.61 and 4.62 possess multiple roots. Using a precision of 2% of a root, find the estimates of all roots using the Newton–Raphson method. The function is given by

$$f(x) = x^4 - 8x^3 + 18x^2 - 16x + 5$$

4.63 Obtain the solution, if any, for the following system of nonlinear equations:

$$x^4 - 8yx^3 + 18yx^2 - 16x = 0$$

$$x^3 + 18xy^2 - 16y = 0$$

4.64   Obtain values of x and y for the following system of equations:

$$x^3 - 3x^2y + 2xy^2 + y - 0.596 = 0$$

$$4x^2 - 6.2y^3 + 5.1xy - 6.4319 = 0$$

Assume the value of $x = 3.7$ and $y = 0.25$ as initial estimates.

Simultaneous Linear Equations

5

5.1 INTRODUCTION

Many engineering and scientific problems can be formulated in terms of systems of simultaneous linear equations. When these systems consist of only a few equations, a solution can be found analytically using the standard methods from algebra, such as substitution. However, complex problems may involve a large number of equations that cannot realistically be solved using analytical methods. In these cases, we will need to find the solution numerically using computers.

Before introducing numerical methods for solving large systems of simultaneous linear equations, we will do two things. First, we will show how systems of simultaneous linear equations arise from physical engineering problems. Second, we will review the substitution method for obtaining an analytical solution in order to illustrate the need for numerical methods. The two example cases introduced in this section are used as motivation for the numerical methods that follow in subsequent sections.

Example 5.1: Material Purchasing for Manufacturing

Let us assume that a manufacturer is marketing a product made of an alloy material meeting a certain specified composition. The three critical ingredients of the alloy are manganese, silicon, and copper. The specifications require 15 pounds of manganese, 22 pounds of silicon, and 39 pounds of copper for each ton of alloy to be produced. This mix of ingredients requires the manufacturer to obtain inputs from three different mining suppliers. Ore from the different suppliers has different concentrations of the alloy ingredients, as detailed in Table 5.1. Given this information, the manufacturer must determine the quantity of ore to purchase from each supplier so that the alloy ingredients are not wasted. A solution to this problem can be found by defining the following variables:

(5.1) X_j = amount of ore purchased from supplier j

(5.2) C_i = amount of ingredient i required per ton of alloy

TABLE 5.1 Concentration of Alloy Ingredients for Three Suppliers

	Supplier 1 (lb/ton of Ore)	Supplier 2 (lb/ton of Ore)	Supplier 3 (lb/ton of Ore)
Manganese	1	3	2
Silicon	2	4	3
Copper	3	4	7

(5.3) a_{ij} = amount of ingredient i contained in each ton of ore shipped from supplier j

Using these notations, we can formulate a general equation that defines (1) the relationships among the compositions of the ore shipped by the different suppliers, (2) the amount of ore needed from each supplier, and (3) the required composition of the final alloy as

$$(5.4) \qquad \sum_{j=1}^{n} a_{ij} X_j = C_i \qquad \text{for } i = 1, 2, \ldots, m$$

in which m is the number of ingredients and n is the number of suppliers. For the case under consideration, both m and n equal 3. More specifically, the problem can be represented by the following system of linear simultaneous equations:

$$(5.5) \qquad X_1 + 3X_2 + 2X_3 = 15$$

$$(5.6) \qquad 2X_1 + 4X_2 + 3X_3 = 22$$

$$(5.7) \qquad 3X_1 + 4X_2 + 7X_3 = 39$$

The solution to these equations determines the amount of ore to purchase from each supplier by the alloy manufacturer.

Example 5.2: Electrical Circuit Analysis

Current flows in circuits are governed by Kirchhoff's laws. Kirchhoff's first law states that the algebraic sum of the currents flowing into a junction of a circuit must equal zero. Kirchhoff's second law states that the algebraic sum of the electromotive forces around a closed circuit must equal the sum of the voltage drops around the circuit, where a voltage drop equals the product of the current and the resistance. In solving circuit problems for which Kirchhoff's laws apply, the circuit network is separated into

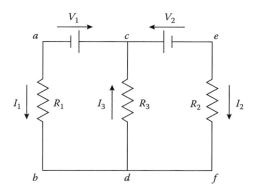

FIGURE 5.1   Example electrical circuit.

individual sections called loops. For the example network shown in Figure 5.1, there are two separate networks: loop *acdb* and loop *aefb*. Applying Kirchhoff's first law at junction *c* yields the linear equation

$$(5.8) \qquad I_1 + I_2 - I_3 = 0$$

Applying Kirchhoff's second law to network loop *acdb* yields the following linear equation:

$$(5.9) \qquad V_1 = R_1 I_1 + R_3 I_3$$

Applying Kirchhoff's second law to network loop *aefb* yields the following linear equation:

$$(5.10) \qquad V_1 - V_2 = R_1 I_1 - R_2 I_2$$

Using the voltages and resistances shown in Figure 5.1 (with assumed $R_1 = 2$, $R_2 = 4$, $R_3 = 5$, $V_1 = 6$, and $V_2 = 2$), the preceding circuit equations governing the current flows in the circuit can be written as the following system of three linear simultaneous equations:

$$(5.11) \qquad I_1 + I_2 - I_3 = 0$$

$$(5.12) \qquad 2I_1 + 5I_3 = 6$$

$$(5.13) \qquad 2I_1 - 4I_2 = 4$$

Solving these three equations would yield the current flows in the network.

5.1.1 GENERAL FORM FOR A SYSTEM OF EQUATIONS

The previous examples are illustrations of the following general form for a set of simultaneous linear equations:

$$a_{11}X_1 + a_{12}X_2 + \cdots + a_{1n}X_n = C_1$$

(5.14)
$$a_{21}X_1 + a_{22}X_2 + \cdots + a_{2n}X_n = C_2$$

$$\vdots$$

$$a_{n1}X_1 + a_{n2}X_2 + \cdots + a_{nn}X_n = C_n$$

in which the a_{ij} terms are the known coefficients of the equations, the X_j terms are the unknown variables, and the C_i terms are the known constants. Since values for both the a_{ij} and C_i terms will be known for any problem, the system of equations in Equation 5.14 represents n linear equations with n unknowns. It is important to emphasize that the solution methods provided in this chapter assume that the number of unknowns equals the number of equations. Also, the equations must be linearly independent; that is, any one equation cannot be produced by a linear combination of any of the other equations or by a multiple of itself.

5.1.2 SOLUTION OF TWO EQUATIONS

If n equals 2, then Equation 5.14 reduces to

(5.15)
$$a_{11}X_1 + a_{12}X_2 = C_1$$

(5.16)
$$a_{21}X_1 + a_{22}X_2 = C_2$$

For simple systems such as the one in Equations 5.15 and 5.16, a solution can be obtained by substitution. Solving Equation 5.15 for X_1 yields

(5.17)
$$X_1 = \frac{C_1 - a_{12}X_2}{a_{11}}$$

which can be substituted into Equation 5.16:

(5.18)
$$a_{21}\frac{C_1 - a_{12}X_2}{a_{11}} + a_{22}X_2 = C_2$$

Equation 5.18 is a single equation with one unknown, X_2. Thus Equation 5.18 can be solved for X_2 as

(5.19)
$$X_2 = \frac{a_{11}C_2 - a_{21}C_1}{a_{11}a_{22} - a_{21}a_{12}}$$

Equation 5.19 can be substituted back into Equation 5.17, producing the following equation for X_1:

(5.20)
$$X_1 = \frac{a_{22}C_1 - a_{12}C_2}{a_{11}a_{22} - a_{21}a_{12}}$$

Thus Equations 5.19 and 5.20 provide a means of determining the values of the unknowns of Equations 5.15 and 5.16 in terms of the known values of the a_{ij} coefficients and the constants C_i.

While the solution procedure based on substitution appears simple to apply, one should try to imagine the effort that would be required to solve a set of 10 simultaneous equations using the substitution procedure that led to Equations 5.19 and 5.20. Many complex engineering problems involve hundreds or even thousands of simultaneous equations. The need for alternative solution procedures is obvious.

Given the availability of computers, numerical solution methods are widely used. Three general types of numerical procedures are available: elimination methods, iteration methods, and the method of determinants. Before discussing the advantages of each, the individual computational methods will be introduced.

5.1.3 CLASSIFICATION OF SYSTEMS OF EQUATIONS BASED ON GRAPHICAL INTERPRETATION

Systems of equations can be classified based on their solutions to the following three types:

1. Systems that have solutions
2. Systems without solutions
3. Systems with an infinite number of solutions

These three types can be illustrated graphically using systems of two equations. For example, the following four systems are shown in Figure 5.2:

(5.21)   System 1: $2X_1 + 3X_2 = 6$

(5.22) $2X_1 + 9X_2 = 12$

(5.23)   System 2: $3X_1 + 9X_2 = 5$

(5.24) $X_1 + 3X_2 = 6$

(5.25)   System 3: $2X_1 + 3X_2 = 4$

(5.26) $4X_1 + 6X_2 = 8$

(5.27)   System 4: $2X_1 + 2.2X_2 = 5.7$

(5.28) $2X_1 + 2X_2 = 5.5$

The first system has a solution according to Equations 5.19 and 5.20 of

(5.29) $X_1 = 1.5$ and $X_2 = 1$

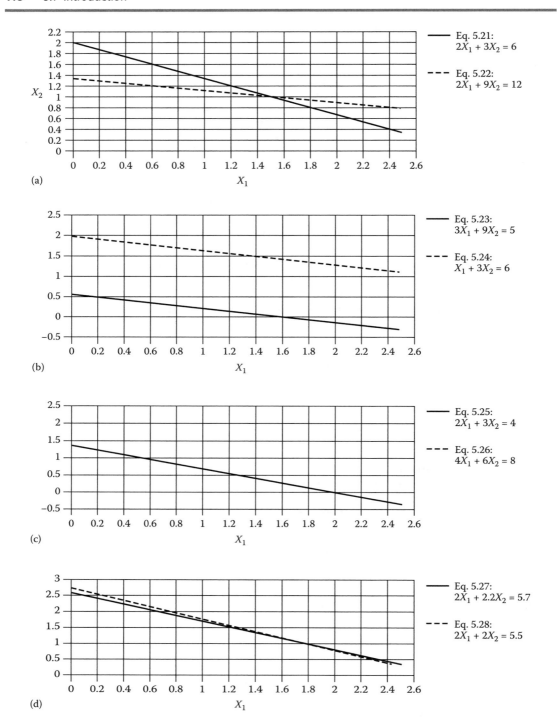

FIGURE 5.2   Three systems of equations: (a) system 1 with a solution, (b) system 2 without a solution, (c) system 3 with an infinite number of solutions, and (d) system 4 with ill-conditioned parameters.

This solution is uniquely defined as the intersection point of Equations 5.21 and 5.22 as shown in Figure 5.2a. System 2 does not have a solution, since the use of Equations 5.19 and 5.20 results in a division by zero. According to Figure 5.2b, the equations of system 2 have the same slope but different intercepts; that is, they are parallel. Therefore, they do not intersect. System 3 results also in a division by zero according to Equations 5.19 and 5.20. The equations in this case are shown in Figure 5.2c as parallel lines with the same slope and intercept—that is, the same line. The intersection of these two lines is defined by the entire line; therefore, there are an infinite number of solutions in this case. The two equations in system 3 are not unique, since one equation can be produced by the other by multiplying it by a constant. System 4 (Equations 5.27 and 5.28) has ill-conditioned parameters that result in two lines with about the same slope as shown in Figure 5.2d. Therefore, the solution defined as the intersection point of the two lines can result in computational difficulties according to numerical methods.

5.2 GAUSSIAN ELIMINATION

Gaussian elimination is similar to the substitution procedure used in deriving expressions for X_1 and X_2 of Equations 5.19 and 5.20, respectively, from Equations 5.15 and 5.16. Generally, the elimination procedure involves performing a set of operations on a system of equations until each equation is reduced to an identity for a single unknown. Details of the procedure are provided following a discussion of the *permissible* operations for a set of equations.

5.2.1 PERMISSIBLE OPERATIONS

Mathematical operations on equations in a set of simultaneous equations that do not alter the solution are termed *permissible* operations. Three permissible operations are of interest here; they are described in the context of simple, two-equation examples.

First, the solution to a set of simultaneous equations does not change if the order (that is, sequence) in which the equations are written is changed. For example, the solution to the equations

(5.30) $2X_1 + 3X_2 = 1$

(5.31) $-4X_1 + X_2 = 5$

can be determined from Equations 5.19 and 5.20; the solution is $X_1 = -1$ and $X_2 = 1$. This solution is identical to the solution for the following rearranged equations:

(5.32) $-4X_1 + X_2 = 5$

(5.33) $2X_1 + 3X_2 = 1$

This can be verified by substituting $X_1 = -1$ and $X_2 = 1$ into Equations 5.32 and 5.33. Although the order of the equations does not have influence on the solution, it is important to note that the order of the equations can in some instances influence both the likelihood of finding a solution or the accuracy of a numerical solution. In cases where the coefficients are of different orders of magnitude, roundoff and truncation errors can lead to inaccurate estimates of the unknowns. This aspect of the solution is discussed and illustrated later.

Second, any one of the simultaneous equations can be multiplied or divided by a nonzero constant without changing the solution. For example, if Equation 5.30 is multiplied by 2, we would have the following set of simultaneous equations:

$$4X_1 + 6X_2 = 2 \qquad (5.34)$$

$$-4X_1 + X_2 = 5 \qquad (5.35)$$

The solutions for Equations 5.30 and 5.31 and Equations 5.34 and 5.35 are identical; this can be verified by substituting the solution for Equations 5.30 and 5.31 into Equations 5.34 and 5.35.

Third, it is permissible to add two equations together and use the resulting equation to replace either of the two original equations. For example, the addition of Equation 5.30 to Equation 5.31 results in

$$-2X_1 + 4X_2 = 6 \qquad (5.36)$$

The solution of the two simultaneous equations as given by Equations 5.30 and 5.36 is the same as for Equations 5.30 and 5.31; also, the solution for Equations 5.31 and 5.36 is the same as for Equations 5.30 and 5.31.

The permissible operations can be applied in any combination to achieve some computational objectives. For example, using the second and third operations, we can multiply an equation by a constant and add it to another equation, with the resulting equation replacing one of the original equations. This type of operation forms the basis for the Gaussian elimination method.

5.2.2 MATRIX REPRESENTATION OF THE SYSTEM OF EQUATIONS

While a solution of two or three simultaneous equations does not present a problem of notation, the solution of a set of n simultaneous equations can be simplified by presenting the equations in terms of matrices. The fundamentals of matrix algebra are provided in Chapter 2. Using the notation of Equation 5.14, we can express the set of simultaneous equations as the following matrix relation:

$$(5.37) \qquad \begin{bmatrix} a_{11} & a_{12} & \cdots & a_{1n} \\ a_{21} & a_{22} & \cdots & a_{2n} \\ \vdots & \vdots & \ddots & \vdots \\ a_{n1} & a_{n2} & \cdots & a_{nn} \end{bmatrix} \begin{bmatrix} X_1 \\ X_2 \\ \vdots \\ X_n \end{bmatrix} = \begin{bmatrix} C_1 \\ C_2 \\ \vdots \\ C_n \end{bmatrix}$$

Equation 5.37 can be expressed in a more compact and convenient form by dropping the unknown X_i terms and incorporating the constant C_i terms as an additional column in the coefficient matrix:

$$
(5.38) \qquad
\begin{bmatrix}
a_{11} & a_{12} & \cdots & a_{1n} & C_1 \\
a_{21} & a_{22} & \cdots & a_{2n} & C_2 \\
\vdots & \vdots & \ddots & \vdots & \vdots \\
a_{n1} & a_{n2} & \cdots & a_{nn} & C_n
\end{bmatrix}
$$

Using this notation, the two simultaneous equations of Equations 5.30 and 5.31 would appear as

$$
(5.39) \qquad
\begin{bmatrix}
2 & 3 & 1 \\
-4 & 1 & 5
\end{bmatrix}
$$

In this notation, the ith row corresponds to the coefficients and constants for the ith equation. The jth column contains the coefficients corresponding to X_j, except for the last column, which contains the vector of the C_i constants. This simplified form is acceptable because row operations act only on the coefficients a_{ij} and constants C_i (but not the X_j values), so the augmented form of Equation 5.38 gives a compact representation by row of only those elements that are actually going to be modified.

5.2.3 GAUSSIAN ELIMINATION PROCEDURE

The Gaussian elimination procedure can be separated into two parts, which are often referred to as the *forward pass* and *back substitution*. The objective of the forward pass is to apply the three permissible operations to transform Equation 5.38 to an upper-triangular matrix having the form

$$
(5.40) \qquad
\begin{bmatrix}
1 & d_{12} & d_{13} & \cdots & d_{1n} & e_1 \\
0 & 1 & d_{23} & \cdots & d_{2n} & e_2 \\
0 & 0 & 1 & \cdots & d_{3n} & e_3 \\
\vdots & \vdots & \vdots & \ddots & \vdots & \vdots \\
0 & 0 & 0 & \cdots & 1 & e_n
\end{bmatrix}
$$

In Equation 5.40, all off-diagonal elements of the lower half of the coefficient matrix equal zero and all the diagonal terms equal 1.

After the forward pass is completed, the unknowns in the equations are found in the back substitution procedure. This procedure can be better illustrated if the matrix of Equation 5.40 is rewritten in the equivalent form in terms of individual equations:

$$(5.41) \quad X_1 + d_{12}X_2 + d_{13}X_3 + \cdots + d_{1,n-2}X_{n-2} + d_{1,n-1}X_{n-1} + d_{1,n}X_n = e_1$$

$$(5.42) \qquad X_2 + d_{23}X_3 + \cdots + d_{2,n-2}X_{n-2} + d_{2,n-1}X_{n-1} + d_{2,n}X_n = e_2$$

(5.43) $$X_3 + \dots + d_{3,n-2}X_{n-2} + d_{3,n-1}X_{n-1} + d_{3,n}X_n = e_3$$
$$\vdots$$

(5.44) $$X_{n-2} + d_{n-2,n-1}X_{n-1} + d_{n-2,n}X_n = e_{n-2}$$

(5.45) $$X_{n-1} + d_{n-1,n}X_n = e_{n-1}$$

(5.46) $$X_n = e_n$$

From Equations 5.41 through 5.46, which represent the system of equations after the forward pass, we can easily obtain the solution for the X_i terms. The last equation (5.46) involves only a single unknown; thus the value of X_n is given by

(5.47) $$X_n = e_n$$

The next to last equation (5.45) involves only the unknowns X_{n-1} and X_n. However, X_n has already been found with Equation 5.46. Thus, the unknown X_{n-1} can be found by rearranging the terms in Equation 5.45 and substituting this known value for X_n as follows:

(5.48) $$X_{n-1} = e_{n-1} - d_{n-1,n}X_n$$

Similarly, Equation 5.44 can be rearranged for X_{n-2} in terms of the now known values of X_{n-1} and X_n. The back substitution step thus involves working upward through the set of reduced equations produced by the forward pass and solving for each unknown in turn.

Example 5.3: Gaussian Elimination Procedure

The Gaussian elimination procedure is best illustrated with an example. Consider the following set of four simultaneous equations:

(5.49) $$2X_1 + 3X_2 - 2X_3 - X_4 = -2$$

(5.50) $$2X_1 + 5X_2 - 3X_3 + X_4 = 7$$

(5.51) $$-2X_1 + X_2 + 3X_3 - 2X_4 = 1$$

(5.52) $$-5X_1 + 2X_2 - X_3 + 3X_4 = 8$$

These equations can be represented in the matrix form of Equation 5.38 as

(5.53) $$\begin{bmatrix} 2 & 3 & -2 & -1 & -2 \\ 2 & 5 & -3 & 1 & 7 \\ -2 & 1 & 3 & -2 & 1 \\ -5 & 2 & -1 & 3 & 8 \end{bmatrix}$$

As the first step in the forward pass, we will convert the element a_{11} of the first transformed matrix to 1 and eliminate—that is, set to zero—all the other elements in the first column. Thus our objective of the first step of the forward pass is to derive a matrix with the following form:

(5.54)

$$
\begin{bmatrix}
1 & d'_{12} & d'_{13} & d'_{14} & e'_1 \\
0 & d'_{22} & d'_{23} & d'_{24} & e'_2 \\
0 & d'_{32} & d'_{33} & d'_{34} & e'_3 \\
0 & d'_{42} & d'_{43} & d'_{44} & e'_4
\end{bmatrix}
$$

Equation 5.54 can be derived by applying the permissible operations to our system of equations. Note that the values of the terms in Equation 5.53 will change after these operations; hence we will use different symbols, d'_{ij} and e'_i, for the modified equation coefficients and constants. In Equation 5.54, the prime notation means that the resulting values after one operation are not the final values as given by Equation 5.40. To obtain Equation 5.54 from Equation 5.53, we first normalize row 1 by dividing the entire row by a_{11}, which equals 2 in our example. Element a_{11} is often termed the *pivot* element for row 1. We then loop over each row below the pivot element and eliminate the elements a_{i1} by multiplying each element of row 1 by $-a_{i1}$ and adding it to the corresponding element a_{ij} in row i. This is shown as follows:

Step 1

| Original Matrix | Operation | Resultant Matrix |

$$
\begin{bmatrix}
2 & 3 & -2 & -1 & -2 \\
2 & 5 & -3 & 1 & 7 \\
-2 & 1 & 3 & -2 & 1 \\
-5 & 2 & -1 & 3 & 8
\end{bmatrix}
\quad
\begin{aligned}
R'_1 &= \dfrac{R_1}{2} \\[4pt]
R'_2 &= R_2 - 2R'_1 \\[4pt]
R'_3 &= R_3 + 2R'_1 \\[4pt]
R'_4 &= R_4 + 5R'_1
\end{aligned}
\quad
\begin{bmatrix}
1 & \frac{3}{2} & -1 & -\frac{1}{2} & -1 \\
0 & 2 & -1 & 2 & 9 \\
0 & 4 & 1 & -3 & -1 \\
0 & \frac{19}{2} & -6 & \frac{1}{2} & 3
\end{bmatrix}
$$

(5.55)

In the preceding notation, the *Operation* column describes the row operations performed on each row R_i where R_i is the vector of row values and R'_i the resulting value. The process is then repeated starting with row 2, with coefficient d'_{22} becoming the pivot element. The objective of this second step of the forward pass is to reduce the equations further to the following form:

(5.56)

$$
\begin{bmatrix}
1 & f_{12} & f_{13} & f_{14} & g_1 \\
0 & 1 & f_{23} & f_{24} & g_2 \\
0 & 0 & f_{33} & f_{34} & g_3 \\
0 & 0 & f_{43} & f_{44} & g_4
\end{bmatrix}
$$

The same procedure is used as in the first step in Equation 5.55. Specifically, the pivot row, row 2, is normalized by dividing the new pivot, d_{22} (which equals 2 in the matrix on the right side of Equation 5.55), and the d_{i2} coefficients are set to zero by performing the following permissible operations on rows 3 and 4:

Step 2

Starting Matrix	Operation	Resultant Matrix

$$
\begin{bmatrix}
1 & \frac{3}{2} & -1 & -\frac{1}{2} & -1 \\
0 & 2 & -1 & 2 & 9 \\
0 & 4 & 1 & -3 & -1 \\
0 & \frac{19}{2} & -6 & \frac{1}{2} & 3
\end{bmatrix}
\qquad
\begin{aligned}
R_1' &= R_1 \\
R_2' &= \frac{R_2}{2} \\
R_3' &= R_3 - 4R_2' \\
R_4' &= R_4 - \frac{19}{2}R_2'
\end{aligned}
\qquad
\begin{bmatrix}
1 & \frac{3}{2} & -1 & -\frac{1}{2} & -1 \\
0 & 1 & -\frac{1}{2} & 1 & \frac{9}{2} \\
0 & 0 & 3 & -7 & -19 \\
0 & 0 & -\frac{5}{4} & -9 & -\frac{159}{4}
\end{bmatrix}
$$

(5.57)

Note that the starting matrix (left side of Equation 5.57) for step 2 of the forward pass is the same as the ending matrix (right side of Equation 5.55) from step 1. The resultant matrix in Equation 5.57 has the same form as the matrix of Equation 5.56, which was the objective of the operations in this step.

In steps 3 and 4 of the forward pass, the third and fourth rows are reduced using the same procedure. The third row is normalized by dividing the row by the pivot element, f_{33}, in the previous Equation 5.56. The revised third row is then used to eliminate f_{43} from row 4. Row 4 is then normalized by dividing the row by its pivot element. The following matrices illustrate these operations:

Step 3

Starting Matrix	Operation	Resultant Matrix

$$
\begin{bmatrix}
1 & \frac{3}{2} & -1 & -\frac{1}{2} & -1 \\
0 & 1 & -\frac{1}{2} & 1 & \frac{9}{2} \\
0 & 0 & 3 & -7 & -19 \\
0 & 0 & -\frac{5}{4} & -9 & -\frac{159}{4}
\end{bmatrix}
\qquad
\begin{aligned}
R_1' &= R_1 \\
R_2' &= R_2 \\
R_3' &= \frac{R_3}{3} \\
R_4' &= R_4 + \frac{5}{4}R_3'
\end{aligned}
\qquad
\begin{bmatrix}
1 & \frac{3}{2} & -1 & -\frac{1}{2} & -1 \\
0 & 1 & -\frac{1}{2} & 1 & \frac{9}{2} \\
0 & 0 & 1 & -\frac{7}{3} & -\frac{19}{3} \\
0 & 0 & 0 & -\frac{143}{12} & -\frac{572}{12}
\end{bmatrix}
$$

(5.58)

Step 4

Starting Matrix   Operation   Resultant Matrix

$$\begin{bmatrix} 1 & \frac{3}{2} & -1 & -\frac{1}{2} & -1 \\ 0 & 1 & -\frac{1}{2} & 1 & \frac{9}{2} \\ 0 & 0 & 1 & -\frac{7}{3} & -\frac{19}{1} \\ 0 & 0 & 0 & -\frac{143}{12} & -\frac{572}{12} \end{bmatrix} \begin{matrix} R_1' = R_1 \\[6pt] R_2' = R_2 \\[6pt] R_3' = R_3 \\[6pt] R_4' = \dfrac{R_4}{-143/12} \end{matrix} \begin{bmatrix} 1 & \frac{3}{2} & -1 & -\frac{1}{2} & -1 \\ 0 & 1 & -\frac{1}{2} & 1 & \frac{9}{2} \\ 0 & 0 & 1 & -\frac{7}{3} & -\frac{19}{3} \\ 0 & 0 & 0 & 1 & \frac{572}{143} \end{bmatrix}$$

(5.59)

The resultant matrix in Equation 5.59 represents a system of four equations with four unknowns:

(5.60)
$$X_1 + \frac{3}{2}X_2 - X_3 - \frac{1}{2}X_4 = -1$$

(5.61)
$$X_2 - \frac{1}{2}X_3 + X_4 = \frac{9}{2}$$

(5.62)
$$X_3 - \frac{7}{3}X_4 = -\frac{19}{3}$$

(5.63)
$$X_4 = \frac{572}{143} = 4$$

Equations 5.60 through 5.63 represent the results from the forward pass in the Gaussian elimination method. The back substitution procedure can now be applied to these equations to solve for the unknown X_i values.

In step 1 of the backward substitution, the objective is to eliminate the known value for X_4 from the other three equations. In matrix form, this step can be expressed as

$$\begin{bmatrix} 1 & \frac{3}{2} & -1 & -\frac{1}{2} & -1 \\ 0 & 1 & -\frac{1}{2} & 1 & \frac{9}{2} \\ 0 & 0 & 1 & -\frac{7}{3} & -\frac{19}{3} \\ 0 & 0 & 0 & 1 & 4 \end{bmatrix} \begin{matrix} R_1' = R_1 + \dfrac{R_4'}{2} \\[6pt] R_2' = R_2 - R_4' \\[6pt] R_3' = R_3 + \dfrac{7R_4'}{3} \\[6pt] R_4' = R_4 \end{matrix} \begin{bmatrix} 1 & \frac{3}{2} & -1 & 0 & 1 \\ 0 & 1 & -\frac{1}{2} & 0 & \frac{1}{2} \\ 0 & 0 & 1 & 0 & 3 \\ 0 & 0 & 0 & 1 & 4 \end{bmatrix}$$

(5.64)

Row 3 of the revised matrix represents the equation $X_3 = 3$. Given the value for X_3, it can be removed from the first and second rows of the matrix; this process represents step 2 of the backward substitution analysis, which has the objective of obtaining zeros in the first and second rows in column 3:

$$
(5.65) \quad
\begin{bmatrix}
1 & \frac{3}{2} & -1 & 0 & 1 \\
0 & 1 & -\frac{1}{2} & 0 & \frac{1}{2} \\
0 & 0 & 1 & 0 & 3 \\
0 & 0 & 0 & 1 & 4
\end{bmatrix}
\begin{array}{l}
R_1' = R_1 + R_3' \\
R_2' = R_2 + \dfrac{R_3'}{2} \\
R_3' = R_3 \\
R_4' = R_4
\end{array}
\begin{bmatrix}
1 & \frac{3}{2} & 0 & 0 & 4 \\
0 & 1 & 0 & 0 & 2 \\
0 & 0 & 1 & 0 & 3 \\
0 & 0 & 0 & 1 & 4
\end{bmatrix}
$$

Step 2 of the backward substitution yields $X_2 = 2$. The solution is completed by eliminating X_2 from the first equation as step 3:

$$
(5.66) \quad
\begin{bmatrix}
1 & \frac{3}{2} & 0 & 0 & 4 \\
0 & 1 & 0 & 0 & 2 \\
0 & 0 & 1 & 0 & 3 \\
0 & 0 & 0 & 1 & 4
\end{bmatrix}
\begin{array}{l}
R_1' = R_1 - \dfrac{3}{2} R_2' \\
R_2' = R_2 \\
R_3' = R_3 \\
R_4' = R_4
\end{array}
\begin{bmatrix}
1 & 0 & 0 & 0 & 1 \\
0 & 1 & 0 & 0 & 2 \\
0 & 0 & 1 & 0 & 3 \\
0 & 0 & 0 & 1 & 4
\end{bmatrix}
$$

This completes the back substitution phase of the Gaussian elimination solution procedure; the solution to the original simultaneous equations of Equations 5.49 through 5.52 is $X_1 = 1$, $X_2 = 2$, $X_3 = 3$, and $X_4 = 4$.

Example 5.4: Gaussian Elimination for Material Purchasing

A problem of material purchasing of ores was discussed in the introduction to this chapter. The statement of the problem was transformed into a solution of three linear equations with three unknowns (Equations 5.5 through 5.7). The equations can be expressed in matrix form, with the forward elimination steps as follows:

Forward Elimination: Step 1

$$
(5.67) \quad
\begin{bmatrix}
1 & 3 & 2 & 15 \\
2 & 4 & 3 & 22 \\
3 & 4 & 7 & 39
\end{bmatrix}
\begin{array}{l}
R_1' = R_1 \\
R_2' = R_2 - 2R_1' \\
R_3' = R_3 - 3R_1'
\end{array}
\begin{bmatrix}
1 & 3 & 2 & 15 \\
0 & -2 & -1 & -8 \\
0 & -5 & 1 & -6
\end{bmatrix}
$$

Forward Elimination: Step 2

$$
(5.68) \quad
\begin{bmatrix}
1 & 3 & 2 & 15 \\
0 & -2 & -1 & -8 \\
0 & -5 & 1 & -6
\end{bmatrix}
\begin{array}{l}
R_1' = R_1 \\
R_2' = \dfrac{R_2}{-2} \\
R_3' = R_3 + 5R_2'
\end{array}
\begin{bmatrix}
1 & 3 & 2 & 15 \\
0 & 1 & \frac{1}{2} & 4 \\
0 & 0 & \frac{7}{2} & 14
\end{bmatrix}
$$

Forward Elimination: Step 3

(5.69)
$$\begin{bmatrix} 1 & 3 & 2 & 15 \\ 0 & 1 & \frac{1}{2} & 4 \\ 0 & 0 & \frac{7}{2} & 14 \end{bmatrix} \begin{array}{c} R_1' = R_1 \\ R_2' = R_2 \\ R_3' = \frac{2\,R_3}{7} \end{array} \begin{bmatrix} 1 & 3 & 2 & 15 \\ 0 & 1 & \frac{1}{2} & 4 \\ 0 & 0 & 1 & 4 \end{bmatrix}$$

The last matrix represents the following set of equations:

(5.70)
$$X_1 + 3X_2 + 2X_3 = 15$$

(5.71)
$$X_2 + \frac{1}{2} X_3 = 4$$

(5.72)
$$X_3 = 4$$

The following backward substitution process can be used to obtain values for the unknowns:

Backward Substitution: Step 1

(5.73)
$$\begin{bmatrix} 1 & 3 & 2 & 15 \\ 0 & 1 & \frac{1}{2} & 4 \\ 0 & 0 & 1 & 4 \end{bmatrix} \begin{array}{c} R_1' = R_1 - 2R_3' \\ R_2' = R_2 - \frac{1}{2}R_3' \\ R_3' = R_3 \end{array} \begin{bmatrix} 1 & 3 & 0 & 7 \\ 0 & 1 & 0 & 2 \\ 0 & 0 & 1 & 4 \end{bmatrix}$$

Backward Substitution: Step 2

(5.74)
$$\begin{bmatrix} 1 & 3 & 0 & 7 \\ 0 & 1 & 0 & 2 \\ 0 & 0 & 1 & 4 \end{bmatrix} \begin{array}{c} R_1' = R_1 - 3R_2' \\ R_2' = R_2 \\ R_3' = R_3 \end{array} \begin{bmatrix} 1 & 0 & 0 & 1 \\ 0 & 1 & 0 & 2 \\ 0 & 0 & 1 & 4 \end{bmatrix}$$

Thus the manufacturer should purchase ore from the three suppliers in the ratio of 1, 2, and 4.

Example 5.5: Gaussian Elimination for an Electrical Circuit

In the introduction to this chapter, the formulation of a set of simultaneous equations was illustrated using an electrical circuit with two loops (Equations 5.11 through 5.13). The resulting equations can be placed in matrix form and solved using the following Gaussian elimination:

Forward Pass: Step 1

(5.75)
$$\begin{bmatrix} 1 & 1 & -1 & 0 \\ 2 & 0 & 5 & 6 \\ 2 & -4 & 0 & 4 \end{bmatrix} \quad \begin{matrix} R_1' = R_1 \\ R_2' = R_2 - 2R_1' \\ R_3' = R_3 - 2R_1' \end{matrix} \quad \begin{bmatrix} 1 & 1 & -1 & 0 \\ 0 & -2 & 7 & 6 \\ 0 & -6 & 2 & 4 \end{bmatrix}$$

Forward Pass: Step 2

(5.76)
$$\begin{bmatrix} 1 & 1 & -1 & 0 \\ 0 & -2 & 7 & 6 \\ 0 & -6 & 2 & 4 \end{bmatrix} \quad \begin{matrix} R_1' = R_1 \\ R_2' = \dfrac{R_2}{-2} \\ R_3' = R_3 + 6R_2' \end{matrix} \quad \begin{bmatrix} 1 & 1 & -1 & 0 \\ 0 & 1 & -\frac{7}{2} & -3 \\ 0 & 0 & -19 & -14 \end{bmatrix}$$

Forward Pass: Step 3

(5.77)
$$\begin{bmatrix} 1 & 1 & -1 & 0 \\ 0 & 1 & -\frac{7}{2} & -3 \\ 0 & 0 & -19 & -14 \end{bmatrix} \quad \begin{matrix} R_1' = R_1 \\ R_2' = R_2 \\ R_3' = \dfrac{R_3}{-19} \end{matrix} \quad \begin{bmatrix} 1 & 1 & -1 & 0 \\ 0 & 1 & -\frac{7}{2} & -3 \\ 0 & 0 & 1 & \frac{14}{19} \end{bmatrix}$$

Backward Substitution: Step 1

(5.78)
$$\begin{bmatrix} 1 & 1 & -1 & 0 \\ 0 & 1 & -\frac{7}{2} & -3 \\ 0 & 0 & 1 & \frac{14}{19} \end{bmatrix} \quad \begin{matrix} R_1' = R_1 + R_3' \\ R_2' = R_2 + \frac{14}{19}R_3' \\ R_3' = R_3 \end{matrix} \quad \begin{bmatrix} 1 & 1 & 0 & \frac{14}{19} \\ 0 & 1 & 0 & -\frac{8}{19} \\ 0 & 0 & 1 & \frac{14}{19} \end{bmatrix}$$

Backward Substitution: Step 2

(5.79)
$$\begin{bmatrix} 1 & 1 & 0 & \frac{14}{19} \\ 0 & 1 & 0 & -\frac{8}{19} \\ 0 & 0 & 1 & \frac{14}{19} \end{bmatrix} \quad \begin{matrix} R_1' = R_1 - R_2' \\ R_2' = R_2 \\ R_3' = R_3 \end{matrix} \quad \begin{bmatrix} 1 & 0 & 0 & \frac{22}{19} \\ 0 & 1 & 0 & -\frac{8}{19} \\ 0 & 0 & 1 & \frac{14}{19} \end{bmatrix}$$

The solution can be verified by substituting the solution (i.e., $I_1 = \dfrac{22}{19}$, $I_2 = -\dfrac{8}{19}$, and $I_3 = \dfrac{14}{19}$) into the original set of three simultaneous equations. The negative value of I_2 indicates that the assumed direction of flow was incorrect and that the current flows in a clockwise direction (that is, from c to e to f to d).

5.3 GAUSS–JORDAN ELIMINATION

The Gaussian elimination procedure requires a forward pass to transform the coefficient matrix into an *upper-triangle* form. That is, the coefficient matrix was transformed to a matrix having all terms below the principal diagonal equal to zero. After the upper-triangular coefficient matrix of Equation 5.40 was obtained, the back substitution pass solved for the unknowns. An alternative process of elimination in which *all* coefficients in a column except for the pivot element are eliminated can also be used to obtain a solution. This alternative procedure is referred to as Gauss–Jordan elimination.

In Gauss–Jordan elimination, the solution is obtained directly after the forward pass; a back substitution phase is not necessary. While the Gauss–Jordan method will involve more computational effort for solving a single problem, it is more efficient when we have several right-side vectors with the coefficient C_i that we need to solve for the same coefficient matrix a_{ij}.

Whereas the rows above the pivot row are not modified during the forward substitution of Gauss elimination, they are transformed in the Gauss–Jordan elimination. The reduction of the rows above the pivot row in the Gauss–Jordan scheme follows the same procedure used to eliminate values from rows below the pivot row in the regular Gauss elimination. The following steps show the Gauss–Jordan procedure for the system of Equations 5.49 through 5.52:

Step 1

$$
\begin{array}{ccc}
\text{Original Matrix} & \text{Operation} & \text{Resultant Matrix} \\
\end{array}
$$

$$
(5.80)\quad
\begin{bmatrix}
2 & 3 & -2 & -1 & -2 \\
2 & 5 & -3 & 1 & 7 \\
-2 & 1 & 3 & -2 & 1 \\
-5 & 2 & -1 & 3 & 8
\end{bmatrix}
\quad
\begin{array}{l}
R_1' = \dfrac{R_1}{2} \\[6pt]
R_2' = R_2 - 2R_1' \\[6pt]
R_3' = R_3 + 2R_1' \\[6pt]
R_4' = R_4 + 5R_1'
\end{array}
\quad
\begin{bmatrix}
1 & \frac{3}{2} & -1 & -\frac{1}{2} & -1 \\[4pt]
0 & 2 & -1 & 2 & 9 \\[4pt]
0 & 4 & 1 & -3 & -1 \\[4pt]
0 & \frac{19}{2} & -6 & \frac{1}{2} & 3
\end{bmatrix}
$$

Step 2

$$
\begin{array}{ccc}
\text{Starting Matrix} & \text{Operation} & \text{Resultant Matrix} \\
\end{array}
$$

$$
(5.81)\quad
\begin{bmatrix}
1 & \frac{3}{2} & -1 & -\frac{1}{2} & -1 \\[4pt]
0 & 2 & -1 & 2 & 9 \\[4pt]
0 & 4 & 1 & -3 & -1 \\[4pt]
0 & \frac{19}{2} & -6 & \frac{1}{2} & 3
\end{bmatrix}
\quad
\begin{array}{l}
R_1' = R_1 - \dfrac{3}{2}R_2' \\[6pt]
R_2' = \dfrac{R_2}{2} \\[6pt]
R_3' = R_3 - 4R_2' \\[6pt]
R_4' = R_4 - \dfrac{19}{2}R_2'
\end{array}
\quad
\begin{bmatrix}
1 & 0 & -\frac{1}{4} & -2 & -\frac{31}{4} \\[4pt]
0 & 1 & -\frac{1}{2} & 1 & \frac{9}{2} \\[4pt]
0 & 0 & 3 & -7 & -19 \\[4pt]
0 & 0 & -\frac{5}{4} & -9 & -\frac{159}{4}
\end{bmatrix}
$$

Step 3

| Starting Matrix | Operation | Resultant Matrix |

$$\begin{bmatrix} 1 & 0 & -\frac{1}{4} & -2 & -\frac{31}{4} \\ 0 & 1 & -\frac{1}{2} & 1 & \frac{9}{2} \\ 0 & 0 & 3 & -7 & -19 \\ 0 & 0 & -\frac{5}{4} & -9 & -\frac{159}{4} \end{bmatrix}$$

$$\begin{aligned} R_1' &= R_1 + \frac{1}{4}R_3' \\ R_2' &= R_2 + \frac{1}{2}R_3' \\ R_3' &= \frac{R_3}{3} \\ R_4' &= R_4 + \frac{5}{4}R_3' \end{aligned}$$

$$\begin{bmatrix} 1 & 0 & 0 & -\frac{31}{12} & -\frac{112}{12} \\ 0 & 1 & 0 & -\frac{1}{6} & \frac{4}{3} \\ 0 & 0 & 1 & -\frac{7}{3} & -\frac{19}{3} \\ 0 & 0 & 0 & -\frac{143}{12} & -\frac{572}{12} \end{bmatrix}$$

(5.82)

Step 4

| Starting Matrix | Operation | Resultant Matrix |

(5.83)

$$\begin{bmatrix} 1 & 0 & 0 & -\frac{31}{12} & -\frac{112}{12} \\ 0 & 1 & 0 & -\frac{1}{6} & \frac{4}{3} \\ 0 & 0 & 1 & -\frac{7}{3} & -\frac{19}{3} \\ 0 & 0 & 0 & -\frac{143}{12} & -\frac{572}{12} \end{bmatrix}$$

$$\begin{aligned} R_1' &= R_1 + \frac{31}{12}R_4' \\ R_2' &= R_2 + \frac{1}{6}R_4' \\ R_3' &= R_3 + \frac{7}{3}R_4' \\ R_4' &= \frac{R_4}{-143/12} \end{aligned}$$

$$\begin{bmatrix} 1 & 0 & 0 & 0 & 1 \\ 0 & 1 & 0 & 0 & 2 \\ 0 & 0 & 1 & 0 & 3 \\ 0 & 0 & 0 & 1 & 4 \end{bmatrix}$$

In each of the four matrix transformations, the pivot row is normalized first and then used to reduce the other rows. For example, rows 1, 2, 3, and 4 were first normalized in the matrices of Equations 5.80 through 5.83, respectively, and then used to reduce the other rows.

Example 5.6: Gauss–Jordan Elimination for Material Purchasing

Gauss–Jordan elimination can be used to find the solution to the three simultaneous equations for the material purchasing example formulated in the introduction to this chapter—that is, Equations 5.5 through 5.7. The equations are placed in matrix form and solved as follows:

Step 1

(5.84)

$$\begin{bmatrix} 1 & 3 & 2 & 15 \\ 2 & 4 & 3 & 22 \\ 3 & 4 & 7 & 39 \end{bmatrix} \quad \begin{aligned} R_1' &= R_1 \\ R_2' &= R_2 - 2R_1' \\ R_3' &= R_3 - 3R_1' \end{aligned} \quad \begin{bmatrix} 1 & 3 & 2 & 15 \\ 0 & -2 & -1 & -8 \\ 0 & -5 & 1 & -6 \end{bmatrix}$$

Step 2

(5.85)
$$
\begin{bmatrix}
1 & 3 & 2 & 15 \\
0 & -2 & -1 & -8 \\
0 & -5 & 1 & -6
\end{bmatrix}
\begin{array}{l}
R_1' = R_1 - 3R_2' \\[4pt]
R_2' = \dfrac{R_2}{-2} \\[4pt]
R_3' = R_3 + 5R_2'
\end{array}
\begin{bmatrix}
1 & 0 & \frac{1}{2} & 3 \\
0 & 1 & \frac{1}{2} & 4 \\
0 & 0 & \frac{7}{2} & 14
\end{bmatrix}
$$

Step 3

(5.86)
$$
\begin{bmatrix}
1 & 0 & \frac{1}{2} & 3 \\
0 & 1 & \frac{1}{2} & 4 \\
0 & 0 & \frac{7}{2} & 14
\end{bmatrix}
\begin{array}{l}
R_1' = R_1 - \frac{1}{2}R_3' \\[4pt]
R_2' = R_2 - \frac{1}{2}R_3' \\[4pt]
R_3' = \dfrac{R_3}{-7/2}
\end{array}
\begin{bmatrix}
1 & 0 & 0 & 1 \\
0 & 1 & 0 & 2 \\
0 & 0 & 1 & 4
\end{bmatrix}
$$

The solution of $X_1 = 1$, $X_2 = 2$, and $X_3 = 4$ indicates that the ore should be purchased from suppliers 1, 2, and 3 in the ratios of $1:2:4$. This material purchase policy will result in the optimum composition without any waste.

5.4 ADDITIONAL CONSIDERATIONS FOR ELIMINATION PROCEDURES

5.4.1 ACCUMULATED ROUNDOFF ERRORS

An important consideration in solving systems of simultaneous equations using either Gaussian or Gauss–Jordan elimination is a loss of precision caused by accumulated roundoff errors in the numerical calculations. Problems with roundoff and truncation are most likely to occur when the coefficients in the equations differ by several orders of magnitude. In the example of Equations 5.49 through 5.52, all the coefficients were of the same order of magnitude; thus the resulting solution was not affected by roundoff and the solution was *exact*, at least within the finite precision of the numerical calculations.

We can minimize roundoff problems by rearranging the equations such that the largest coefficient in each equation is placed on the principal diagonal of the coefficient matrix, using one of the permissible operations that allows equations to be rearranged without changing the solution. Additionally, the equations should be ordered such that the equation having the largest pivot is reduced first, followed by the equation having the next largest pivot, and so on. These manipulations of equations have been shown by experience to minimize the effect of accumulated roundoff errors on the solution. These manipulations in pivoting are called partial pivoting.

5.4.2 ZERO PIVOT ELEMENT

In many problems, one or more of the coefficients will equal zero in the initial matrix. If any of the pivot elements equal zero, a *division by zero* error will occur when that pivot equation is normalized during the forward pass of the elimination procedure. However, in most cases the equations can be rearranged to eliminate the division by zero. The fact that a matrix begins with a zero element on the principal diagonal does not mean that a solution does not exist.

In some cases, the system of equations may include linearly dependent equations—that is, equations that are a linear transformation or combination of one or more other equations in the system. This condition produces what is termed a *singular* coefficient matrix. In this case, one or more of the pivot elements in the system will always equal zero regardless of how the equations are written or rearranged. For example, the following pair of equations is singular:

(5.87) $$2X_1 + 3X_2 = 7$$

(5.88) $$4X_1 + 6X_2 = 14$$

The matrix solution is

(5.89) $$\begin{bmatrix} 2 & 3 & 7 \\ 4 & 6 & 14 \end{bmatrix} \quad \begin{matrix} R_1' = \dfrac{R_1}{2} \\ R_2' = R_2 - 4R_1' \end{matrix} \quad \begin{bmatrix} 1 & \frac{3}{2} & \frac{7}{2} \\ 0 & 0 & 0 \end{bmatrix}$$

Thus more than one solution to the equations exist. In fact, any pair of values of X_1 and X_2 that satisfies either equation is a valid solution. For example, each of the following three pairs of values satisfies Equations 5.87 and 5.88: $(2, 1)$, $(1/2, 2)$, and $\left(0, \dfrac{7}{3}\right)$.

5.4.3 CONSIDERATIONS IN PROGRAMMING

Computers are most efficient when repetition exists in the solution procedure. It should be evident from Equations 5.55, 5.57 through 5.59, and 5.64 through 5.66 that the procedure underlying Gaussian elimination (and also Gauss–Jordan elimination) involves much repetition. In the forward pass, the algorithm loops over all the rows, with each row successively being designated as the pivot row. Operations are not performed on rows above the pivot row. The elements of the pivot row are normalized by dividing the row by the pivot element a_{ii}. The rows below the pivot row are then reduced so that the column elements below the pivot are zero. The forward pass can thus be programmed using the following steps:

1. Loop over each row i, making each row i in turn the pivot row.
2. Normalize the elements of the pivot row (row i) by dividing each element in the row by a_{ii} as follows:

(5.90) $$a_{ij} = \frac{a_{ij}}{a_{ii}} \quad \text{for } j = (i+1),(i+2),\ldots,(n+1)$$

(5.91) $$C_i = \frac{C_i}{a_{ii}}$$

(5.92) $$a_{ii} = 1$$

3. Loop over rows $(i + 1)$ to n below the pivot row and reduce the elements in each row as follows:

(5.93) $$a_{kj} = a_{kj} - a_{ki}a_{ij} \quad \text{for } j = i,\ldots, n$$

(5.94) $$C_k = C_k - a_{ki}C_i \quad \text{for } k = (i + 1),\ldots, n$$

The back substitution loops over the results from the forward pass, working from the last equation to the first as follows:

1. For the last row n:

(5.95) $$X_n = C_n$$

2. For rows $(n - 1)$ through 1,

(5.96) $$X_i = C_i - \sum_{j=i+1}^{n} a_{ij}X_j \quad \text{for } j = (i+1),\ldots,n \text{ and } i = (n-1),\ldots,1$$

The outlined procedure in Equations 5.90 to 5.96 can be programmed. In developing a computer program for this method, the programmer needs to consider the recommendations in Sections 5.4.1 and 5.4.2.

5.5 LU DECOMPOSITION

5.5.1 GENERAL CASE

As discussed in Section 5.2, the Gaussian elimination method consists of two stages: forward pass and back substitution. The forward pass results in the upper triangular matrix given by Equation 5.40. This matrix can be denoted as **U** and is given by

(5.97) $$\mathbf{U} = \begin{bmatrix} 1 & u_{12} & u_{13} & \cdots & u_{1n} \\ 0 & 1 & u_{23} & \cdots & u_{2n} \\ 0 & 0 & 1 & \cdots & u_{3n} \\ \vdots & \vdots & \vdots & \ddots & \vdots \\ 0 & 0 & 0 & \cdots & 1 \end{bmatrix}$$

The values of this matrix can be related to the values in Equation 5.40 as

$$(5.98) \qquad u_{ij} = d_{ij} \quad \text{for } i = 1,2,\ldots,n \text{ and } j = 1,2,\ldots,n$$

This upper-triangular matrix $\mathbf{U}$ can be related to the coefficient matrix $\mathbf{A}$ as

$$(5.99) \qquad\qquad \mathbf{LU} = \mathbf{A}$$

where $\mathbf{L}$ is a lower-triangular matrix that can be viewed to represent the back substitution aspect of the Gaussian elimination method. The general expression for $\mathbf{L}$ is

$$(5.100) \qquad \mathbf{L} = \begin{bmatrix} l_{11} & 0 & 0 & \cdots & 0 \\ l_{21} & l_{22} & 0 & \cdots & 0 \\ l_{31} & l_{32} & l_{33} & \cdots & 0 \\ \vdots & \vdots & \vdots & \ddots & \vdots \\ l_{n1} & l_{n2} & l_{n3} & \cdots & l_{nn} \end{bmatrix}$$

Therefore, Equation 5.99 can be expressed as

$$(5.101)$$
$$\begin{bmatrix} l_{11} & 0 & 0 & \cdots & 0 \\ l_{21} & l_{22} & 0 & \cdots & 0 \\ l_{31} & l_{32} & l_{33} & \cdots & 0 \\ \vdots & \vdots & \vdots & \ddots & \vdots \\ l_{n1} & l_{n2} & l_{n3} & \cdots & l_{nn} \end{bmatrix} \begin{bmatrix} 1 & u_{12} & u_{13} & \cdots & u_{1n} \\ 0 & 1 & u_{23} & \cdots & u_{2n} \\ 0 & 0 & 1 & \cdots & u_{3n} \\ \vdots & \vdots & \vdots & \ddots & \vdots \\ 0 & 0 & 0 & \cdots & 1 \end{bmatrix} = \begin{bmatrix} a_{11} & a_{12} & a_{13} & \cdots & a_{1n} \\ a_{21} & a_{22} & a_{23} & \cdots & a_{2n} \\ a_{31} & a_{32} & a_{33} & \cdots & a_{3n} \\ \vdots & \vdots & \vdots & \ddots & \vdots \\ a_{n1} & a_{n2} & a_{n3} & \cdots & a_{nn} \end{bmatrix}$$

Equation 5.101 states that the coefficient matrix $\mathbf{A}$ can be decomposed to two triangular matrices $\mathbf{L}$ and $\mathbf{U}$. These matrices can be determined, without use of the Gaussian elimination method, by performing the matrix multiplication $\mathbf{LU}$ in Equation 5.101 and equating the resulting terms with the corresponding coefficients in the matrix $\mathbf{A}$. For example, the multiplication of the first row of $\mathbf{L}$ with the first column of $\mathbf{U}$ results in the following value that is equal to a_{11}:

$$(5.102) \qquad\qquad l_{11}(1) = a_{11}$$

Performing other multiplication operations of rows and columns produces the following for the elements' values of $\mathbf{L}$ and $\mathbf{U}$:

$$(5.103) \qquad\qquad l_{i1} = a_{i1} \quad \text{for } i = 1,2,\ldots,n$$

(5.104) $$u_{1j} = \frac{a_{1j}}{l_{11}} \quad \text{for } j = 2,3,\ldots,n$$

(5.105) $$l_{ij} = a_{ij} - \sum_{k=1}^{j-1} l_{ik} u_{kj} \quad \text{for } j = 2,3,\ldots,n-1 \text{ and } i = j, j+1,\ldots,n$$

(5.106) $$u_{ji} = \frac{a_{ji} - \sum_{k=1}^{j-1} l_{jk} u_{ki}}{l_{jj}} \quad \text{for } j = 2,3,\ldots,n-1 \text{ and } i = j+1, j+2,\ldots,n$$

and

(5.107) $$l_{nn} = a_{nn} - \sum_{k=1}^{n-1} l_{nk} u_{kn}$$

Once the matrix **A** is decomposed into **L** and **U**, these matrices can be used to solve a system of equations with constants C_i, $i = 1, 2,\ldots, n$. The solution can be obtained in two steps: forward pass and back substitution. The forward pass produces the e_i values, $i = 1, 2,\ldots, n$, of Equation 5.40 as follows:

(5.108) $$e_1 = \frac{C_1}{l_{11}}$$

(5.109) $$e_i = \frac{C_i - \sum_{j=1}^{i-1} l_{ij} e_j}{l_{ii}} \quad \text{for } i = 2,3,\ldots,n$$

The back substitution results in the X_i values, $i = 1, 2,\ldots, n$, as follows:

(5.110) $$X_n = e_n$$

(5.111) $$X_i = e_i - \sum_{j=i+1}^{n} u_{ij} X_j \quad \text{for } i = n-1, n-2,\ldots,1$$

The considerations discussed in Section 5.4 for partial pivoting are applicable for **LU** decomposition and should be used. Since partial pivoting results in rearranging the rows of the **A** matrix, the constants in the **C** vector need to be arranged to maintain their correspondence with the rows of **A**. The rearranged **C** vector should be used in Equations 5.108 through 5.111 to obtain the solution of a system of linear equations.

Example 5.7: LU Decomposition for Material Purchasing

The **LU** decomposition method can be applied to the three simultaneous equations for the material purchasing example formulated according to Equations 5.5 through 5.7. The coefficient matrix is given by

$$(5.112) \qquad \mathbf{A} = \begin{bmatrix} 1 & 3 & 2 \\ 2 & 4 & 3 \\ 3 & 3 & 7 \end{bmatrix}$$

Using Equations 5.103 through 5.107, the **L** and **U** matrices can be obtained. The first column of the **L** matrix can be computed using Equation 5.103 as follows:

$$(5.113) \qquad l_{11} = a_{11} = 1$$

$$(5.114) \qquad l_{21} = a_{21} = 2$$

$$(5.115) \qquad l_{31} = a_{31} = 3$$

Equation 5.104 can be used to compute the first row of the **U** matrix as

$$(5.116) \qquad u_{12} = \frac{a_{12}}{l_{11}} = \frac{3}{1} = 3$$

$$(5.117) \qquad u_{13} = \frac{a_{13}}{l_{11}} = \frac{2}{1} = 2$$

The second column of the **L** matrix can now be computed using Equation 5.105 as follows:

$$(5.118) \qquad l_{22} = a_{22} - \sum_{k=1}^{2-1} l_{2k} u_{k2} = 4 - 2(3) = -2$$

$$(5.119) \qquad l_{32} = a_{32} - \sum_{k=1}^{2-1} l_{3k} u_{k2} = 4 - 3(3) = -5$$

Using Equation 5.106, the value of the last element of **U** can be computed as

$$(5.120) \qquad u_{23} = \frac{a_{23} - \sum_{k=1}^{2-1} l_{2k} u_{k3}}{l_{22}} = \frac{3 - 2(2)}{-2} = 0.5$$

The last element of the **L** matrix can be computed using Equation 5.107 as

$$(5.121) \quad l_{33} = a_{33} - \sum_{k=1}^{3-1} l_{3k} u_{k3} = 7 - (3)(2) - (-5)(0.5) = 3.5$$

Therefore, the **L** and **U** matrices are

$$(5.122) \quad \mathbf{L} = \begin{bmatrix} 1 & 0 & 0 \\ 2 & -2 & 0 \\ 3 & -5 & 3.5 \end{bmatrix}$$

$$(5.123) \quad \mathbf{U} = \begin{bmatrix} 1 & 3 & 2 \\ 0 & 1 & 0.5 \\ 0 & 0 & 1 \end{bmatrix}$$

The forward substitution can be performed using Equations 5.108 and 5.109 as follows:

$$(5.124) \quad e_1 = \frac{C_1}{l_{11}} = \frac{15}{1} = 15$$

$$(5.125) \quad e_2 = \frac{C_2 - \sum_{j=1}^{2-1} l_{2j} e_j}{l_{22}} = \frac{22 - 2(15)}{-2} = 4$$

$$(5.126) \quad e_3 = \frac{C_3 - \sum_{j=1}^{3-1} l_{3j} e_j}{l_{33}} = \frac{39 - 3(15) - (-5)(4)}{3.5} = 4$$

The back substitution can be performed using Equations 5.110 and 5.111 as follows:

$$(5.127) \quad X_3 = e_3 = 4$$

$$(5.128) \quad X_2 = e_2 - \sum_{j=2+1}^{3} u_{2j} X_j = 4 - 0.5(4) = 2$$

$$(5.129) \quad X_1 = e_1 - \sum_{j=1+1}^{3} u_{1j} X_j = 15 - 3(2) - (2)(4) = 1$$

The result obtained with the **LU** decomposition method is the same as what was previously obtained.

5.5.2 BANDED MATRICES

In Section 2.1.3 (Chapter 2), a banded matrix was defined as a square matrix with elements of zero except on the principal diagonal and the values in the positions adjacent to the principal diagonal. For example, a tridiagonal matrix is the special case of a banded matrix that has zeros, except in the three diagonals, as follows:

$$(5.130) \qquad \begin{bmatrix} a_{11} & a_{12} & 0 & 0 & 0 \\ a_{21} & a_{22} & a_{23} & 0 & 0 \\ 0 & a_{32} & a_{33} & a_{34} & 0 \\ 0 & 0 & a_{43} & a_{44} & a_{45} \\ 0 & 0 & 0 & a_{54} & a_{55} \end{bmatrix}$$

This matrix is described to have a band width of 3, in reference to the three nonzero diagonals. Alternatively, the matrix is described to have a half-band width of 1, in reference to the number of nonzero diagonals on one side of the principal diagonal—that is, not including the diagonal where $i = j$ for a_{ij}. Therefore, the following relationship between the band width (b_{w1}) and half-band width (b_{w2}) can be obtained:

$$(5.131) \qquad b_{w1} = 2b_{w2} + 1$$

The elimination methods previously described can be used to solve simultaneous equations with banded coefficient matrices. The efficiency of these methods can be enhanced by recognizing the non-necessity of pivoting for the zero elements that do not fall within the bands of the matrices. The computational algorithm can, therefore, be developed to reduce the number of pivotings by performing the pivotings for element a_{ij} within the bands. Also, banded matrices can be stored more efficiently by storing the banded elements only, thereby reducing the storage requirements for a solution.

5.5.3 SYMMETRIC MATRICES

A symmetric square matrix is defined as a matrix where $a_{ij} = a_{ji}$. In engineering, it is common to deal with symmetric matrices. For example, the stiffness properties of a structure can be described using a symmetric stiffness matrix, and correlations among the structural variables can be described using a symmetric correlation matrix. A symmetric matrix **A** has the following property:

$$(5.132) \qquad \mathbf{A} = \mathbf{A}^T$$

where $\mathbf{A}^T$ = the transpose of the matrix. For a symmetric matrix, the decomposition of **A** can be expressed as

$$(5.133) \qquad \mathbf{A} = \mathbf{LL}^T$$

As a result, the **LU** decomposition can be computed more effectively using a recurrence method called the *Cholesky decomposition method*. According to this method, l_{11} is given by

(5.134)
$$l_{11} = \sqrt{a_{11}}$$

The rest of the ith row of the **L** matrix is

(5.135)
$$l_{ii} = \sqrt{a_{ii} - \sum_{k=1}^{i-1} l_{ik}^2} \quad \text{for } i = 2,3,\ldots,n$$

(5.136)
$$l_{ij} = \frac{a_{ij} - \sum_{k=1}^{j-1} l_{ik} l_{jk}}{l_{jj}} \quad \text{for } j = 2,3,\ldots,i-1, \text{ and } j < i$$

These equations can be used to replace Equations 5.103 through 5.107 in the **LU** decomposition method.

Example 5.8: Cholesky Decomposition Method

The computational aspects of the Cholesky decomposition method are illustrated in this example using the following symmetric matrix:

(5.137)
$$\mathbf{A} = \begin{bmatrix} 1 & 2 & 3 \\ 2 & 8 & 10 \\ 3 & 10 & 22 \end{bmatrix}$$

Using Equation 5.136, the first row ($i = 1$) can be computed as follows:

(5.138)
$$l_{11} = \sqrt{a_{11}} = \sqrt{1} = 1$$

The second row ($i = 2$) can be computed using Equations 5.135 and 5.136 as

(5.139)
$$l_{21} = \frac{a_{21}}{l_{11}} = \frac{2}{1} = 2$$

(5.140)
$$l_{22} = \sqrt{a_{22} - \sum_{k=1}^{2-1} l_{2k}^2} = \sqrt{8 - 2^2} = 2$$

The third row ($i = 3$) is

(5.141) $$l_{31} = \frac{a_{31}}{l_{11}} = \frac{3}{1} = 3$$

(5.142) $$l_{32} = \frac{a_{32} - \sum_{k=1}^{2-1} l_{3k} l_{2k}}{l_{22}} = \frac{10 - (3)(2)}{2} = 2 \quad \text{for } j = 1,2,3,\ldots,i-1$$

(5.143) $$l_{33} = \sqrt{a_{33} - \sum_{k=1}^{3-1} l_{3k}^2} = \sqrt{22 - 3^2 - (2)^2} = 3$$

Therefore, the **L** matrix is

(5.144) $$\mathbf{L} = \begin{bmatrix} 1 & 0 & 0 \\ 2 & 2 & 0 \\ 3 & 2 & 3 \end{bmatrix}$$

The validity of Equation 5.133 can be verified as

(5.145) $$\mathbf{A} = \mathbf{LL}^T = \begin{bmatrix} 1 & 0 & 0 \\ 2 & 2 & 0 \\ 3 & 2 & 3 \end{bmatrix} \cdot \begin{bmatrix} 1 & 2 & 3 \\ 0 & 2 & 2 \\ 0 & 0 & 3 \end{bmatrix} = \begin{bmatrix} 1 & 2 & 3 \\ 2 & 8 & 10 \\ 3 & 10 & 22 \end{bmatrix}$$

5.6 ITERATIVE EQUATION-SOLVING METHODS

Elimination methods like the Gaussian elimination procedure are often called *direct* equation-solving methods because the solution is found after a fixed, predictable number of operations. As an alternative, however, we can solve sets of simultaneous equations by using a *trial-and-error* procedure. In this procedure, we assume a solution—that is, a set of estimates for the unknowns—and successively refine our estimate of the solution through some set of rules. This approach is the basis for *iterative* methods for solving simultaneous equations. Iterative methods, in general, produce an exact solution for the problem; the precision of the solution depends on our patience in performing the iterations—that is, on the number of trial-and-error or iteration cycles that we perform. Obviously, the number of operations required to obtain a given solution using iterative methods is not fixed; this, too, is a function of the number of iteration cycles that we choose to perform. However, a major advantage of iterative methods is that they can be used to solve nonlinear simultaneous equations, a task that is not possible using direct elimination methods. The use of these methods for solving nonlinear simultaneous equations is discussed in Section 4.10 in Chapter 4.

A number of iteration methods are available for solving simultaneous equations. Two of the most common methods, the Jacobi and Gauss–Seidel procedures, are presented herein.

5.6.1 JACOBI ITERATION

The general procedure for Jacobi iteration is illustrated using the general notation for the simultaneous equations in Equation 5.14. As a first step, each equation is rearranged to produce an expression for a single unknown. For example, if we solve each row i for the unknown X_i, then Equation 5.14 transforms to

(5.146)
$$X_1 = \frac{C_1 - a_{12}X_2 - a_{13}X_3 - \ldots - a_{1n}X_n}{a_{11}}$$

(5.147)
$$X_2 = \frac{C_2 - a_{21}X_1 - a_{23}X_3 - \cdots - a_{2n}X_n}{a_{22}}$$

(5.148)
$$X_n = \frac{C_n - a_{n1}X_1 - a_{n2}X_2 - \cdots - a_{n-1,n}X_{n-1}}{a_{nn}}$$

To start the iterative calculations, an initial solution estimate for the X_i unknowns is required. The initial estimates for all the X_i are substituted into the right sides of Equations 5.146 through 5.148 to obtain a new set of calculated (left side) values for the X_i's. These new values are substituted into the right side of Equations 5.146 to 5.148 and another new set of values is obtained for the X_i's. This *iteration* process continues until the calculated values for the X_i terms converge to the true solution. Convergence can be defined in many ways. For now, it is sufficient to say that convergence is obtained when the changes in all the X_i values between successive iteration cycles are acceptably small; that is, we can assume that the solution has converged when the values of the X_i for one iteration are not much different from the values of the previous iteration. The acceptable difference is set by the user and influenced by the need for accuracy.

Example 5.9: Jacobi Iteration

Given the following set of equations, we wish to solve for values of the unknowns using Jacobi iteration:

(5.149)
$$3X_1 + X_2 - 2X_3 = 9$$

(5.150)
$$-X_1 + 4X_2 - 3X_3 = -8$$

(5.151)
$$X_1 - X_2 + 4X_3 = 1$$

As a first step, we will rearrange Equations 5.149 through 5.151 such that each equation is an expression for a single, different unknown:

$$(5.152) \qquad X_1 = \frac{9 - X_2 + 2X_3}{3}$$

$$(5.153) \qquad X_2 = \frac{-8 + X_1 + 3X_3}{4}$$

$$(5.154) \qquad X_3 = \frac{1 - X_1 + X_2}{4}$$

Next, we must assume an initial estimate for the solution; values of $X_1 = X_2 = X_3 = 1$ for this initial estimate are assumed. We can now substitute these values into the appropriate locations on the right side of Equations 5.152 through 5.154 to obtain a new set of values for the unknowns, as follows:

$$(5.155) \qquad X_1 = \frac{9 - 1 + 2(1)}{3} = \frac{10}{3}$$

$$(5.156) \qquad X_2 = \frac{-8 + 1 + 3(1)}{4} = -1$$

$$(5.157) \qquad X_3 = \frac{1 - 1 + 1}{4} = \frac{1}{4}$$

These new values for X_1, X_2, and X_3 are then used as the new solution estimate. These values are substituted into the right side of Equations 5.152 through 5.154 to obtain a second set of computed values for the unknowns:

$$(5.158) \qquad X_1 = \frac{9 - (-1) + 2\left(\dfrac{1}{4}\right)}{3} = \frac{7}{2}$$

$$(5.159) \qquad X_2 = \frac{-8 + \dfrac{10}{3} + 3\left(\dfrac{1}{4}\right)}{4} = -\frac{47}{48}$$

$$(5.160) \qquad X_3 = \frac{1 - \dfrac{10}{3} + (-1)}{4} = -\frac{5}{6}$$

This process is repeated until the differences between the previous values and the new values are small. The values for the first 13 iteration cycles are shown in Table 5.2. The accuracy of the estimates

TABLE 5.2   Example of Jacobi Iteration

Iteration	X_1	$|\Delta X_1|$	X_2	$|\Delta X_2|$	X_3	$|\Delta X_3|$
0	1	–	1	–	1	–
1	3.333	2.333	−1.000	2.000	0.250	0.750
2	3.500	0.167	−0.979	0.021	−0.833	1.083
3	2.771	0.729	−1.750	0.771	−0.870	0.036
4	3.003	0.233	−1.960	0.210	−0.880	0.010
5	3.066	0.063	−1.909	0.050	−0.991	0.111
6	2.976	0.090	−1.976	0.067	−0.994	0.003
7	2.996	0.020	−2.001	0.025	−0.988	0.006
8	3.008	0.012	−1.992	0.009	−0.999	0.011
9	2.998	0.011	−1.997	0.005	−1.000	0.001
10	2.999	0.001	−2.001	0.003	−0.999	0.001
11	3.001	0.002	−1.999	0.001	−1.000	0.001
12	3.000	0.001	−2.000	0.000	−1.000	0.001

of the unknowns depends on the number of iteration cycles performed. To illustrate the convergence of the iterative calculations, the change in absolute value of the calculated X_i's from each iteration cycle to the next is also shown in Table 5.2. If we had chosen a convergence criterion for the value of any unknown based on a maximum absolute change of less than 0.05, then the iteration would have been discontinued after the seventh cycle, when the value of $|\Delta X_1|$ (the largest $|\Delta X_i|$) equals 0.049. At this point, the true errors for X_1, X_2, and X_3 (that is, the differences between the seventh cycle estimates and the true solution of $X_1 = 3$, $X_2 = -2$, and $X_3 = -1$) are 0.028, 0.025, and 0.004, respectively. The true solution, or at least the estimated solution matching the true solution to three decimal places, is not obtained until the 13th iteration cycle.

5.6.2 GAUSS–SEIDEL ITERATION

A careful examination of the calculation steps used in our Jacobi iteration example suggests that we may not have been as efficient as possible in our iterative solution. For example, in Equation 5.156 we used our initial estimate of $X_1 = 1$ on the right side when, in fact, we had already computed an updated estimate of $X_1 = \dfrac{10}{3}$ in Equation 5.155. Similarly, for Equation 5.157 the original solution estimates for X_1 and X_2 were used, rather than the revised estimates obtained from Equations 5.155 and 5.166. One could reasonably argue that, if supposedly better estimates are available, then they should be used as soon as possible. This argument is the basis of the Gauss–Seidel iteration method.

In the Jacobi iteration procedure, we always complete a full iteration cycle over all the equations before updating our solution estimates. In the Gauss–Seidel iteration procedure, however, we update each unknown as soon as a new estimate of that unknown is computed. The philosophy here is that the most recent estimate of the unknown is usually the best estimate and, therefore, should be used as soon as it is available. However, in practice, the most recent estimate of any one unknown is not always the best. For example, the calculations in Table 5.2 show that the initial estimate of X_1 from iteration 1 is better than the revised estimate from iteration 2. Nevertheless, we can show that, in general, the procedure converges much faster using the most recent estimate (i.e., Gauss–Seidel iteration), rather than the values from the previous Jacobi iteration. Thus the Gauss–Seidel iteration method is more efficient than the Jacobi method.

Example 5.10: Gauss–Seidel Iteration

The following system of equations as provided in Equations 5.149 through 5.151 was solved using the Gauss–Seidel iteration method:

$$3X_1 + X_2 - 2X_3 = 9$$

$$-X_1 + 4X_2 - 3X_3 = -8$$

$$X_1 - X_2 + 4X_3 = 1$$

Again, we will assume an initial solution estimate of $X_1 = X_2 = X_3 = 1$. The rearranged Equations 5.152 through 5.154 can be used to perform the first iteration cycle, expressed as

(5.161)
$$X_1 = \frac{9 - 1 + 2(1)}{3} = 3.333$$

(5.162)
$$X_2 = \frac{-8 + 3.333 + 3(1)}{4} = -0.417$$

(5.163)
$$X_3 = \frac{1 - 3.333 + (-0.417)}{4} = -0.688$$

Note that in Equation 5.161 the initial assumed estimates for X_2 and X_3 are used in the right side of the equation. However, in Equation 5.162 the initial assumed estimate for X_1 is not used; instead, the just-calculated updated estimate of X_1 from Equation 5.161 is substituted in the right side. The initial assumed estimate for X_3 is still used in Equation 5.162 because a more recent estimate has not been computed yet. Similarly, in Equation 5.163, the updated values for X_1 and X_2 computed in Equations 5.161 and 5.162 are used in the right side of the equation.

The second iteration cycle is computed as

(5.164)
$$X_1 = \frac{9 - (-0.417) + 2(-0.688)}{3} = 2.680$$

TABLE 5.3   Example of Gauss–Seidel Iteration

Iteration	X_1	$|\Delta X_1|$	X_2	$|\Delta X_2|$	X_3	$|\Delta X_3|$
0	1	–	1	–	1	–
1	3.333	2.333	−0.417	1.417	−0.688	1.688
2	2.680	0.348	−1.845	1.428	−0.882	0.194
3	3.027	0.346	−1.904	0.059	−0.983	0.101
4	2.979	0.048	−1.992	0.088	−0.993	0.010
5	3.002	0.023	−1.994	0.002	−0.999	0.006
6	2.999	0.003	−2.000	0.006	−1.000	0.001
7	3.000	0.001	−2.000	0.000	−1.000	0.000
8	3.000	0.000	−2.000	0.000	−1.000	0.000

(5.165) $$X_2 = \frac{-8 + 2.680 + 3(-0.688)}{4} = -1.845$$

(5.166) $$X_3 = \frac{1 - 2.680 + (-1.845)}{4} = -0.882$$

Again, note that the most recently computed values for the X_i terms are always used in the right side of Equations 5.164 through 5.166. The calculations for the remaining iteration cycles are summarized in Table 5.3.

For a convergence criterion requiring a maximum absolute difference between successive estimates to be less than 0.05, the Gauss–Seidel procedure (Table 5.3) converges in five cycles, versus seven cycles for the Jacobi iteration method (Table 5.2). Whereas the Jacobi method requires 13 iteration cycles before all values of the unknowns are accurate to three decimal places, the Gauss–Seidel method requires only seven iteration cycles.

5.6.3 CONVERGENCE CONSIDERATIONS OF THE ITERATIVE METHODS

Although both the Jacobi and Gauss–Seidel iterative methods converged for our example problem, in the general case either one or both of the methods may diverge; that is, the differences between successive estimates of the solution become *larger* with each iteration cycle. In fact, the example problem can be made divergent by interchanging the order of the starting Equations 5.149 and 5.150. That is, Equation 5.149 is used to solve for X_2 and Equation 5.150 is used to solve for X_1. In this case, the following set of equations is obtained:

(5.167) $$X_1 = 8 + 4X_2 - 3X_3$$

(5.168) $$X_2 = 9 - 3X_1 + 2X_3$$

(5.169) $$X_3 = \frac{1 - X_1 + X_2}{4}$$

TABLE 5.4   Divergence of Gauss–Seidel Iteration

Iteration	X_1	$|\Delta X_1|$	X_2	$|\Delta X_2|$	X_3	$|\Delta X_3|$
0	1	–	1	–	1	–
1	9	8	–16	17	–6	7
2	–38	47	111	127	37.5	43.5
3	339.5	377.5	–934.5	1045.5	–318.25	355.75

The computations for three iteration cycles are shown in Table 5.4 using the Gauss–Seidel method with the same initial estimates as used for the calculations in Tables 5.2 and 5.3. It is evident that the solution is diverging and that, even with more iteration cycles, the solution would not converge. Since the solution converged when the equations were written in the order of Equations 5.152 through 5.154 but diverged for the order in Equations 5.167 through 5.169, we must conclude that convergence can depend on the order in which the equations are processed.

It is important to note that the divergence of the iterative calculations in Table 5.4 does not imply that a solution to Equations 5.167 through 5.169 does not exist. Recall that it is a permissible operation to interchange the order in which a set of equations is written; by the definition of a permissible operation, the solution should be unaffected by the reordering. However, although the solution may still exist, the method for calculating this solution may not work after the equations have been reordered. This is exactly the result in this example problem.

A means of ordering the equations so that the solution procedure converges can be demonstrated using two simultaneous equations ($n = 2$ in Equation 5.14):

$$(5.170) \qquad a_{11}X_1 + a_{12}X_2 = C_1$$

$$(5.171) \qquad a_{21}X_1 + a_{22}X_2 = C_2$$

If we solve both equations individually for X_1 we get

$$(5.172) \qquad X_1 = \frac{C_1}{a_{11}} - \frac{a_{12}}{a_{11}}X_2$$

$$(5.173) \qquad X_1 = \frac{C_2}{a_{21}} - \frac{a_{22}}{a_{21}}X_2$$

If we refer to Equations 5.172 and 5.173 as f_1 and f_2, respectively, and plot Equations 5.172 and 5.173 on an (X_1, X_2) axis system as shown in Figure 5.3, the solution occurs at the intersection of the two functions f_1 and f_2. It is easy to show by equating Equations 5.172 and 5.173 and solving for X_2 that the intersection of f_1 and f_2 is given by Equations 5.19 and 5.20. But, since we do not know the values of X_1 and X_2, we must make an initial estimate. If we select a value of X_2, say $X_1^{(1)}$, and substitute it into f_1, we get a value for X_1, say $X_1^{(1)}$. For cycle 2, we substitute $X_1^{(1)}$ into f_2 and get a revised value of X_2, which is denoted as $X_2^{(2)}$. The value $X_2^{(2)}$ is closer to

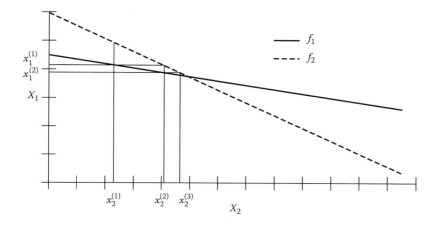

FIGURE 5.3   Converging order of equations.

the intersection of f_1 and f_2 than $X_1^{(1)}$. If we continue this iterative process, we will converge on the intersection of f_1 and f_2. This convergence occurs whether $X_2^{(1)}$ is greater than or less than the true value of X_2.

If the slope of the line of f_1 is greater than the slope of the line of f_2, then we have the situation shown in Figure 5.4. We begin with an estimate of X_2, say $X_2^{(1)}$, and use f_1 to obtain a value $X_1^{(1)}$. This value is then used with f_2 to get a value of X_2, say $X_2^{(2)}$. It is evident from Figure 5.4 that the initial estimate of X_2 was a better estimate of X_2 than is $X_2^{(2)}$. The same result occurs when we use f_1 and $X_2^{(2)}$ to obtain $X_1^{(2)}$. This process, if continued, results in a diverging solution, with the differences between successive estimates of X_1 and X_2 changing sign and increasing in magnitude. Again, divergence occurs regardless of the initial value $X_2^{(1)}$, unless the value of $X_2^{(1)}$ exactly equals the true value of X_2.

In comparing the configurations in Figures 5.3 and 5.4, it should be evident that for two simultaneous equations the iterative solution converges only when the initial value of $X_2^{(1)}$ is substituted into the equation that has

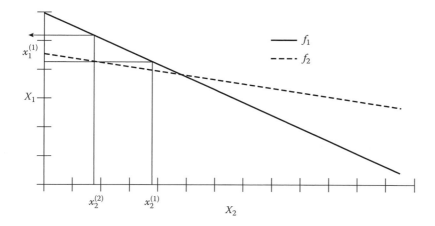

FIGURE 5.4   Diverging order of equations.

the smaller slope—that is, when the slope of f_1 is less than the slope of f_2. Thus the equations should be arranged such that X_1 is expressed in terms of X_2 with the following conditions:

$$(5.174) \qquad \left| \frac{a_{12}}{a_{11}} \right| < \left| \frac{a_{22}}{a_{21}} \right|$$

It should be evident from Equation 5.174 that this inequality holds if we make both a_{11} and a_{22} as large as possible. The term on the left side of the inequality of Equation 5.174 most probably is smaller if the equation having the largest value of a_{11} is selected as the first equation and the other equation as the second equation. A system of equations in which the coefficients with the largest absolute values appear on the diagonal is referred to as being diagonally dominant.

While it is not practical to try to show the convergence–divergence analysis for systems that include more than two equations, the underlying concepts are the same. We can increase the likelihood of convergence of our iterative solution scheme by selecting the equation with the largest coefficient for X_1 as the first equation, the equation with the largest coefficient for X_2 in the remaining equations as the second equation, and so on.

Example 5.11: Gauss–Seidel Solution for an Electrical Circuit

A system of three linear equations was developed to represent the distribution of current flow in the circuit of Figure 5.1. These equations were rearranged so that the matrix is diagonally dominant:

$$(5.175) \qquad I_1 + I_2 - I_3 = 0$$

$$(5.176) \qquad 2I_1 - 4I_2 = 4$$

$$(5.177) \qquad 2I_1 + 5I_3 = 6$$

Solving these equations for the variable on the diagonal yields

$$(5.178) \qquad I_1 = I_3 - I_2$$

$$(5.179) \qquad I_2 = \frac{2I_1 - 4}{4}$$

$$(5.180) \qquad I_3 = \frac{6 - 2I_1}{5}$$

Initial estimates of $I_1 = 1$ A (ampere), $I_2 = 0$ A, and $I_3 = 1$ A were selected; these initial estimates satisfy Kirchhoff's first law, which is represented by the first of the three linear equations. The Gauss–Seidel method was applied, with a partial listing shown in Table 5.5. Because of the slopes of the planes representing the equation, convergence is a slow process. It takes 20 cycles before the largest error in absolute value is 0.05 or less. After 31 cycles, the process indicates

TABLE 5.5   Solution for Electrical Circuit

Iteration	I_1	$|\Delta I_1|$	I_2	$|\Delta I_2|$	I_3	$|\Delta I_3|$
0	1	–	0	–	1	–
1	1	0	-0.5	0.50	0.8	0.20
2	1.3	0.30	-0.35	0.15	0.68	0.12
3	1.03	0.27	-0.49	0.14	0.47	0.11
4	1.27	0.24	-0.36	0.13	0.69	0.10
5	1.05	0.22	-0.47	0.11	0.78	0.09
6	1.25	0.20	-0.37	0.10	0.70	0.08
7	1.07	0.18	-0.46	0.09	0.77	0.07
10	1.22	0.13	-0.39	0.07	0.71	0.05
15	1.12	0.08	-0.44	0.04	0.75	0.03
16	1.19	0.07	-0.40	0.04	0.72	0.03
20	1.18	0.05	-0.41	0.02	0.73	0.02
21	1.14	0.04	-0.43	0.02	0.74	0.01
25	1.15	0.02	-0.43	0.02	0.74	0.01
26	1.17	0.02	-0.42	0.01	0.73	0.01
30	1.17	0.02	-0.42	0.01	0.73	0.01
31	1.15	0.02	-0.42	0.00	0.74	0.01

the following solution: I_1 = 1.15 A, I_2 = −0.42 A, and I_3 = 0.74 A. This solution results in absolute value errors of 0.01, 0.02, and 0.00 for the three unknown currents.

5.6.4 CONSIDERATIONS IN PROGRAMMING

Both the Jacobi and the Gauss–Seidel iteration methods are based on Equations 5.146 through 5.148. Each of these has the same general form:

(5.181)
$$X_i = \frac{C_i - \sum_{\substack{j=1 \\ j \neq i}}^{n} a_{ij} X_j}{a_{ii}} \quad \text{for } i = 1, \ldots, n$$

The summation includes all the n terms except for the case where $i = j$. One iteration cycle requires n evaluations of Equation 5.181, with i varying from 1 to n. Thus the general computational procedure consists of the following steps:

1. For iteration 1, assume initial estimates for all X_i.
2. For each subsequent iteration cycle, perform the following steps:
   a. Loop over all values of X_i from i = 1 to n, solving Equation 5.181 for each new value of X_i.
   b. Check for convergence, and if the solution has converged, exit from the loop; otherwise, perform another cycle starting with step 2a.

These steps can be used to solve a system of equations by either Jacobi or Gauss–Seidel iteration. The only difference between the two methods is the handling of the X_i. For the Jacobi method, the values computed from Equation 5.181 should be stored in a vector labeled differently than the X_i, say Y_i. Thus Equation 5.181 becomes

$$(5.182) \qquad Y_i = \frac{C_i - \displaystyle\sum_{\substack{j=1 \\ j\neq i}}^{n} a_{ij} X_j}{a_{ii}} \qquad \text{for } i = 1,\ldots,n$$

After completing each iteration cycle, the values of Y_i must be transferred into the X_i vector for the start of the next iteration cycle.

5.7 USE OF DETERMINANTS

A determinant is a unique Fiscalar number that can be used to represent a square matrix. For the matrix $\mathbf{A}$, the determinant is denoted as $|\mathbf{A}|$. To solve a system of linear equations using determinants, it is necessary to view the equations in Equation 5.14 as a coefficient matrix $\mathbf{A}$ and the column vector of constants $\mathbf{C}$:

$$(5.183) \qquad \mathbf{A} = \begin{bmatrix} a_{11} & a_{12} & \cdots & a_{1n} \\ a_{21} & a_{22} & \cdots & a_{2n} \\ \vdots & \vdots & \vdots & \vdots \\ a_{n1} & a_{n2} & \cdots & a_{nn} \end{bmatrix}$$

and

$$(5.184) \qquad \mathbf{C} = \begin{bmatrix} C_1 \\ C_2 \\ \vdots \\ C_n \end{bmatrix}$$

The value of X_i is obtained using Cramer's rule as

$$(5.185) \qquad X_i = \frac{|\mathbf{A}_i|}{|\mathbf{A}|}$$

where $|\mathbf{A}|$ is the determinant of matrix $\mathbf{A}$ and $|\mathbf{A}_i|$ is the determinant of a matrix formed by replacing column i of $\mathbf{A}$ with the column vector of Equation 5.184. For example, $|\mathbf{A}_2|$ would be given by

$$(5.186) \qquad |\mathbf{A}_2| = \begin{bmatrix} a_{11} & C_1 & a_{13} & \cdots & a_{1n} \\ a_{21} & C_2 & a_{23} & \cdots & a_{2n} \\ \vdots & \vdots & \vdots & \vdots & \vdots \\ a_{n1} & C_n & a_{n3} & \cdots & a_{nn} \end{bmatrix}$$

Example 5.12: Three Linear Equations

The system of three linear equations given by Equations 5.149 to 5.151 is easily solved using Equation 5.185. The determinant of the coefficient matrix is given by

$$|\mathbf{A}| = \begin{vmatrix} -1 & 4 & -3 \\ 3 & 1 & -2 \\ 1 & -1 & 4 \end{vmatrix} = (-1)[1(4)-(-2)(-1)] - 4[(3(4)-(-2)(1)]$$
$$(5.187) \qquad\qquad\qquad -3[3(-1)-(1)(1)] = -46$$

The determinants for the matrices for the numerator of Equation 5.185 are given by

$$(5.188) \qquad |\mathbf{A}_1| = \begin{vmatrix} -8 & 4 & -3 \\ 9 & 1 & -2 \\ 1 & -1 & 4 \end{vmatrix} = -138$$

$$(5.189) \qquad |\mathbf{A}_2| = \begin{vmatrix} -1 & -8 & -3 \\ 3 & 9 & -2 \\ 1 & 1 & 4 \end{vmatrix} = 92$$

$$(5.190) \qquad |\mathbf{A}_3| = \begin{vmatrix} -1 & 4 & -8 \\ 3 & 1 & 9 \\ 1 & -1 & 1 \end{vmatrix} = 46$$

Thus the solution is

$$(5.191) \qquad X_1 = \frac{|\mathbf{A}_1|}{|\mathbf{A}|} = \frac{-138}{-46} = 3$$

$$(5.192) \qquad X_2 = \frac{|\mathbf{A}_2|}{|\mathbf{A}|} = \frac{92}{-46} = -2$$

$$(5.193) \qquad X_3 = \frac{|\mathbf{A}_3|}{|\mathbf{A}|} = \frac{46}{-46} = -1$$

Thus the solution of Equations 5.191 through 5.193 is the same solution that was determined for Equations 5.149 through 5.151 using the Jacobi iteration method.

Example 5.13: Method of Determinants for Material Purchasing

The method of determinants could be used to obtain a solution to the manufacturing problem (Equations 5.5 through 5.7). Using the system of linear equations and Cramer's rule, we get the following solution:

(5.194)
$$|\mathbf{X}| = \begin{vmatrix} 1 & 3 & 3 \\ 2 & 4 & 3 \\ 3 & 4 & 7 \end{vmatrix} = -7$$

and

(5.195)
$$X_1 = \dfrac{\begin{vmatrix} 15 & 3 & 2 \\ 22 & 4 & 3 \\ 39 & 4 & 7 \end{vmatrix}}{-7}$$

(5.196)
$$X_2 = \dfrac{\begin{vmatrix} 1 & 15 & 2 \\ 2 & 22 & 3 \\ 3 & 39 & 7 \end{vmatrix}}{-7}$$

(5.197)
$$X_3 = \dfrac{\begin{vmatrix} 1 & 3 & 15 \\ 2 & 4 & 22 \\ 3 & 4 & 39 \end{vmatrix}}{-7}$$

Therefore, the solution is

(5.198)
$$X_1 = \frac{-7}{-7} = 1$$

(5.199)
$$X_2 = \frac{-14}{-7} = 2$$

(5.200)
$$X_3 = \frac{-28}{-7} = 4$$

The solution is the same as that obtained using Gaussian elimination (Example 5.4).

Example 5.14: Method of Determinants for Electrical Circuit

The Gauss–Seidel solution for Example 5.11 required a large number of iterations to achieve convergence. The method of determinants would thus be easier to apply. The solution is as follows:

(5.201)
$$|\mathbf{I}| = \begin{vmatrix} 1 & 1 & -1 \\ 2 & 0 & 5 \\ 2 & -4 & 0 \end{vmatrix} = 38$$

and

(5.202)
$$I_1 = \frac{\begin{vmatrix} 0 & 1 & -1 \\ 6 & 0 & 5 \\ 4 & -4 & 0 \end{vmatrix}}{38}$$

(5.203)
$$I_2 = \frac{\begin{vmatrix} 1 & 0 & -1 \\ 2 & 6 & 5 \\ 2 & 4 & 0 \end{vmatrix}}{38}$$

(5.204)
$$I_3 = \frac{\begin{vmatrix} 1 & 1 & 0 \\ 2 & 0 & 6 \\ 2 & -4 & 4 \end{vmatrix}}{38}$$

Therefore, the solution is

(5.205)
$$I_1 = \frac{44}{38} = 1.158 \text{ A}$$

(5.206)
$$I_2 = \frac{-16}{38} = -0.421 \text{ A}$$

(5.207)
$$I_3 = \frac{28}{38} = 0.737 \text{ A}$$

The solution is exact for the number of significant digits shown. Comparing this solution with the results of the Gauss–Seidel solution, the errors for the Gauss–Seidel solution were less than 0.01 A in absolute value.

5.7.1 CONSIDERATIONS IN PROGRAMMING

The solution of Equation 5.185 for each X_i requires a subprogram that determines the determinant of a matrix. Given such a subprogram, the determinant $|\mathbf{A}|$ can be computed. Then it is only necessary to iterate over the X_i values from $i = 1$ to n. For each step, the vector of constants $\mathbf{C}$ is used to replace column i in matrix $\mathbf{A}$, which is the iteration loop counter, of the coefficient matrix; this replacement is accomplished using another iteration loop. The determinant $|\mathbf{A}_i|$ is then determined using the subprogram, and the value of X_i is computed.

5.8 MATRIX INVERSION

Matrix inversion is frequently necessary in the solution of engineering problems, as was discussed in Chapter 2. For example, in fitting a linear multiple regression equation (see Chapter 10), the vector of unknowns ($\mathbf{t}$) is related to the correlation matrix ($\mathbf{R}$) by

$$(5.208) \qquad \mathbf{t} = \mathbf{R}_{11}^{-1}\,\mathbf{R}_{12}$$

where $\mathbf{R}_{11}$ is the matrix of correlation coefficients between the predictor (independent) variables, and $\mathbf{R}_{12}$ is the vector of correlation coefficients between the dependent variable and each predictor variable. To solve Equation 5.208, it is necessary to find the inverse of $\mathbf{R}_{11}$—that is, $\mathbf{R}_{11}^{-1}$.

The inverse of a matrix $\mathbf{P}$ is defined by the following equation (see Section 2.2, Chapter 2):

$$(5.209) \qquad \mathbf{P}^{-1}\mathbf{P} = \mathbf{I}$$

in which $\mathbf{I}$ is the identity or unit matrix and both $\mathbf{P}$ and $\mathbf{I}$ are square matrices. The unit matrix is a square matrix with values of 1.0 on the principal diagonal and values of zero for all other elements. In Equation 5.209, we consider $\mathbf{P}$ and $\mathbf{I}$ to be given, with the elements of $\mathbf{P}^{-1}$ the unknowns. The values of $\mathbf{P}^{-1}$ can be computed by expanding Equation 5.209 into a set of n^2 simultaneous equations, where n is the order of the matrix. The solution of Equation 5.209 for the elements of $\mathbf{P}^{-1}$ can be demonstrated using the case where $n = 3$. Letting the elements of $\mathbf{P}^{-1}$ and $\mathbf{P}$ be denoted as q_{ij} and p_{ij}, respectively, Equation 5.209 becomes

$$(5.210) \qquad \begin{bmatrix} q_{11} & q_{12} & q_{13} \\ q_{21} & q_{22} & q_{23} \\ q_{31} & q_{32} & q_{33} \end{bmatrix} \cdot \begin{bmatrix} p_{11} & p_{12} & p_{13} \\ p_{21} & p_{22} & p_{23} \\ p_{31} & p_{32} & p_{33} \end{bmatrix} = \begin{bmatrix} 1 & 0 & 0 \\ 0 & 1 & 0 \\ 0 & 0 & 1 \end{bmatrix}$$

Expanding Equations 5.210 and 5.211 yields the following set of $n^2 = 9$ simultaneous equations:

$$p_{11}q_{11} + p_{21}q_{12} + p_{31}q_{13} = 1$$
$$p_{12}q_{11} + p_{22}q_{12} + p_{32}q_{13} = 0$$
$$p_{13}q_{11} + p_{23}q_{12} + p_{33}q_{13} = 0$$
$$p_{11}q_{21} + p_{21}q_{22} + p_{31}q_{23} = 0$$
$$p_{12}q_{22} + p_{22}q_{22} + p_{32}q_{23} = 1$$
$$p_{13}q_{23} + p_{23}q_{22} + p_{33}q_{23} = 0$$
$$p_{11}q_{31} + p_{21}q_{32} + p_{31}q_{33} = 0$$
$$p_{12}q_{31} + p_{22}q_{32} + p_{32}q_{33} = 0$$
$$p_{13}q_{31} + p_{23}q_{32} + p_{33}q_{33} = 1$$

(5.211)

Equation 5.210 includes nine simultaneous equations with nine q_{ij} unknowns (that is, q_{ij} for $i = 1, 2,$ and 3 and $j = 1, 2,$ and 3). The solution yields the values of the inverse $\mathbf{P}^{-1}$.

It should be evident from Equation 5.211 that the solution could be found by considering the set of nine equations to consist of three sets of three equations. The unknowns for row i of $\mathbf{P}^{-1}$ are independent in Equations 5.210 and 5.211 of the unknowns for every other row. The solution could be made by solving n sets of n simultaneous equations as given by

(5.212) $$\mathbf{P}^T\mathbf{R}_i = I_i \quad \text{for } i = 1,2,\ldots,n$$

where $\mathbf{P}^T$ is the transpose of the $(n \times n)$ $\mathbf{P}$ matrix of Equations 5.210 and 5.211, $\mathbf{R}_i$ is the nth-order column vector that consists of the elements of the ith row of the matrix $\mathbf{P}^{-1}$, and $\mathbf{I}_i$ is the nth-order column vector of the identity matrix.

Example 5.15: Inverse of a Matrix

To illustrate the estimation of the inverse of a matrix, consider the following correlation matrix $\mathbf{P}$:

(5.213) $$\mathbf{P} = \begin{bmatrix} 1 & -0.42 \\ -0.42 & 1 \end{bmatrix}$$

If we were interested in developing an equation to predict X_3 for values of X_1 and X_2, then we would be interested in the inverse of the correlation matrix for X_1 and X_2, which is

(5.214) $$\begin{bmatrix} q_{11} & q_{12} \\ q_{21} & q_{22} \end{bmatrix} \begin{bmatrix} 1 & -0.42 \\ -0.42 & 1 \end{bmatrix} = \begin{bmatrix} 1 & 0 \\ 0 & 1 \end{bmatrix}$$

Based on Equation 5.212, the values of q_{11} and q_{12} can be found by solving the following two simultaneous equations:

(5.215)
$$q_{11} - 0.42q_{12} = 1$$

(5.216)
$$-0.42q_{11} + q_{12} = 0$$

Equations 5.215 and 5.216 can be solved using any of the methods previously introduced. The method of determinants yields

(5.217)
$$q_{11} = \frac{\begin{vmatrix} 1 & -0.42 \\ 0 & 1 \end{vmatrix}}{\begin{vmatrix} 1 & -0.42 \\ -0.42 & 1 \end{vmatrix}} = 1.214$$

and

(5.218)
$$q_{12} = \frac{\begin{vmatrix} 1 & 1 \\ -0.42 & 0 \end{vmatrix}}{\begin{vmatrix} 1 & -0.42 \\ -0.42 & 1 \end{vmatrix}} = 0.510$$

For the elements of the second row of Equation 5.214, we get the following two simultaneous equations:

(5.219)
$$q_{21} - 0.42q_{22} = 0$$

(5.220)
$$-0.42q_{21} + q_{22} = 1$$

The method of determinants yields

(5.221)
$$q_{21} = \frac{\begin{vmatrix} 0 & -0.42 \\ 1 & 1 \end{vmatrix}}{\begin{vmatrix} 1 & -0.42 \\ -0.42 & 1 \end{vmatrix}} = 0.510$$

and

(5.222)
$$q_{22} = \frac{\begin{vmatrix} 1 & 0 \\ -0.42 & 1 \end{vmatrix}}{\begin{vmatrix} 1 & -0.42 \\ -0.42 & 1 \end{vmatrix}} = 1.214$$

Thus the inverse matrix is

(5.223)
$$\mathbf{P}^{-1} = \begin{bmatrix} 1.214 & 0.510 \\ 0.510 & 1.214 \end{bmatrix}$$

5.9 APPLICATIONS

5.9.1 FLEXIBILITY AND STIFFNESS ANALYSES OF A BEAM

The cantilever beam shown in Figure 5.5 can be loaded with a concentrated force (P) or moment (M) at the free end. The loads result in deformations at the free end in the form of a vertical deflection (y) and rotation (θ). The deformations due to a unit concentrated force applied at the free end are

(5.224)
$$y_P = \frac{L^3}{3EI}$$

(5.225)
$$\theta_P = \frac{L^2}{2EI}$$

where E = modulus of elasticity, L = span length of the beam, and I = moment of inertia of the beam's cross-sectional area. The deformations due to a unit concentrated moment applied at the free end are

(5.226)
$$y_M = \frac{L^2}{2EI}$$

(5.227)
$$\theta_M = \frac{L}{EI}$$

The deformation expressions provided by Equations 5.224 through 5.227 are called the flexibility coefficients. They can be used to assemble the flexibility matrix ($\mathbf{F}$) as follows:

(5.228)
$$\mathbf{F} = \begin{bmatrix} \dfrac{L^3}{3EI} & \dfrac{L^2}{2EI} \\ \dfrac{L^2}{2EI} & \dfrac{L}{EI} \end{bmatrix}$$

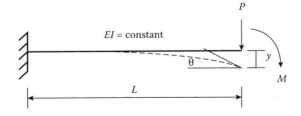

FIGURE 5.5   Cantilever beam.

The flexibility matrix has the symmetry property and can be used to compute the deformations due to any combination of P and M as

(5.229)
$$
\begin{bmatrix} y \\ \theta \end{bmatrix} = \begin{bmatrix} \dfrac{L^3}{3EI} & \dfrac{L^2}{2EI} \\[2mm] \dfrac{L^2}{2EI} & \dfrac{L}{EI} \end{bmatrix} \begin{bmatrix} P \\ M \end{bmatrix}
$$

The inverse of Equation 5.229 is sometimes needed for the analysis of statically indeterminate structures; that is,

(5.230)
$$
\begin{bmatrix} P \\ M \end{bmatrix} = \begin{bmatrix} \dfrac{L^3}{3EI} & \dfrac{L^2}{2EI} \\[2mm] \dfrac{L^2}{2EI} & \dfrac{L}{EI} \end{bmatrix}^{-1} \begin{bmatrix} y \\ \theta \end{bmatrix}
$$

where the inverse of the flexibility matrix is called the stiffness matrix ($\mathbf{S}$). The relationship between $\mathbf{F}$ and $\mathbf{S}$ according to Equations 5.210 and 5.211 can be expressed as

(5.231)
$$
\begin{bmatrix} S_{11} & S_{12} \\ S_{21} & S_{22} \end{bmatrix} \begin{bmatrix} \dfrac{L^3}{3EI} & \dfrac{L^2}{2EI} \\[2mm] \dfrac{L^2}{2EI} & \dfrac{L}{EI} \end{bmatrix} = \begin{bmatrix} 1 & 0 \\ 0 & 1 \end{bmatrix}
$$

Therefore, the following four equations can be obtained:

(5.232)
$$
\frac{L^3}{3EI} S_{11} + \frac{L^2}{2EI} S_{12} = 1
$$

(5.233)
$$
\frac{L^2}{2EI} S_{11} + \frac{L}{EI} S_{12} = 0
$$

(5.234)
$$
\frac{L^3}{3EI} S_{21} + \frac{L^2}{2EI} S_{22} = 0
$$

(5.235)
$$
\frac{L^2}{2EI} S_{21} + \frac{L}{EI} S_{22} = 1
$$

Equations 5.232 and 5.233 can be simultaneously solved for S_{11} and S_{12} using Equations 5.19 and 5.20 to obtain

(5.236)
$$
S_{11} = \frac{12EI}{L^3}
$$

(5.237)
$$S_{12} = -\frac{6EI}{L^2}$$

Equations 5.234 and 5.235 can be simultaneously solved for S_{21} and S_{22} using Equations 5.19 and 5.20 to obtain

(5.238)
$$S_{21} = -\frac{6EI}{L^2}$$

(5.239)
$$S_{22} = \frac{4EI}{L}$$

The results of Equations 5.236 through 5.239 can be used to assemble the stiffness matrix as

(5.240)
$$\mathbf{S} = \begin{bmatrix} \dfrac{12EI}{L^3} & -\dfrac{6EI}{L^2} \\ -\dfrac{6EI}{L^2} & \dfrac{4EI}{L} \end{bmatrix}$$

The stiffness matrix is also symmetric, similarly to the flexibility matrix. The stiffness matrix can be used to obtain the forces (e.g., reactions) due to known deformations (e.g., boundary conditions) according to Equation 5.230.

In practice, the analysis of real indeterminate structures can result in flexibility and stiffness matrices of relatively large sizes requiring numerical evaluations rather than analytical ones, as was performed in Equations 5.236 through 5.239. The concepts behind the development of these matrices is similar to what is covered herein; however, a complete development of a case study would require more advanced background in the stiffness analysis of structures.

5.9.2 CONCRETE MIX DESIGN

A concrete mix was designed to have the weight proportions of [cement: fine aggregates: coarse aggregates] as [1 : 1.9 : 2.8] with a water–cement ratio of 7 gallons of water per sack of cement. A sack of cement weighs 94 pounds. Table 5.6 shows the unit weights for the components of the

TABLE 5.6   Unit Weights for the Components of the Mix

Component	Unit Weight
Cement	195 pcf
Fine aggregates	165 pcf
Coarse aggregates	165 pcf
Water	62.4 pcf

mix. How much of each constituent is needed to produce 50 cubic yards of concrete?

Using the notations of C = cement weight (lb), F = weight of fine aggregates (lb), G = weight of coarse aggregates (lb), and W = water volume (gal), the following conditions can be developed based on the proportionality requirements:

(5.241)
$$F = 1.9C$$

(5.242)
$$G = 2.8C$$

Using the water–cement ratio, the following condition can be used:

(5.243)
$$W = \frac{7C}{94}$$

Using the conversion factor of 1 cubic yard = 27 ft³, the requirement for the total amount of concrete can be expressed as

(5.244)
$$50(27) = \frac{C}{195} + \frac{F+G}{165} + \frac{W}{7.48}$$

where the factor 7.48 is to convert gallons to cubic feet. Equations 5.241 through 5.244 can be written as

(5.245)
$$\begin{bmatrix} 1.9 & -1 & 0 & 0 \\ 2.8 & 0 & -1 & 0 \\ \frac{7}{94} & 0 & 0 & -1 \\ \frac{1}{195} & \frac{1}{165} & \frac{1}{165} & \frac{1}{7.48} \end{bmatrix} \begin{bmatrix} C \\ F \\ G \\ W \end{bmatrix} = \begin{bmatrix} 0 \\ 0 \\ 0 \\ 1350 \end{bmatrix}$$

The system of equations can be arranged such that nonzero elements are on the diagonal of the matrix, resulting in

(5.246)
$$\begin{bmatrix} \frac{1}{195} & \frac{1}{165} & \frac{1}{165} & \frac{1}{74.8} \\ 1.9 & -1 & 0 & 0 \\ 2.8 & 0 & -1 & 0 \\ \frac{7}{94} & 0 & 0 & -1 \end{bmatrix} \begin{bmatrix} W \\ C \\ F \\ G \end{bmatrix} = \begin{bmatrix} 1350 \\ 0 \\ 0 \\ 0 \end{bmatrix}$$

The Gauss–Jordan elimination can be used on the following coefficient matrix:

(5.247)
$$\begin{bmatrix} \frac{1}{195} & \frac{1}{165} & \frac{1}{165} & \frac{1}{7.48} & 1350 \\ 1.9 & -1 & 0 & 0 & 0 \\ 2.8 & 0 & -1 & 0 & 0 \\ \frac{7}{94} & 0 & 0 & -1 & 0 \end{bmatrix}$$

The result of the Gauss–Jordan elimination is

(5.248)
$$\begin{bmatrix} 1 & 0 & 0 & 0 & 2307.4 \\ 0 & 1 & 0 & 0 & 30{,}985.6 \\ 0 & 0 & 1 & 0 & 58{,}872.6 \\ 0 & 0 & 0 & 1 & 86{,}759.6 \end{bmatrix}$$

Therefore, the amounts of the components of the mix are

(5.249)   Water = 2307.4 gal

(5.250)   Cement = 30,985.6 lb = 330 sacks

(5.251)   Fine aggregates = 58,872.6 lb = 29.4 tons

(5.252)   Coarse aggregates = 86,759.6 lb = 43.4 tons

PROBLEMS

Find the solutions to Problems 5.1 through 5.6 using Gaussian elimination.

5.1
$$X_1 + 2X_2 - X_3 = 3.1$$
$$-3X_1 + X_2 + X_3 = 1.4$$
$$-X_1 - X_2 + 4X_3 = 7.3$$

5.2
$$-2X_1 + 3X_2 - X_3 = -0.6$$
$$3X_1 + X_2 - 2X_3 = -3.3$$
$$X_1 + 2X_2 + X_3 = 1.9$$

5.3
$$2X_1 - 3X_2 - X_3 = -3.7$$
$$-3X_1 + 3X_2 - X_3 = -1.0$$
$$4X_1 - 2X_2 + 3X_3 = 4.4$$

5.4
$$3X_1 + 2X_2 - 4X_3 = -5.9$$
$$7X_1 - 4X_2 + 3X_3 = -5.7$$
$$-2X_1 + 2X_2 - 3X_3 = 1.2$$

5.5
$$2X_1 - 3X_2 + 2X_3 - 4X_4 = 2.3$$
$$-X_1 - 2X_2 + 3X_3 + 2X_4 = -3.0$$
$$2X_1 + X_2 - 4X_3 - 2X_4 = 7.7$$
$$-3X_1 + 4X_2 - X_3 + X_4 = -7.0$$

5.6
$$4X_1 - 2X_2 + 6X_3 - 2X_4 = 11.0$$
$$-3X_1 + 2X_2 + 2X_3 - 4X_4 = -9.9$$
$$X_1 + 5X_2 - 2X_3 - 2X_4 = -3.3$$
$$-2X_1 + X_2 - 3X_3 + X_4 = -5.5$$

5.7 $8X_1 + 7X_2 + 6X_3 = 21$
$18X_1 + 18X_2 + 16X_3 = 52$
$9X_1 + 8X_2 + 7X_3 = 24$

5.8 $2X_1 - 4X_2 + 3X_3 = 0$
$-X_1 + 2X_2 + 3X_3 = -9$
$4X_1 + X_2 - 5X_3 = 13$

Find the solutions to Problems 5.9 through 5.16 using Gauss–Jordan elimination.

5.9 $3X_1 + 3X_2 - 2X_3 = 7.6$
$2X_1 - 4X_2 + X_3 = 1.4$
$-X_1 - 2X_2 + 5X_3 = -6.3$

5.10 $2X_1 - 4X_2 + 6X_3 = 1.4$
$3X_1 + X_2 - 3X_3 = -9.8$
$-X_1 + 2X_2 - 3X_3 = -0.7$

5.11 $5X_1 - 3X_2 - X_3 = 20.3$
$2X_1 + 2X_2 - 3X_3 = -4.0$
$-3X_1 - 4X_2 + 5X_3 = 9.6$

5.12 $2X_1 - 3X_2 - X_3 = 5.2$
$-3X_1 - X_2 + 4X_3 = 3.9$
$5X_1 + 3X_2 - 4X_3 = -11.3$

5.13 $2X_1 - 3X_2 - X_3 + 2X_4 = 11.7$
$-3X_1 + X_2 + 2X_3 - X_4 = -11.5$
$-X_1 - 2X_2 + 3X_3 + 5X_4 = -0.4$
$3X_1 + 3X_2 - X_3 + X_4 = 3.6$

5.14 $-X_1 + 2X_2 + 3X_3 - 4X_4 = 1.5$
$2X_1 - 3X_2 - X_3 - X_4 = -12.0$
$-X_1 - 3X_2 - X_3 + 5X_4 = -1.5$
$3X_1 + 4X_2 + X_3 - 2X_4 = 5.1$

5.15 $3X_1 + 5X_2 - 6X_3 = -41$
$-2X_1 + 2X_2 - 2X_3 = -24$
$-X_1 - 4X_2 + 6X_3 = 43$

5.16 $0.6X_1 - 0.4X_2 + 11.1X_3 = 2.8$
$-0.5X_1 - 0.7X_2 - 0.9X_3 = 0.8$
$-0.6X_1 - 1.3X_2 + 0.2X_3 = 3.7$

Find the solutions to Problems 5.17 through 5.22 using the **LU** decomposition method.

5.17 $X_1 + X_2 = 3$
$4X_1 - 2X_2 = 2$

5.18 $2X_1 + 5X_2 = 3$
$-3X_1 - X_2 = 2$

5.19   The set of equations in Problem 5.1.

5.20   The set of equations in Problem 5.2.

5.21   The set of equations in Problem 5.8.

5.22   The set of equations in Problem 5.16.

Find the solutions to Problems 5.23 through 5.28 using the Jacobi iteration method.

5.23
$$3X_1 - X_2 + 2X_3 = -3.7$$
$$2X_1 + 3X_2 - 2X_3 = 17.9$$
$$-X_1 + 2X_2 + 4X_3 = -6.2$$

5.24
$$4X_1 - 3X_2 - X_3 = -2.1$$
$$3X_1 - 2X_2 + 5X_3 = -12.3$$
$$-2X_1 + 2X_2 + 3X_3 = 12.9$$

5.25
$$2X_1 + X_2 - 3X_3 + 2X_4 = 3.7$$
$$-3X_1 - X_2 - 4X_3 + 5X_4 = 3.3$$
$$X_1 + 5X_2 - 3X_3 + 4X_4 = 4.1$$
$$5X_1 - 3X_2 + 2X_3 - 2X_4 = 6.4$$

5.26
$$3X_1 - 3X_2 + 2X_3 - 4X_4 = 7.9$$
$$-2X_1 - X_2 + 3X_3 - X_4 = -12.5$$
$$5X_1 - 2X_2 - 3X_3 + 2X_4 = 18.0$$
$$-2X_1 + 4X_2 + X_3 + 2X_4 = -8.1$$

5.27
$$6X_1 + 4X_2 - 5X_3 = -2.6$$
$$2X_1 - 7X_2 + 9X_3 = 9.7$$
$$-4X_1 + 8X_2 - 7X_3 = -9.8$$

5.28
$$2.6X_1 - 3.1X_2 + 4.5X_3 - 1.7X_4 = 1.46$$
$$-1.4X_1 + 4.1X_2 + 3.7X_3 - 0.9X_4 = 9.86$$
$$-3.3X_1 - 1.6X_2 + 5.4X_3 + 3.6X_4 = 10.41$$
$$4.4X_1 + 2.8X_2 - 3.3X_3 + 4.7X_4 = -3.63$$

Find the solutions to Problems 5.29 through 5.36 using the Gauss–Seidel iteration method.

5.29
$$5X_1 - 2X_2 + 3X_3 = 1.3$$
$$-2X_1 + 4X_2 - X_3 = -1.4$$
$$2X_1 - X_2 + 3X_3 = 1.3$$

5.30
$$-2X_1 + 3X_2 - X_3 = -5.7$$
$$3X_1 - 2X_2 - 5X_3 = 2.1$$
$$X_1 + 4X_2 + 2X_3 = 17.1$$

5.31
$$3X_1 - 3X_2 + 6X_3 = -5.7$$
$$2X_1 + 4X_2 - 2X_3 = 11.2$$
$$-X_1 + X_2 - 2X_3 = 1.9$$

5.32
$$5X_1 + 2X_2 - 3X_3 = -25.2$$
$$-3X_1 + 4X_2 + 4X_3 = 19.2$$
$$2X_1 - 2X_2 + 3X_3 = 2.8$$

5.33
$$2X_1 - 3X_2 + 3X_3 - X_4 = -7.9$$
$$-5X_1 + X_2 + X_3 - 2X_4 = -8.1$$
$$2X_1 + 3X_2 + 4X_3 - 3X_4 = -9.2$$
$$-X_1 - 2X_2 + 2X_3 - 3X_4 = -11.8$$

5.34
$$3X_1 + 4X_2 - 2X_3 - X_4 = 29.9$$
$$4X_1 - 2X_2 + 2X_3 + 3X_4 = 2.4$$
$$-2X_1 - 3X_2 + 4X_3 - X_4 = -29.6$$
$$2X_1 + 2X_3 - 3X_4 = 2.6$$

5.35
$$2.6X_1 - 3.8X_2 + 4.1X_3 = 4.62$$
$$6.1X_1 + 1.9X_2 = 2.79$$
$$-3.4X_1 - 4.6X_2 - 2.7X_3 = 3.42$$

5.36
$$6.1X_1 - 2.8X_2 - 3.3X_3 + 4.4X_4 = 1.09$$
$$-2.5X_1 + 3.6X_2 + 1.7X_3 + 5.1X_4 = 5.69$$
$$-1.9X_1 + 4.0X_2 + 3.6X_3 - 1.8X_4 = 3.65$$
$$2.8X_1 - 1.6X_2 + 1.4X_3 - 1.7X_4 = 0.38$$

For Problems 5.37 through 5.42, find the solutions using both the Jacobi and Gauss–Seidel iteration methods and compare the rates of convergence.

5.37
$$2X_1 - 3X_2 = 6.5$$
$$5X_1 + 2X_2 = 9.6$$

5.38
$$6X_1 - 5X_2 = -36.3$$
$$3X_1 + 4X_2 = -4.5$$

5.39
$$3X_1 + X_2 - 3X_3 = -4.2$$
$$-2X_1 + 4X_2 - X_3 = -14.1$$
$$4X_1 - 2X_2 + 3X_3 = 15.7$$

5.40
$$4X_1 - 2X_2 + 3X_3 = -10.8$$
$$3X_1 + 5X_2 - 2X_3 = 11.9$$
$$-2X_1 - 2X_2 + 5X_3 = -0.2$$

5.41
$$2.3X_1 - 1.4X_2 = -0.93$$
$$-0.8X_1 + 1.8X_2 = 4.00$$

5.42
$$3.2X_1 - 1.7X_2 - 1.6X_3 = -8.57$$
$$-1.1X_1 + 0.6X_2 + 3.4X_3 = 9.23$$
$$0.9X_1 - 0.2X_2 + 3.7X_3 = 6.97$$

Find the solutions to Problems 5.43 through 5.50 using the method of determinants.

5.43
$$2X_1 + 3X_2 = -0.3$$
$$-3X_1 + 4X_2 = -12.3$$

5.44
$$6X_1 - 5X_2 = -19.6$$
$$3X_1 + 4X_2 = 11.0$$

5.45
$$3X_1 - 2X_2 + 4X_3 = 1.2$$
$$-2X_1 + 5X_2 = 10.7$$
$$2X_1 + 3X_2 - 4X_3 = 8.5$$

5.46
$$6X_1 - 5X_2 + 2X_3 = -20.3$$
$$3X_1 - 2X_2 + 7X_3 = 2.1$$
$$4X_2 - 3X_3 = 1.1$$

5.47
$$2X_1 - 3X_2 + X_3 - 2X_4 = -10.3$$
$$3X_1 + 4X_2 - 5X_3 + X_4 = 4.3$$
$$-2X_1 - 2X_2 + 4X_4 = -5.4$$
$$X_1 + X_2 - 3X_3 + X_4 = -1.8$$

5.48
$$5X_1 - X_2 + X_3 + 3X_4 = 11.9$$
$$-2X_1 + 3X_2 - X_3 + 2X_4 = -2.0$$
$$3X_1 + X_2 - 3X_3 + 5X_4 = 14.1$$
$$2X_1 - 4X_2 + X_3 + 2X_4 = 4.4$$

5.49   Solve the equations in Problem 5.8.

5.50   Solve the equations in Problem 5.28.

For Problems 5.51 through 5.56, find the inverse of the coefficient matrix shown.

5.51
$$\begin{bmatrix} 1 & -0.45 \\ -0.45 & 1 \end{bmatrix}$$

5.52
$$\begin{bmatrix} 2.1 & -1.4 \\ 0.7 & 1.8 \end{bmatrix}$$

5.53
$$\begin{bmatrix} 1 & -0.62 & 0.37 \\ -0.62 & 1 & -0.51 \\ 0.37 & -0.51 & 1 \end{bmatrix}$$

5.54
$$\begin{bmatrix} 7 & -4 & 3 \\ -2 & 2 & -3 \\ 3 & 2 & -4 \end{bmatrix}$$

5.55
$$\begin{bmatrix} 1 & 0.62 \\ 0.62 & 1 \end{bmatrix}$$

5.56
$$\begin{bmatrix} 1.0 & -0.2 & 0.3 \\ -0.2 & 1.0 & 0.6 \\ 0.3 & 0.6 & 1.0 \end{bmatrix}$$

In the statistical fitting of a linear equation of the form $Y = a + bX$, values for the fitting coefficients a and b are obtained by solving the two simultaneous equations

$$an + b\sum_{i=1}^{n} X_i = \sum_{i=1}^{n} Y_i$$

and

$$a\sum_{i=1}^{n} X_i + b\sum_{i=1}^{n} X_i^2 = \sum_{i=1}^{n} X_i Y_i$$

where n is the number of pairs of values of the two variables X and Y, and each summation is made over the n values. For Problems 5.57 through 5.60, solve for the values of a and b using any of the numerical algorithms to solve simultaneous equations.

5.57

I	X_i	Y_i
1	1	2
2	2	1
3	3	1
4	3	3
5	6	4

5.58

I	X_i	Y_i
1	0	5
2	1	3
3	2	4
4	4	3
5	6	1
6	7	2

5.59

I	X_i	Y_i
1	4	-2
2	9	-3
3	13	2
4	21	1
5	24	4
6	28	7

5.60

I	X_i	Y_i
1	1	12
2	3	8
3	4	9
4	7	5
5	9	1
6	12	4

Formulate a set of linear equations that represent the relationship between voltage, current, and resistances for the circuits shown in Problems 5.61 through 5.66. Solve the simultaneous equations using any of the methods discussed in this chapter.

5.61

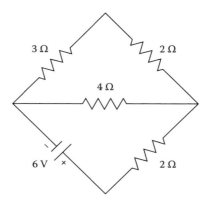

5.62

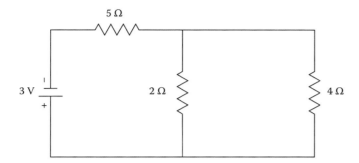

5.63

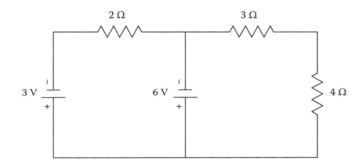

5.64

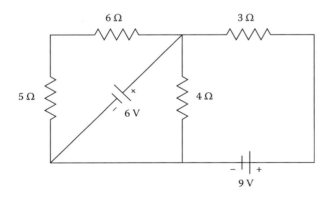

5.65 Write a computer program for the Gauss elimination method as outlined in Section 5.4.3 (in Equations 5.90 through 5.96). Test the program using Examples 5.3 through 5.5.

5.66 Write a computer program for the Gauss–Jordan elimination method as outlined in Equations 5.90 through 5.96 in Section 5.4.3. Test the program using Example 5.6.

Numerical Interpolation

6

6.1 INTRODUCTION

Problems requiring interpolation between individual data points occur frequently in science and engineering. For example, the design of a computerized energy control system for a building may require the typical temperature variation occurring in the building each day as input data. Sample temperature values would be measured in the building at discrete time points. However, the computer program for the energy control system may require temperature at times other than those at which the sample measurements were taken, such as on an hourly increment. One way to overcome this problem is to have the program fit a curve to the measured temperature points and interpolate for values between the sample measurements.

Interpolation is required in many engineering applications that use tabular data as input. The basis of all interpolation algorithms is the fitting of some type of curve or function to a subset of the tabular data; linear interpolation uses a straight line. Interpolation algorithms differ in the form of their interpolation functions.

6.2 METHOD OF UNDETERMINED COEFFICIENTS

The method of undetermined coefficients is conceptually the simplest interpolation algorithm, and it illustrates many of the key points that hold for all interpolation schemes. In this method, an nth-order polynomial is used as the interpolation function, $f(x)$:

$$(6.1) \qquad f(x) = b_0 + b_1 x + b_2 x^2 + b_3 x^3 + \ldots + b_n x^n$$

The constants in Equation 6.1, b_0, b_1, b_2, b_3, ..., b_n, are determined using the measured data points.

As an example of the method of undetermined coefficients, consider the easiest case: linear interpolation—that is, an interpolation function consisting of a straight line. Table 6.1 contains numbers and their cubes for the purpose of illustration.

The method of undetermined coefficients is used to estimate the cube of 2.2. Of course, in this case, the answer can be calculated exactly, if we desired; however, linear interpolation is used instead

TABLE 6.1   Data for Linear Interpolation

x	x^3
0	0
1	1
2	8
3	27
4	64

to illustrate the general concepts. By truncating all except the first two terms of Equation 6.1, the linear interpolation function has the form

(6.2) $$f(x) = b_0 + b_1 x$$

Since there are two unknown coefficients (b_0 and b_1 in Equation 6.2), two data points are necessary to obtain values for the coefficients. Any two data points could be used, but common sense dictates that the best approximation is obtained when the two data points closest to the value we are trying to determine are used. As shown in Figure 6.1, the data points corresponding to $x = 2$ and $x = 3$ are used to estimate the value for $x = 2.2$; our interpolation function $f(x)$ is then the straight line connecting these two points from the table. Using concepts from algebra, the equation of the linear interpolation function is

(6.3) $$f(x) = f(x_i) + \frac{x - x_i}{x_{i+1} - x_i}[f(x_{i+1}) - f(x_i)]$$

The ratio $[f(x_{i+1}) - f(x_i)]/(x_{i+1} - x_i)$ is the slope of the line between the two points. The difference $(x - x_i)$ represents a deviation from point x_i so that the product of the deviation and slope represents a correction to the base point value $f(x_i)$. Substituting values from Table 6.1 yields

(6.4) $$f(x) = 8 + \frac{x - 2}{3 - 2}(27 - 8)$$

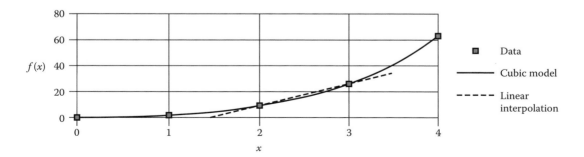

FIGURE 6.1   Linear interpolation.

Rearranging yields

(6.5) $f(x) = -30 + 19x$

At the two observation points, the values for $f(x)$ from Equation 6.5 exactly equal the cubes of the x values; at values for x other than the observation points, the values for $f(x)$ from Equation 6.5 only approximate the true values for the cubes of x. From Figure 6.1, it is evident that the accuracy of the approximation varies with the value of x, with the poorest accuracy within the range $2 < x < 3$ near the center of the interval.

Equation 6.3 is an algebraic form for estimating $f(x)$. Algebraic manipulation of Equation 6.3 yields the following, which has the form of Equation 6.2:

(6.6) $$f(x) = \frac{x_{i+1} f(x_i) - x_i f(x_{i+1})}{x_{i+1} - x_i} + \frac{f(x_{i+1}) - f(x_i)}{x_{i+1} - x_i} x$$

It can be noted that the manipulation produced the slope in the second term of Equation 6.6. Using the preceding values of x and x^3 with Equation 6.6 yields

(6.7) $f(x) = -30 + 19x$

This is the same interpolation function derived with Equation 6.3.

Although we could write the interpolation function in Equations 6.4 and 6.5 by inspection for this simple case, we can alternatively approach the problem in a more methodical fashion. Since the interpolation passed through the two data points spanning $x = 2.2$, Equation 6.2 can be used with the values of the two points nearest to $x = 2.2$ to write the following two equations:

(6.8) $8 = b_0 + 2b_1$

(6.9) $27 = b_0 + 3b_1$

Equations 6.8 and 6.9 represent a system of two equations for the two unknown coefficients. These equations can be solved for b_0 and b_1 using the usual methods from algebra or the equation-solving algorithms covered in Chapter 5 to yield the solution $b_0 = -30$ and $b_1 = 19$. These values can be substituted into Equation 6.2, producing the same interpolation function as we obtained in Equation 6.7.

The linear interpolation function predicts a value of 11.8 for the cube of $x = 2.2$. Since the actual value for the cube of 2.2 is 10.648, the discrepancy or error in the interpolation is 1.152, or 10.8% of the true value at this point. As shown in Figure 6.1, the reason for this discrepancy is the fact that the interpolation function only approximately matches the actual function between the two end points; in this case, the interpolation function always lies above the actual function. The error in the interpolation will always be zero at the observation end points and a maximum at some location between the end points.

The interpolation function could also be used to extrapolate values for the cube of x outside the range of our two observation points. However, as shown in Figure 6.1, the errors in the approximation quickly grow very large. The extrapolated value from Equation 6.7 for the cube of 4.0 is 46, which differs from the true value of 64 by over 28%. Extrapolation, in general, gives larger errors than interpolation.

Using a function that is more flexible than a straight line, the accuracy of an estimate can be improved. For example, a nonlinear function that has more coefficients than the linear interpolation function should provide greater accuracy. However, since a nonlinear function may contain more unknown coefficients than the straight line, it may be necessary to use more of the data points to estimate the coefficients. For example, by truncating the terms of Equation 6.1 beyond the x^2 term, the following second-order polynomial could be used as an interpolation function:

$$(6.10) \qquad f(x) = b_0 + b_1 x + b_2 x^2$$

Determining the three unknown coefficients of Equation 6.10 requires three pairs of values of x and its cube. To estimate the value of x^3 for $x = 2.2$ using the data of Table 6.1, coefficients for Equation 6.10 can be fitted using x values of either 1, 2, and 3 or 2, 3, and 4. Since 2.2 is closer to 2 than to 3, 2 will be used as the center point. Evaluating Equation 6.10 at the three data points yields the following set of simultaneous equations:

$$(6.11) \qquad 1 = b_0 + b_1 + b_2$$

$$(6.12) \qquad 8 = b_0 + 2b_1 + 4b_2$$

$$(6.13) \qquad 27 = b_0 + 3b_1 + 9b_2$$

Solving these three simultaneous equations yields the following values for the coefficients of Equation 6.10: $b_0 = 6$, $b_1 = -11$, and $b_2 = 6$. Therefore, the interpolation equation is

$$(6.14) \qquad f(x) = 6 - 11x + 6x^2$$

The estimated value of $(2.2)^3$ from Equation 6.14 is 10.840, which differs from the true value (10.648) by 1.8%. Thus, using a second-order function of Equation 6.14 instead of the first-order function of Equation 6.7 reduced the interpolation error by nearly an order of magnitude. The cost of this increased precision is the additional effort required to determine the coefficients of the second-order equation—that is, Equation 6.10 rather than Equation 6.2.

The method of undetermined coefficients can be generalized to interpolation polynomials of any order. The general nth-order polynomial contains $n + 1$ undetermined coefficients and therefore requires $n + 1$ distinct data points for its solution. Each of the $n + 1$ data points is substituted into the interpolation polynomial in turn, generating $n + 1$ simultaneous equations that are solved to obtain the polynomial coefficients.

6.3 GREGORY–NEWTON INTERPOLATION METHOD

The previous analysis indicated that the second-order polynomial provides a significant improvement in accuracy when compared with the linear, or first-order, equation. Thus it may seem reasonable to try an interpolation polynomial of higher order. However, we need to recognize that the solution procedure becomes more complex as the order of the polynomial increases when the solution procedure based on solving a set of simultaneous equations is used. The Gregory–Newton method provides a means of developing an nth order interpolation polynomial without requiring the solution of a set of simultaneous equations. The nth-order Gregory–Newton interpolation formula has the form

$$f(x) = a_1 + a_2(x - x_1) + a_3(x - x_1)(x - x_2) + a_4(x - x_1)(x - x_2)(x - x_3) + \dots$$
$$+ a_n(x - x_1)(x - x_2)\cdots(x - x_{n-1}) + a_{n+1}(x - x_1)(x - x_2)\dots(x - x_n)$$

(6.15)

in which
   x is the independent variable
   x_i for $i = 1, 2, \dots, n$ are n known values of the independent variable
   a_i for $i = 1, 2, \dots, (n + 1)$ are the unknown coefficients
   $f(x)$ is the value of the dependent variable for a value x of the independent variable

The $(n + 1)$ coefficients of Equation 6.15 can be evaluated from the set of $(n + 1)$ values of the dependent and independent variables. Assuming that the $(n + 1)$ values of the independent variable are placed in increasing order (that is, $x_j < x_{j+1}$), the value of a_1 can be computed from Equation 6.15 using $f(x_j)$ and x_1 For this pair of values, every term in Equation 6.15 except the constant a_1 is zero because $(x - x_1)$ equals $(x_1 - x_1)$, which equals zero. Thus the value of a_1 equals $f(x_j)$. Given the just-computed value of a_1 and the pair of values x_2 and $f(x_2)$, we can compute a_2. For this case, all terms in Equation 6.15 that contain $(x - x_2)$ are zero because $(x_2 - x_2)$ is zero. Thus a_2 is derived from

(6.16) $$f(x_2) = a_1 + a_2(x_2 - x_1)$$

which can be solved for a_2:

(6.17) $$a_2 = \frac{f(x_2) - a_1}{x_2 - x_1} = \frac{f(x_2) - f(x_1)}{x_2 - x_1}$$

The coefficient a_3 can be computed using the values of a_1 and a_2, and the pair $[x_3, f(x_3)]$. In this case, all terms containing $(x_3 - x_3)$ are zero, and therefore

$$(6.18) \qquad f(x_3) = a_1 + a_2(x_3 - x_1) + a_3(x_3 - x_1)(x_3 - x_2)$$

Solving for a_3 yields

$$(6.19) \qquad a_3 = \frac{f(x_3) - a_1 - a_2(x_3 - x_1)}{(x_3 - x_1)(x_3 - x_2)}$$

The process can be repeated to solve for the remaining coefficients. At each step a new pair of points $[x_i$ and $f(x_i)]$ is used with the values of a_i computed in previous steps. The Gregory–Newton method yields the same polynomial as the solution of the simultaneous equations used for the method of undetermined coefficients. Thus it will provide the same interpolated value and have the same accuracy.

Example 6.1: Cubic Function

The Gregory–Newton method can be demonstrated using the three pairs of values of x and x^3 from Table 6.1. Using the pair (1, 1), we get $f(x_1) = a_1 = 1$. Using the second pair (2, 8) and Equation 6.17, the value of a_2 can be computed as

$$(6.20) \qquad a_2 = \frac{8 - 1}{2 - 1} = 7$$

The third data pair (3, 27) and Equation 6.19 can be used to compute the value of a_3:

$$(6.21) \qquad a_3 = \frac{27 - 1 - 7(3 - 1)}{(3 - 1)(3 - 2)} = 6$$

This produces the following interpolation polynomial, which has the form of Equation 6.15:

$$(6.22) \qquad f(x) = 1 + 7(x - 1) + 6(x - 1)(x - 2)$$

Thus, for $x = 2.2$ we get

$$(6.23) \qquad f(2.2) = 1 + 7(2.2 - 1) + 6(2.2 - 1)(2.2 - 2) = 10.84$$

which is the same interpolated value as was computed before using Equations 6.11 through 6.14. Equation 6.22 can be rearranged to produce

(6.24) $$f(x) = 6 - 11x + 6x^2$$

This is the same interpolation function derived previously using the method of undetermined coefficients (Equation 6.14).

Example 6.2: Vapor Pressure of Water

The vapor pressure of water is an important element in many engineering problems, including chemical reactions involving steam, evapotranspiration estimation problems in agricultural engineering design, and cavitation problems in hydraulic engineering. Let us assume that the vapor pressure of water between temperatures of 40°C and 72°C are of interest. Table 6.2 provides five pairs of values at 8° increments.

The interpolation polynomial can be formed by applying Equation 6.15 for each pair of values in increasing order of T. The calculations in this example are summarized to four significant figures in Table 6.3. This yields the following interpolating polynomial:

TABLE 6.2 Vapor Pressure Data

T (°C)	P (mm Hg)
40	55.3
48	83.7
56	123.8
64	179.2
72	254.5

TABLE 6.3 Interpolation of Vapor Pressure

i	T (°C)	P (mm Hg)	Equation 6.15	a_i
1	40	55.3	$55.3 = a_1$	55.3
2	48	83.7	$83.7 = a_1 + a_2(48 - 40)$	3.55
3	56	123.8	$123.8 = a_1 + a_2(56 - 40)$ $+ a_3(56 - 40)(56 - 48)$	0.0914063
4	64	179.2	$179.2 = a_1 + a_2(64 - 40)$ $+ a_3(64 - 40)(64 - 48)$ $+ a_4(64 - 40)(64 - 48)(64 - 56)$	0.001172
5	72	254.5	$254.5 = a_1 + a_2(72 - 40)$ $+ a_3(72 - 40)(72 - 48)$ $+ a_4(72 - 40)(72 - 48)(72 - 56)$ $+ a_5(72 - 40)(72 - 48)(72 - 56)(72 - 64)$	0.00001017

$$P(T) = 55.3 + 3.55(T - 40) + 0.0914063(T - 40)(T - 48)$$
$$+ 0.001172(T - 40)(T - 48)(T - 56)$$
$$+ 0.00001017(T - 40)(T - 48)(T - 56)(T - 64)$$

(6.25)

The interpolating polynomial of Equation 6.25 is used to estimate the vapor pressure for a temperature of 52°C:

$$P(52) = 55.3 + 3.55(52 - 40) + 0.0914063(52 - 40)(52 - 48)$$
$$+ 0.001172(52 - 40)(52 - 48)(52 - 56)$$
$$+ 0.00001017(52 - 40)(52 - 48)(52 - 56)(52 - 64) = 102.0 \, \text{mm} \, | \, \text{Hg}$$

(6.26)

The true value of P for a T of 52°C is 102.1; thus the interpolated value is in error by approximately 0.1, which is about 0.1% of the true value.

6.4 FINITE-DIFFERENCE INTERPOLATION

A finite-difference scheme can be used to develop an interpolation polynomial when the known values of the independent variable are equally spaced. If Δx is the increment on which the values of the independent variable are recorded, then the first finite difference of the dependent variable $y = f(x)$ is

(6.27)
$$\Delta f = f(x + \Delta x) - f(x)$$

in which $f(x)$ is the value of the dependent variable for the value of the independent variable x. The second finite difference is the following difference of the first finite difference:

(6.28) $\Delta^2 f = \Delta[\Delta f] = [f(x + 2\Delta x) - f(x + \Delta x)] - [f(x + \Delta x) - f(x)]$

Continuing for successive differences, it can be shown that the nth finite difference is the following first difference of the $(n - 1)$th difference:

(6.29)
$$\Delta^n f = \Delta[\Delta^{n-1} f]$$

The calculation of the interpolating polynomial is based on the following rule: The ith finite difference, where i is less than or equal to n, of an nth-degree polynomial is a polynomial of degree $(n - i)$. This rule implies that the nth difference would be a constant. Assume the following nth-order polynomial:

(6.30)
$$f(x) = b_0 + b_1 x + b_2 x^2 + \cdots + b_{n-1} x^{n-1} + b_n x^n$$

in which b_n does not equal zero. For $(x + \Delta x)$, the polynomial of Equation 6.30 becomes

(6.31) $\quad f(x + \Delta x) = b_0 + b_1(x + \Delta x) + b_2(x + \Delta x)^2 + \cdots + b_n(x + \Delta x)^n$

which can be rearranged into the form

(6.32) $\qquad\qquad f(x + \Delta x) = b_n x^n + (b_{n-1} + n\Delta x b_n)x^{n-1} + \cdots$

The first finite difference can be formed by subtracting Equation 6.30 from Equation 6.32 as follows:

(6.33) $\qquad\qquad \Delta f = f(x + \Delta x) - f(x)$

or

(6.34) $\qquad\qquad \Delta f = (b_{n-1} + n\Delta x b_n)x^{n-1} + \cdots$

where Equation 6.34 is a polynomial of order $(n - 1)$. For $(x + 2\Delta x)$, Equation 6.31 becomes

(6.35) $\qquad f(x + 2\Delta x) = b_n(x + 2\Delta x)^n + b_{n-1}(x + 2\Delta x)^{n-1} + \cdots$

Equations 6.35, 6.32, and 6.30 can be substituted into Equation 6.28 to compute the second finite difference, with the following result:

(6.36) $\qquad\qquad \Delta^2 f = n(n - 1)(\Delta x)^2 b_n x^{n-2} + \cdots$

which is a polynomial of order $(n - 2)$. By induction, the ith finite difference is given by

(6.37) $\qquad \Delta^i f = n(n - 1)(n - 2) \cdots (n - i + 1)(\Delta x)^i b_n x^{n-i} + \cdots$

which is a polynomial of order $(n - i)$. The nth difference equals the following constant:

(6.38) $\qquad \Delta^n f = n(n - 1)(n - 2) \cdots (1)(\Delta x)^n b_n = n!\,(\Delta x)^n b_n$

For the case where n equals 2, Equation 6.30 becomes

(6.39) $\qquad\qquad f(x) = b_2 x^2 + b_1 x + b_0$

For $(x + \Delta x)$ and $(x + 2\Delta x)$, we can define the following:

(6.40) $\qquad\qquad f(x + \Delta x) = b_2(x + \Delta x)^2 + b_1(x + \Delta x) + b_0$
(6.41) $\qquad\qquad\qquad\quad = b_2(x^2 + 2x\Delta x + (\Delta x)^2) + b_1(x + \Delta x) + b_0$

$$\text{(6.42)} \qquad = b_2x^2 + (2\Delta x b_2 + b_1)x + (b_2(\Delta x)^2 + b_1\Delta x + b_0)$$

$$\text{(6.43)} \qquad f(x + 2\Delta x) = b_2(x + 2\Delta x)^2 + b_1(x + 2\Delta x) + b_0$$

$$\text{(6.44)} \qquad = b_2(x^2 + 4x\Delta x + (2\Delta x)^2) + b_1(x + 2\Delta x) + b_0$$

$$\text{(6.45)} \qquad = b_2x^2 + (4b_2\Delta x + b_1)x + (b_2(2\Delta x)^2 + 2b_1\Delta x + b_0)$$

The first difference is computed using Equations 6.27, 6.39, and 6.40 through 6.42 as follows:

$$\text{(6.46)} \qquad \Delta f = f(x + \Delta x) - f(x)$$

$$\text{(6.47)} \qquad = 2b_2(\Delta x)x + (b_2(\Delta x)^2 + b_1\Delta x)$$

which is a first-order polynomial. The second finite difference is computed for Equations 6.42 and 6.39 with Equation 6.28 as

$$\text{(6.48)} \qquad \Delta^2 f = [f(x + 2\Delta x) - f(x + \Delta x)] - [f(x + \Delta x) - f(x)]$$

$$\text{(6.49)} \qquad = 2b_2(\Delta x)^2$$

which is a constant.

Example 6.3: Finite-Difference Interpolation Polynomial

The equations for the second-order polynomial (Equations 6.39 through 6.49) can be used to develop an interpolation polynomial for the data in Table 6.4. The first finite difference from Equation 6.27 is

$$\text{(6.50)} \qquad \Delta f = f(x + \Delta x) - f(x) = 5 - 3 = 2$$

This can be equated to Equation 6.47 with $\Delta x = 1$ to obtain the following:

$$\text{(6.51)} \qquad 5 - 3 = 2b_2(\Delta x)x + b_2(\Delta x)^2 + b_1\Delta x$$

$$\text{(6.52)} \qquad 2 = 2b_2(1)(1) + b_2(1)^2 - b_1(1)$$

$$\text{(6.53)} \qquad 2 = 3b_2 + b_1$$

The second finite difference from Equation 6.28 is

$$\text{(6.54)} \qquad \Delta^2 f = [8 - 5] - [5 - 3] = 1$$

TABLE 6.4   Data for Finite-Difference Interpolation

x	f(x)
1	3
2	5
3	8

This can be equated to Equation 6.49 to produce the following:

(6.55) $$1 = 2b_2(\Delta x)^2 = 2b_2$$

Solving Equation 6.55 yields $b_2 = 0.5$. Substituting this result into Equation 6.53 yields $b_1 = 0.5$. Using these values of b_2 and b_1 and any one of the three points, a value of b_0 can be computed from Equation 6.39. Using the point $f(1) = 3$, we get the following equation:

(6.56) $$f(x) = 3 = b_2(1)^2 + b_1(1) + b_0 = 1 + b_0$$

Therefore, b_0 equals 2. Thus the interpolating polynomial is

(6.57) $$f(x) = 0.5x^2 + 0.5x + 2$$

For any of the three points used to derive the interpolating polynomial, Equation 6.57 yields an exact value. The equation can be used to interpolate a value of $f(x)$ for any x in the range of the data. For example, the value of $f(x)$ at $x = 2.3$ can be computed as

$$f(2.3) = 0.5(2.3)^2 + 0.5(2.3) + 2 = 5.795$$

6.4.1 FINITE-DIFFERENCE TABLE

The finite-differences computed previously can be organized into a table, as shown in Table 6.5. The finite-difference table can be used to derive an interpolating polynomial. The finite-difference table of Table 6.6 shows the differences when integers from 10 to 14 are the independent variable x and their cubes are the dependent variable $f(x)$.

TABLE 6.5   Finite-Difference Table

x	$f(x)$	Δf	$\Delta^2 f$	$\Delta^3 f$	...	$\Delta^n f$
x	$f(x)$					
		$\Delta f(x)$				
$x + \Delta x$	$f(x + \Delta x)$		$\Delta^2 f(x)$			
		$\Delta f(x + \Delta x)$		$\Delta^3 f(x)$		
$x + 2\Delta x$	$f(x + 2\Delta x)$		$\Delta^2 f(x + \Delta x)$			
		$\Delta f(x + 2\Delta x)$		$\Delta^3 f(x + \Delta x)$		
$x + 3\Delta x$	$f(x + 3\Delta x)$		$\Delta^2 f(x + 2\Delta x)$			
		$\Delta f(x + 3\Delta x)$		$\Delta^3 f(x + 2\Delta x)$		
			$\Delta^2 f(x + 3\Delta x)$			
$\vdots$	$\vdots$	$\vdots$	$\vdots$	$\vdots$	$\vdots\vdots\vdots$	$\Delta^n f(x)$
$x + (n - 2)\Delta x$	$f[x + (n - 2)\Delta x]$		$\Delta^2 f[x + (n - 3)\Delta x]$			
		$\Delta f[x + (n - 2)\Delta x]$		$\Delta^3 f[x + (n - 3)\Delta x]$		
$x + (n - 1)\Delta x$	$f[x + (n - 1)\Delta x]$		$\Delta^2 f[x + (n - 2)\Delta x]$			
		$\Delta f[x + (n - 1)\Delta x]$				
$x + n\Delta x$	$f(x + n\Delta x)$					

TABLE 6.6   Finite-Difference Table for Cubes

x	f(x)	Δf	Δ²f	Δ³f	Δ⁴f
10	1000				
		331			
11	1331		66		
		397		6	
12	1728		72		0
		469		6	
13	2197		78		
		547			
14	2744				

Example 6.4: Interpolation of the Standard Normal Distribution Table

It is frequently necessary to interpolate values from probabilities tables—for example, the standard normal distribution. The finite-difference scheme provides a convenient method of developing an interpolating polynomial. Consider the values of the standard normal deviate z and the cumulative probability, $f(z)$, which are given in the finite-difference table of Table 6.7. To fit the coefficients of the second-order polynomial, Equations 6.47, 6.49, and 6.39 can be applied with the top diagonal row of the finite-difference table. For computing b_2, $\Delta z = -1.6 - (-1.5) = -0.1$ and Equation 6.49 can be stated as

$$\Delta^2 f = 2b_2(\Delta z)^2 \tag{6.58}$$

$$0.0018 = 2b_2(-0.1)^2 \tag{6.59}$$

$$b_2 = 0.09 \tag{6.60}$$

For computing b_1, Equation 6.47 produces

$$\Delta f = 2b_2(\Delta z)z + b_2(\Delta z)^2 + b_1(\Delta z) \tag{6.61}$$

TABLE 6.7   Finite-Difference Table for Standard Normal Distribution

z	f(z)	Δf	Δ²f
−1.5	0.0668		
		−0.0120	
−1.6	0.0548		0.0018
		−0.0102	
−1.7	0.0446		

(6.62) $-0.0120 = 2(0.09)(-0.1)(-1.5) + 0.09(-0.1)^2 + b_1(-0.1)$

(6.63) $b_1 = 0.399$

For computing b_0, Equation 6.39 produces

(6.64) $f(z) = b_2z^2 + b_1z + b_0$

(6.65) $0.0668 = 0.09(-1.5)^2 + 0.399(-1.5) + b_0$

(6.66) $b_0 = 0.4628$

Thus the interpolating polynomial is

(6.67) $f(z) = 0.096z^2 + 0.399z + 0.4628$

Substituting the values for z of -1.5, -1.6, and -1.7 yields the exact values of $f(z)$ shown in the finite-difference table. The interpolating polynomial can be used to estimate the probability f for any z. For example, given $z = -1.55$, the polynomial yields 0.0606, which equals the tabulated value in probability tables. The polynomial can be used to extrapolate; for example, for $z_0 = -1.8$, the polynomial yields 0.0362, which is about 1% greater than the actual value of 0.0359.

6.5 NEWTON'S METHOD

The finite-difference table of Table 6.5 provides a means for deriving an interpolation function. It can be shown that the finite differences of the first diagonal row are given by

(6.68) $f(x + \Delta x) = f(x) + \Delta f(x)$
(6.69) $\Delta f(x + \Delta x) = \Delta f(x) + \Delta^2 f(x)$
(6.70) $\Delta^2 f(x + \Delta x) = \Delta^2 f(x) + \Delta^3 f(x)$
$$\vdots$$
(6.71) $\Delta^{n-1} f(x + \Delta x) = \Delta^{n-1} f(x) + \Delta^n f(x)$

The same procedure could be used for any diagonal row. For example, the equations for the next diagonal row are

(6.72) $f(x + 2\Delta x) = f(x) + \Delta f(x) + \Delta f(x + \Delta x)$
                $= f(x) + \Delta f(x) + \Delta f(x) + \Delta^2 f(x)$
(6.73) $= f(x) + 2\Delta f(x) + \Delta^2 f(x)$
(6.74) $\Delta f(x + 2\Delta x) = \Delta f(x) + 2\Delta^2 f(x) + \Delta^3 f(x)$
(6.75) $\Delta^2 f(x + 2\Delta x) = \Delta^2 f(x) + 2\Delta^3 f(x) + \Delta^4 f(x)$
$$\vdots$$
(6.76) $\Delta^{n-3} f(x + 2\Delta x) = \Delta^{n-3} f(x) + 2\Delta^{n-2} f(x) + \Delta^{n-1} f(x)$

All these expansions have the form of a binomial expansion; thus they can be rewritten as

$$\text{(6.77)} \quad \Delta f(x+m\Delta x) = \sum_{i=0}^{m} b_i \Delta^i f(x)$$

$$\text{(6.78)} \quad \Delta f(x+m\Delta x) = f(x) + m\Delta f(x) + \frac{m(m-1)}{2}\Delta^2 f(x) + \ldots + \Delta^m f(x)$$

where b_i = constant as defined by Equations 6.77 and 6.78. In general,

$$\text{(6.79)} \quad \Delta^r f(x+m\Delta x) = \sum_{i=0}^{m} b_i \Delta^{r+1} f(x)$$

$$\text{(6.80)} \quad \Delta^r f(x+m\Delta x) = \Delta^r f(x) + m\Delta^{r+1} f(x) + \ldots + \Delta^{r+m} f(x)$$

These equations yield the following Newton's interpolation formula:

$$\text{(6.81)} \quad \begin{aligned} f(x) &= f(x_0) + n\Delta f(x_0) + \frac{n(n-1)}{2!}\Delta^2 f(x_0) + \ldots \\ &+ \frac{n(n-1)\ldots(n-m+1)}{m!}\Delta^m f(x_0) \end{aligned}$$

in which

$$\text{(6.82)} \quad x = x_0 + n\Delta x$$

or

$$\text{(6.83)} \quad n = \frac{x - x_0}{\Delta x}$$

Thus, given the finite-difference table of Table 6.5, Newton's formula can be used to derive an interpolation polynomial.

Example 6.5: Tangents of Angles

Assume that we are interested in the values of the tangent of angles (θ) between 30° and 40° and a table with an increment of 2° is available. The accuracy of estimated values can be increased by developing a nonlinear interpolation polynomial, rather than using linear interpolation. The data are given in columns 1 and 2 of Table 6.8. Using the first diagonal row, an interpolation formula can be formed as follows:

$$\tan(\theta) = 0.5774 + n(0.0475) + \frac{n(n-1)}{2!}(0.0021) + \frac{n(n-1)(n-2)}{3!}(0.0003)$$

$$+ \frac{n(n-1)(n-2)(n-3)}{4!}(0.0001) + \frac{n(n-1)(n-2)(n-3)(n-4)}{5!}(-0.0003)$$

$$\text{(6.84)}$$

TABLE 6.8   Finite-Difference Table for tan θ Where 30° ≤ θ ≤ 40°

θ	tan θ	Δ tan θ	Δ² tan θ	Δ³ tan θ	Δ⁴ tan θ	Δ⁵ tan θ
30°	0.5774					
		0.0475				
32°	0.6249		0.0021			
		0.0496		0.0003		
34°	0.6745		0.0024		0.0001	
		0.0520		0.0004		−0.0003
36°	0.7265		0.0028		−0.0002	
		0.0548		0.0002		
38°	0.7813		0.0030			
		0.0578				
40°	0.8391					

For example, the tangent for θ equal to 36.5° can be computed using Equation 6.83 as

$$(6.85) \qquad n = \frac{36.5 - 30}{2} = 3.25$$

Thus Equation 6.84 yields

$$\tan(36.5) = 0.5774 + 3.25(0.0475) + \frac{3.25(2.25)}{2}(0.0021)$$
$$+ \frac{3.25(2.25)(1.25)}{6}(0.0003) + \frac{3.25(2.25)(1.25)(0.25)}{24}(0.0001)$$
$$+ \frac{3.25(2.25)(1.25)(0.25)(-0.75)}{120}(-0.0003) = 0.73992$$

(6.86)

which is in error by 0.00004 since the true value is 0.73996. Given that the original values were recorded with four significant digits, the error is not significantly different from zero. Linear interpolation gives a value of 0.8402, which is in error by 0.00024 or approximately six times greater than the error resulting from the fifth-order polynomial of Equation 6.86.

Example 6.6: Dissolved Oxygen Concentration

The dissolved oxygen concentration is an important element in some water-quality analyses. The saturation values are a function of temperature. Table 6.9 gives the saturated values (D) for temperatures (T) from 0°C to 25°C in increments of 5°. Table 6.9 also shows the finite-difference table, with D as the dependent variable. The following is Newton's interpolation polynomial for the data of Table 6.9:

TABLE 6.9 Finite-Difference Table for Dissolved Oxygen Concentrations (D) as a Function of Temperature (T)

T (°C)	D (mg/L)	ΔD	Δ²D	Δ²D	Δ³D	Δ⁵D
0	14.62					
		−1.82				
5	12.80		0.35			
		−1.47		−0.06		
10	11.33		0.29		−0.03	
		−1.18		−0.09		0.11
15	10.15		0.20		0.08	
		−0.98		−0.01		
20	9.17		0.19			
		−0.79				
25	8.38					

$$D = 14.62 + n(-1.82) + \frac{n(n-1)}{2}(0.35) + \frac{n(n-1)(n-2)}{6}(-0.06)$$
$$+ \frac{n(n-1)(n-2)(n-3)}{24}(-0.03) + \frac{n(n-1)(n-2)(n-3)(n-4)}{120}(0.11)$$

(6.87)

Since the increment of the independent variable is 5°, then Equation 6.83 becomes

(6.88)
$$n = \frac{T-0}{5} = \frac{T}{5}$$

Thus, if it is necessary to predict the saturation concentration of D at 13°C, n equals 2.6 and Equation 6.87 results in

$$D = 14.62 + 2.6(-1.82) + \frac{2.6(1.6)}{2}(0.35) + \frac{2.6(1.6)(0.6)}{6}(-0.06)$$
$$+ \frac{2.6(1.6)(0.6)(-0.4)}{24}(-0.03) + \frac{2.6(1.6)(0.6)(-0.4)(-1.4)}{120}(0.11)$$

(6.89)

Therefore,

(6.90) $D = 10.593569$ mg/L

The true value is 10.60 mg/L, which yields an error of 0.006 mg/L. Linear interpolation yields an estimated value of 10.622 mg/L, which yields an error of 0.022 mg/L.

The interpolating polynomial of Equations 6.89 and 6.90 can also be used for extrapolation. For example, for T equal to 30°, and n

equal to 6, the computed value of D would be 7.41 mg/L. This represents an error of 0.22 mg/L since the true value is 7.63 mg/L, or 2.9% of the true value.

6.6 LAGRANGE POLYNOMIALS

For many problems in engineering analysis, the data collection cannot be controlled, and the resulting data are for measurements taken at unequal intervals. In this case, data are collected for one variable, $f(x)$, which is called the dependent variable, as a function of a second variable, x, which is called the independent variable. For the previously discussed methods of interpolation, the x variable was assumed to be measured at a constant interval, Δx. A method that can be used with data for which the independent variable is measured on a variable interval Δx is introduced here. The method is referred to as Lagrangian interpolation.

Before introducing the method, it is necessary to provide the notations that are used in this section. The data sample is assumed to consist of n pairs of values measured on x and $f(x)$, with x_i being the ith measured value of the independent variable. The method provides an estimate of the value of $f(x)$ at a specific value of x, which is denoted as x_0; the estimated value is denoted as $f(x_0)$. A measured value of the dependent variable is denoted as $f(x_i)$, where $i = 1, 2, ..., n$. The method involves a weighting function, with the weight given to the ith value of f for x_0 denoted as $w_i(x_0)$.

The Lagrange interpolating equation for estimating the value of $f(x_0)$ is

(6.91)
$$f(x_0) = \sum_{i=1}^{n} w_i(x_0) f(x_i)$$

The values of $f(x_j)$ are the measured values of the dependent variable, and the weights are a function of x_0 and the n values of the independent variable x. The weights are given by

(6.92)
$$w_i(x_0) = \frac{\prod_{\substack{j=1 \\ j \neq i}}^{n} (x_0 - x_j)}{\prod_{\substack{j=1 \\ j \neq i}}^{n} (x_i - x_j)}$$

in which the products (Π) include $(n - 1)$ terms, with the term where $j = i$ omitted. Equations 6.91 and 6.92 can be illustrated using the case where $n = 3$. In this case, Equation 6.91 becomes

(6.93)
$$f(x_0) = w_1(x_0)f(x_1) + w_2(x_0)f(x_2) + w_3(x_0)f(x_3)$$

and the three weights are given by:

(6.94)
$$w_1(x_0) = \frac{(x_0 - x_2)(x_0 - x_3)}{(x_1 - x_2)(x_1 - x_3)}$$

(6.95)
$$w_2(x_0) = \frac{(x_0 - x_1)(x_0 - x_3)}{(x_2 - x_1)(x_2 - x_3)}$$

(6.96)
$$w_3(x_0) = \frac{(x_0 - x_1)(x_0 - x_2)}{(x_3 - x_1)(x_3 - x_2)}$$

Example 6.7: Sine Function

This method is illustrated using the sine as the dependent variable $f(x)$ and the angle x in degrees as the independent variable; that is, $f(x) = \sin(x)$. The data in Table 6.10 are applied with Equations 6.93 and 6.94 through 6.96. If we wish to estimate the sine of 12°, then, using the Equations 6.94 through 6.96, the weights are

(6.97)
$$w_1(12) = \frac{(12-11)(12-13)}{(10-11)(10-13)} = -\frac{1}{3}$$

(6.98)
$$w_2(12) = \frac{(12-10)(12-13)}{(11-10)(11-13)} = 1.0$$

(6.99)
$$w_3(12) = \frac{(12-10)(12-11)}{(13-10)(13-11)} = \frac{1}{3}$$

Substituting the weights and the tabulated values of the sine of x into Equation 6.93 yields

(6.100) $f(12) = w_1(12)f(x_1) + w_2(12)f(x_2) + w_3(12)f(x_3)$

(6.101) $f(12) = -\dfrac{1}{3}(0.17365) + 1.0(0.19081) + \dfrac{1}{3}(0.22495) = 0.20791$

The computed value equals the true value to five significant digits.

Example 6.8: Lagrange Polynomial for Well-Water Elevation

A well pumping at 250 gallons per minute has observation wells located at 15, 42, 128, 317, and 433 feet away along a straight line

TABLE 6.10   Sine Function Data

i	x_i	$f(x_i)$
1	10	0.17365
2	11	0.19081
3	13	0.22495

from the well. After 3 hours of pumping, the following drawdowns in the five wells were observed: 14.6, 10.7, 4.8, 1.7, and 0.3 feet, respectively. If an estimate of the drawdown (f) at a distance of 25 feet is needed, the interpolating polynomial is

(6.102) $f(25) = 14.6w_1 + 10.7w_2 + 4.8w_3 + 1.7w_4 + 0.3w_5$

with the weights given by:

$$w_1(25) = \frac{(25-42)(25-128)(25-317)(25-433)}{(15-42)(15-128)(15-317)(15-433)} = 0.5416$$

(6.103)

$$w_2(25) = \frac{(25-15)(25-128)(25-317)(25-433)}{(42-15)(42-128)(42-317)(42-433)} = 0.4915$$

(6.104)

$$w_3(25) = \frac{(25-15)(25-42)(25-317)(25-433)}{(128-15)(128-42)(128-317)(128-433)} = -0.0362$$

(6.105)

$$w_4(25) = \frac{(25-15)(25-42)(25-128)(25-433)}{(317-15)(317-42)(317-128)(317-433)} = 0.0039$$

(6.106)

$$w_5(25) = \frac{(25-15)(25-42)(25-128)(25-317)}{(433-15)(433-42)(433-128)(433-317)} = -0.0008$$

(6.107)

Thus, substituting the weights of Equations 6.103 through 6.107 into Equation 6.102 yields $f(25) = 13.0$ feet.

6.7 INTERPOLATION USING SPLINES

For a given data set of the type $(x_i, f(x_i))$, $i = 1, 2, ..., n$, the highest order interpolating polynomial that can be used is of the $(n - 1)$ order. Generally, the accuracy of interpolation can be improved by using higher order polynomials. However, sometimes the interpolation accuracy can decrease, especially in data sets that include local abrupt changes in the values of $f(x)$ for a steady change in the value of x. The resulting high-order interpolating polynomial includes oscillations around the abrupt changes. Lower order polynomials in this case can be more accurate than higher order polynomials. The lower order polynomials are called *splines*. The splines can be classified depending on the order of these polynomials. The most commonly used splines are linear, quadratic, and cubic. In this section these three types are described. Other higher order splines can be similarly developed with a large increase in computational difficulty.

6.7.1 LINEAR SPLINES

A linear spline used for interpolation with a data set of the type $(x_i, f(x_i))$, $i = 1, 2, ..., n$, results in connecting adjacent data points with straight-line segments. The resulting spline is sometimes called a first-order spline. The interpolation function can be expressed as

$$(6.108) \qquad f_1(x) = f(x_1) + \frac{f(x_2) - f(x_1)}{x_2 - x_1}(x - x_1) \qquad \text{for } x_1 \leq x \leq x_2$$

$$(6.109) \qquad f_2(x) = f(x_2) + \frac{f(x_3) - f(x_2)}{x_3 - x_2}(x - x_2) \qquad \text{for } x_2 \leq x \leq x_3$$

$$\vdots$$

$$(6.110) \qquad f_i(x) = f(x_i) + \frac{f(x_{i+1}) - f(x_i)}{x_{i+1} - x_i}(x - x_i) \qquad \text{for } x_i \leq x \leq x_{i+1}$$

$$\vdots$$

$$(6.111) \quad f_{n-1}(x) = f(x_{n-1}) + \frac{f(x_n) - f(x_{n-1})}{x_n - x_{n-1}}(x - x_{n-1}) \qquad \text{for } x_{n-1} \leq x \leq x_n$$

These equations result in linear interpolation between the points. The resulting overall interpolation function satisfies the following condition:

$$(6.112) \qquad f_i(x_i) = f_{i+1}(x_i) \quad \text{for } i = 1, 2, ..., n - 1$$

However, the function is discontinuous at the data points, which might make it visually or physically unacceptable.

Example 6.9: Linear Spline for Well-Water Elevation

The data set that was previously used in Example 6.8 can be utilized to illustrate the linear spline. Using Equations 6.108 through 6.111, Table 6.11 can be constructed. The resulting linear spline is shown in Figure 6.2. The figure shows that the linear spline results in exact predictions at the data points, but has discontinuities of slope at the data points.

TABLE 6.11   Linear Spline for Well-Water Elevation

i	x_i (ft)	$f(x_i)$ (ft)	Linear Spline $f_i(x)$
1	15	14.6	
			$14.6 + \dfrac{10.7 - 14.6}{14 - 15}(x - 15) = 14.6 - 0.1444(x - 15)$
2	42	10.7	
			$10.7 - 0.0686046(x - 42)$
3	128	4.8	
			$4.8 - 0.0164021(x - 128)$
4	317	1.7	
			$1.7 - 0.0120689(x - 317)$
5	433	0.3	

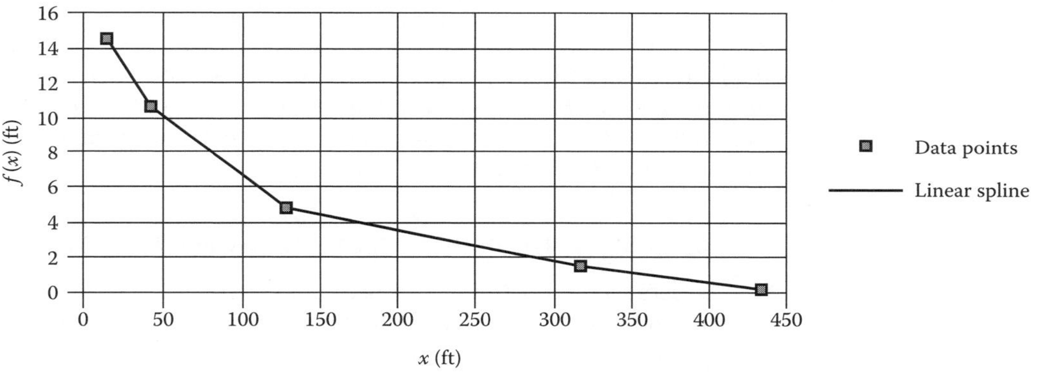

FIGURE 6.2   Linear spline for well-water elevation.

6.7.2 QUADRATIC SPLINES

A quadratic spline provides a quadratic equation connecting any two adjacent data points within a data set of the type $(x_i, f(x_i))$, $i = 1, 2, …, n$. The general form of a quadratic equation between point $(x_i, f(x_i))$ and $(x_{i+1}, f(x_{i+1}))$ is

$$(6.113) \qquad f_i(x) = a_i x^2 + b_i x + c_i \quad \text{for } i = 1, 2, …, n - 1$$

Therefore, every two adjacent data points have an interpolation equation as given by Equation 6.113, with three constants: a_i, b_i, and c_i. As a result, there are $3(n - 1)$ unknowns that need to be determined, using the given data set, requiring $3(n - 1)$ conditions. The conditions in this case are given by the following:

1. The splines must pass through the data points. For the ith spline (where $i = 1, 2, …, n - 1$), this condition can be expressed as

$$(6.114) \qquad f_i(x_i) = a_i x_i^2 + b_i x + c_i = f(x_i) \quad \text{for } i = 1, 2, 3, …, n - 1$$

$$(6.115) \quad f_i(x_{i+1}) = a_i x_{i+1}^2 + b_i x_{i+1} + c_i = f(x_{i+1}) \quad \text{for } i = 1, 2, 3, \ldots, n-1$$

2. The splines must be continuous at the interior data points. This condition can be expressed using first derivatives of the quadratic splines as

$$(6.116) \quad 2a_i x_i + b_i = 2a_{i+1} x_{i+1} + b_{i+1} \quad \text{for } i = 1, 2, 3, \ldots, n-2$$

3. The last needed condition can be arbitrarily set as the second derivative for the spline between the first two data points to be zero. Since the second derivative for the first spline is $2a_1$, this condition can be expressed as

$$(6.117) \quad a_1 = 0$$

Equations 6.114 through 6.117 provide the needed $3(n-1)$ conditions to solve for the $3(n-1)$ unknowns, a_i, b_i, and c_i, $i = 1, 2, \ldots, n-1$. The resulting quadratic splines provide the desired continuity while passing through the data points.

Example 6.10: Quadratic Spline for Well-Water Elevation

The data set that was previously used in Examples 6.8 and 6.9 can be utilized to illustrate the quadratic spline. Using Equation 6.114, the following equations can be developed:

$$(6.118) \quad 225a_1 + 15b_1 + c_1 = 14.6$$

$$(6.119) \quad 1764a_2 + 42b_2 + c_2 = 10.7$$

$$(6.120) \quad 16{,}384a_3 + 128b_3 + c_3 = 4.8$$

$$(6.121) \quad 100{,}489a_4 + 317b_4 + c_4 = 1.7$$

Equation 6.115 results in the following conditions:

$$(6.122) \quad 1764a_1 + 42b_1 + c_1 = 10.7$$

$$(6.123) \quad 16{,}384a_2 + 128b_2 + c_2 = 4.8$$

$$(6.124) \quad 100{,}489a_3 + 317b_3 + c_3 = 1.7$$

$$(6.125) \quad 187{,}489a_4 + 433b_4 + c_4 = 0.3$$

Equation 6.116 results in the following conditions:

$$(6.126) \quad 2a_1(42) + b_1 = 2a_2(42) + b_2$$

(6.127) $2a_2(128) + b_2 = 2a_3(128) + b_3$

(6.128) $2a_3(317) + b_3 = 2a_4(317) + b_4$

The last condition comes from Equation 6.117 as

(6.129) $a_1 = 0$

Since $a_1 = 0$, the resulting system of 11 equations can be summarized in a matrix format as

$$
\begin{bmatrix}
15 & 1 & 0 & 0 & 0 & 0 & 0 & 0 & 0 & 0 & 0 \\
0 & 0 & 1764 & 42 & 1 & 0 & 0 & 0 & 0 & 0 & 0 \\
0 & 0 & 0 & 0 & 0 & 16{,}384 & 128 & 1 & 0 & 0 & 0 \\
0 & 0 & 0 & 0 & 0 & 0 & 0 & 0 & 100{,}489 & 317 & 1 \\
42 & 1 & 0 & 0 & 0 & 0 & 0 & 0 & 0 & 0 & 0 \\
0 & 0 & 16{,}384 & 128 & 1 & 0 & 0 & 0 & 0 & 0 & 0 \\
0 & 0 & 0 & 0 & 0 & 100{,}489 & 317 & 1 & 0 & 0 & 0 \\
0 & 0 & 0 & 0 & 0 & 0 & 0 & 0 & 187{,}489 & 433 & 1 \\
1 & 0 & -84 & -1 & 0 & 0 & 0 & 0 & 0 & 0 & 0 \\
0 & 0 & 256 & 1 & 0 & -256 & -1 & 0 & 0 & 0 & 0 \\
0 & 0 & 0 & 0 & 0 & 634 & 1 & 0 & -634 & -1 & 0
\end{bmatrix}
\begin{bmatrix}
b_1 \\ c_1 \\ a_2 \\ b_2 \\ c_2 \\ a_3 \\ b_3 \\ c_3 \\ a_4 \\ b_4 \\ c_4
\end{bmatrix}
=
\begin{bmatrix}
14.6 \\ 10.7 \\ 4.8 \\ 1.7 \\ 10.7 \\ 4.8 \\ 1.7 \\ 0.3 \\ 0 \\ 0 \\ 0
\end{bmatrix}
$$

(6.130)

The solution of this system of equations can be determined numerically to be

(6.131)
$$
\begin{bmatrix}
b_1 \\ c_1 \\ a_2 \\ b_2 \\ c_2 \\ a_3 \\ b_3 \\ c_3 \\ a_4 \\ b_4 \\ c_4
\end{bmatrix}
=
\begin{bmatrix}
-0.144444 \\
16.766667 \\
8.818583 \times 10^{-4} \\
-0.2185206 \\
18.32227 \\
-1.2506 \times 10^{-4} \\
3.925179 \times 10^{-2} \\
1.824836 \\
2.411243 \times 10^{-4} \\
-0.1929122 \\
38.62283
\end{bmatrix}
$$

The resulting quadratic splines are shown in Figure 6.3. For example, the first spline is $f(x) = (0)x^2 + (-0.14444)x + (16.766667)$, which is valid for $15 \leq x \leq 42$. The first spline is shown in the figure over

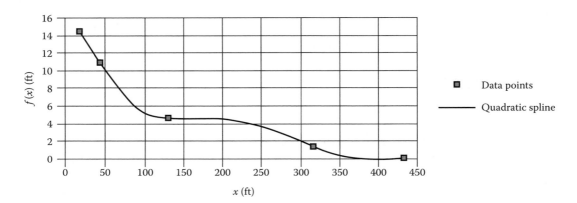

FIGURE 6.3   Quadratic spline for well pump.

its range of applicability (i.e., $15 \leq x \leq 42$). Other splines can be similarly expressed. Plotting each spline over its respective range of applicability produces Figure 6.3. The figure shows that the quadratic spline results in predicting the data points with the appearance of continuity in both magnitude and slope at the measured data points.

6.7.3 CUBIC SPLINES AND OTHER HIGHER ORDER SPLINES

The cubic splines provide a cubic equation connecting any two adjacent data points within a data set of the type $(x_i, f(x_i))$, $i = 1, 2, ..., n$. The general form of a cubic equation between point $(x_i, f(x_i))$ and $(x_{i+1}, f(x_{i+1}))$ is

(6.132) $f_i(x) = a_i x^3 + b_i x^2 + c_i x + d_i$ for $i = 1, 2, ..., n - 1$

Therefore, every two adjacent data points have an interpolation equation as given by Equation 6.132, with four constants: a_i, b_i, c_i, and d_i. As a result, there are $4(n - 1)$ unknowns that need to be determined using the given data set, requiring $4(n - 1)$ conditions. The conditions in this case are given by the following:

1. The splines must pass through the data points. This condition can be expressed as

(6.133)
$$f_i(x_i) = a_i x_i^3 + b_i x_i^2 + c_i x_i + d_i = f_i(x_i)$$
$$\text{for } i = 1, 2, 3, ..., n-1$$

(6.134)
$$f_i(x_{i+1}) = a_i x_{i+1}^3 + b_i x_{i+1}^2 + c_i x_{i+1} + d_i = f_i(x_{i+1})$$
$$\text{for } i = 1, 2, 3, ..., n-1$$

2. The splines must be continuous at the interior data points. This condition can be expressed using first derivatives of the cubic splines as

(6.135)
$$3a_i x_{i+1}^2 + b_i x_{i+1} + c_i = 3a_{i+1} x_{i+1}^2 + b_{i+1} x_{i+1} + c_{i+1}$$
$$\text{for } i = 1, 2, 3, \ldots, n-2$$

3. The splines need to satisfy the second derivative continuity at the interior points. This condition can be expressed as

(6.136)
$$6a_i x_{i+1} + b_i = 6a_{i+1} x_{i+1} + b_{i+1}$$
$$\text{for } i = 1, 2, 3, \ldots, n-2$$

4. The last needed two conditions can be set as the second derivative at the first and last data points to be zero. This condition can be expressed as

(6.137)
$$6a_1 x_i + b_1 = 0$$

(6.138)
$$6a_{n-1} x_n + b_{n-1} = 0$$

Equations 6.133 through 6.138 provide the needed $4(n - 1)$ conditions to solve for the $4(n - 1)$ unknowns: a_i, b_i, c_i, and d_i, $i = 1, 2, \ldots, n - 1$. The resulting cubic splines provide the desirable continuity while passing through the data points.

The cubic splines are sometimes referred to as the third-order splines. Higher order splines can be developed using similar conditions on higher order derivatives. However, the computational difficulty greatly increases by needing to solve larger systems of simultaneous equations. The gained visual or physical benefits might be too negligible to justify their use.

6.8 GUIDELINES FOR CHOICE OF INTERPOLATION METHOD

The best interpolation method to use for any problem often depends on the details of the particular problem. However, the following general guidelines can be offered:

1. The method of undetermined coefficients is perhaps the easiest method to understand conceptually. However, the method does require the solution of a set of $n + 1$ simultaneous equations for an nth-order interpolation polynomial. One advantage of this method is that the interpolation polynomial can be *customized* for the particular problem at hand; for example, some of

the terms of the general nth-order interpolation polynomial can be dropped, if desired. This is not possible in most of the other interpolation methods.

2. Both the Gregory–Newton and Lagrange polynomial interpolation methods eliminate the need to solve a set of simultaneous equations. Of the two methods, the Lagrange polynomial scheme is perhaps the easier to program. Although both of these methods (as well as the method of undetermined coefficients) work with unequally spaced data points, the best accuracy is usually achieved when the differences in the spacing of the data points is minimal.

3. The finite-difference interpolation scheme requires equally spaced data points for its application.

4. Interpolation using splines can be more accurate than other methods, especially in cases where the data show large, sudden changes.

All of the interpolation schemes can also be used for extrapolation. Always remember, though, that extrapolation of data usually produces less accurate estimates than interpolated values within the range of the data and that the accuracy gets worse as the distance between the extrapolated point and the bounds of the data points increases.

6.9 MULTIDIMENSIONAL INTERPOLATION

So far the discussion has been limited to one independent variable, x. In this section, the objective is to generalize the previous concepts to several independent variables. In general, the function of interest can be of the following form:

$$(6.139) \qquad f(x_1, x_2, x_3, ..., x_n)$$

where $x_1, x_2, x_3, ..., x_n$ are n independent variables. The objective herein is similar to the objective in previous sections for one-dimensional interpolation. The objective can be stated as follows: We are given data in the form of the function $f(x_1, x_2, x_3, ..., x_n)$ value at specified sets of $x_1, x_2, x_3, ..., x_n$ values, and the value of the function at some other values of $x_1, x_2, x_3, ..., x_n$ is needed.

The concepts behind multidimensional interpolation can be shown using two-dimensional interpolation. Once understanding is established for the two-dimensional case, the concepts can be extended to other higher dimensions. In this section, the discussion is limited to two-dimensional interpolation. Also, the discussion is limited to one interpolation method, which is linear interpolation. The development of other methods is similar, but with an increased computational difficulty. In most engineering applications that require multidimensional interpolation, linear interpolation is adequate.

6.9.1 LINEAR INTERPOLATION IN TWO DIMENSIONS

Two-dimensional interpolation involves one dependent variable and two independent variables. Thus a given data set consists of the triads x_1, x_2, and $f(x_1, x_2)$. The data set can have the following form for two independent variables, x_1 and x_2:

	x_{11}	x_{12}	x_{1i}	$\cdots$	x_{1n}
x_{21}	$f(x_{11}, x_{21})$	$f(x_{12}, x_{21})$	$f(x_{1i}, x_{21})$	$\cdots$	$f(x_{1n}, x_{21})$
x_{22}	$f(x_{11}, x_{22})$	$f(x_{12}, x_{22})$	$f(x_{1i}, x_{22})$	$\cdots$	$f(x_{1n}, x_{22})$
x_{2j}	$f(x_{11}, x_{2j})$	$f(x_{12}, x_{2j})$	$f(x_{1i}, x_{2j})$	$\cdots$	$f(x_{1n}, x_{2j})$
$\vdots$	$\vdots$	$\vdots$	$\vdots$	$\vdots$	$\vdots$
x_{2m}	$f(x_{11}, x_{2m})$	$f(x_{12}, x_{2m})$	$f(x_{1i}, x_{2m})$	$\cdots$	$f(x_{1n}, x_{2m})$

(6.140)

These values can be viewed as a three-dimensional graph, with the values of $f(x_1, x_2)$ as the surface. The objective is to estimate the value on the surface that corresponds to values of x_1 and x_2. For given values of x_1 and x_2, the value of $f(x_1, x_2)$ needs to be determined. Assume that the values of x_1 and x_2 fall between x_{1i} and $x_{1(i+1)}$ and x_{2j} and $x_{2(j+1)}$, respectively, as follows:

(6.141)

	x_{1i}	x_1	$x_{1(i+1)}$
x_2	$f(x_{1i}, x_{2j})$	$f(x_1, x_{2j})$	$f(x_{1(i+1)}, x_{2j})$
x_2	$f(x_{1i}, x_2)$	$f(x_1, x_2)$	$f(x_{1(i+1)}, x_2)$
$x_{2(j+1)}$	$f(x_{1i}, x_{2(j+1)})$	$f(x_1, x_{2(j+1)})$	$f(x_{1(i+1)}, x_{2(j+1)})$

All of the values in Equation 6.141 are known except $f(x_1, x_{2j})$, $f(x_{1i}, x_2)$, $f(x_{1(i+1)}, x_2)$, $f(x_1, x_{2(j+1)})$, and $f(x_1, x_2)$. The objective of the interpolation is to determine $f(x_1, x_2)$ by performing the following one-dimensional interpolations:

1. A one-dimensional interpolation is performed for x_1 at x_{2j} to obtain $f(x_1, x_{2j})$ as follows:

(6.142) $$f(x_1, x_{2j}) = f(x_{1i}, x_{2j}) + \frac{f(x_{1(i+1)}, x_{2j}) - f(x_{1i}, x_{2j})}{x_{1(i+1)} - x_{1i}}(x_1 - x_{1i})$$

2. A one-dimensional interpolation is performed for x_1 at $x_{2(j+1)}$ to obtain $f(x_1, x_{2(j+1)})$ as follows:

$$f(x_1, x_{2(j+1)}) = f(x_{1i}, x_{2(j+1)}) + \frac{f(x_{1(i+1)}, x_{2(j+1)}) - f(x_{1i}, x_{2(j+1)})}{x_{1(i+1)} - x_{1i}}(x_1 - x_{1i})$$

(6.143)

3. Using the results of the previous one-dimensional interpolations, a one-dimensional interpolation is performed for x_2 at x_1
to obtain $f(x_1, x_2)$ as follows:

$$(6.144) \quad f(x_1, x_2) = f(x_{1i}, x_{2j}) + \frac{f(x_1, x_{2(j+1)}) - f(x_1, x_{2j})}{x_{2(j+1)} - x_{2j}}(x_2 - x_{2j})$$

Alternatively, the two-dimensional interpolation can be performed
using one-dimensional interpolation over x_2 first, followed by interpolation over x_1. The computational procedure is similar to Equations 6.142
through 6.144. The data of Equation 6.141 can be viewed as a planar surface defined by the four corner points.

Example 6.11: Gravitational Acceleration
as a Function of Latitude and Elevation

The gravitational acceleration (g) in feet per second squared is a
function of latitude (L) in degrees and elevation (E) in feet. This relationship is commonly tabulated with L and E as the independent
variables and g as the dependent variable. Data that cover the range
of latitude from 0° to 20° and elevation from 0 to 2000 ft are shown
in Equation 6.145. For the purpose of illustration, a problem might
require the value of g at $L = 12.5°$ and $E = 500$. The relevant portion
of the table that defines this relationship is given by four values with
$L = 10°$ and 15° and 1000 ft as follows:

	$E = 0$ ft	$E = 1000$ ft	$E = 2000$ ft
$L = 0°$	32.0877	32.0847	32.0816
$L = 5°$	32.0890	32.0859	32.0829
$L = 10°$	32.0928	32.0897	32.0867
$L = 15°$	32.0991	32.0960	32.0929
$L = 20°$	32.1075	32.1044	32.1013

(6.145)

Using the four values and $L = 12.5°$ and 1000 ft, the two-dimensional
interpolation can be performed as follows:

1. A one-dimensional interpolation is performed for E at $L =$
10° to obtain $g(10°, 500$ ft$)$ as follows:

$$g(10,500) = 32.0928 + \frac{32.0897 - 32.0928}{1000 - 0}(500 - 0) = 32.09125 \text{ ft/sec}^2$$

(6.146)

2. A one-dimensional interpolation is performed for E at $L =$
15° to obtain $g(15°, 500$ ft$)$ as follows:

$$g(15,500) = 32.0991 + \frac{32.0960 - 32.0991}{1000 - 0}(500 - 0) = 32.09755 \text{ ft/sec}^2$$

(6.147)

3. Using the results of the previous one-dimensional interpolations, a one-dimensional interpolation is performed for L at $E = 500$ ft to obtain $g(12.5°, 500$ ft$)$ as follows:

$$g(12.5, 500) = 32.09125 + \frac{32.09755 - 32.09125}{15 - 10}(12.5 - 10) = 32.0944 \text{ ft/sec}^2$$

(6.148)

Therefore, the estimated value of g at $L = 12.5°$ and $E = 500$ ft is 32.0944 ft/sec^2.

6.10 APPLICATIONS

6.10.1 PROBABILITY OF WIND LOADING

A structural engineer is designing a temporary structure that will be in place for 1 year. To account for wind load in the design, the engineer obtains past records for wind speeds in the area. Using the 34-year record of the annual maximum wind speeds, the mean value $(\bar{Y})$ and standard deviation (S_Y) are 90 and 20 km/hr, respectively. The engineer decides to use a wind speed of 125 km/hr for the design wind loading. The 34-year record also suggests that the annual maximum wind speed is normally distributed, so the engineer estimates the probability that the design loading will exceed 125 km/hr using the normal distribution. The standard normal deviate can be defined as

(6.149)
$$z = \frac{125 - \bar{Y}}{S_y} = \frac{125 - 90}{20} = 1.75$$

The table of normal probabilities that the engineer has available includes probabilities for z values to the nearest tenth. Thus an interpolating polynomial to estimate the probability must be developed.

For linear interpolation, values of z of 1.7 and 1.8 will be used, as follows:

$$f(z = 1.75) = f(1.7) + \frac{1.75 - 1.70}{1.80 - 1.70} = [f(1.8) - f(1.7)]$$

(6.150)
$$= 0.0446 + \frac{0.05}{0.10}(0.0359 - 0.0446)$$

$$= 0.04025$$

The true value is 0.0401, so the estimate is in error by 0.00015, or 0.4%.

To obtain a more accurate estimate, Newton's finite-difference equation is used. The finite-difference table is shown in Table 6.12 with the following interpolation formula:

TABLE 6.12   Finite-Difference Table for Normal Probabilities

z	f(z)	Δf	$\Delta^2 f$	$\Delta^3 f$	$\Delta^4 f$	$\Delta^5 f$
1.5	0.0668					
		−0.0120				
1.6	0.0548		0.0018			
		−0.0102		−0.0003		
1.7	0.0446		0.0015		0.0003	
		−0.0087		0.0000		−0.0005
1.8	0.0359		0.0015		−0.0002	
		−0.0072		−0.0002		
1.9	0.0289		0.0013			
		−0.0059				
2.0	0.0228					

$$f(z) = 0.0668 - 0.0120n + 0.0018\frac{n(n-1)}{2} - 0.0003\frac{n(n-1)(n-2)}{6}$$

(6.151)

$$+ 0.0003\frac{n(n-1)(n-2)(n-3)}{24} - 0.0005\frac{n(n-1)(n-2)(n-3)(n-4)}{120}$$

For $z = 1.75$, the value of n is

(6.152)
$$n = \frac{1.75 - 1.50}{0.1} = 2.5$$

Thus, the estimate is

$$f(1.75) = 0.0668 - 0.0120(2.5) + 0.0018\frac{2.5(1.5)}{2} - 0.0003\frac{2.5(1.5)(0.5)}{6}$$

$$+ 0.0003\frac{2.5(1.5)(0.5)(-0.5)}{24}$$

$$- 0.0005\frac{2.5(1.5)(0.5)(-0.5)(-1.5)}{120}$$

$$= 0.0401$$

(6.153)

Thus, the Newton method gives an exact value.

6.10.2 SHEAR STRESS OF OIL BETWEEN TWO PARALLEL PLATES

In the development of a new piece of equipment, it is necessary to measure the shear stress (τ) between two parallel plates, where the lower plate is stationary and the upper plate moves at a velocity (V) of 2 m/sec. The plates are 0.025 m apart (h). The shear stress τ is estimated by

$$(6.154) \qquad \tau = \mu \frac{V}{h} = \mu \frac{2\,\text{m/sec}}{0.025\,\text{m}} = 80\,\mu$$

The viscosity (μ) in N-sec/m^2 is a function of temperature. Laboratory measurements have determined μ at the temperatures shown in Table 6.13. Since the measurements are at unequal intervals, the method of Lagrange will be used to develop the interpolating polynomial. For a temperature of 38°C, the weights for the values of μ are

$$(6.155) \qquad w_1(38) = \frac{(38-20)(38-30)(38-50)(38-55)}{(5-20)(5-30)(5-50)(5-55)} = 0.03482$$

$$(6.156) \qquad w_2(38) = \frac{(38-5)(38-30)(38-50)(38-55)}{(20-5)(20-30)(20-50)(20-55)} = -0.34194$$

$$(6.157) \qquad w_3(68) = \frac{(38-5)(38-20)(38-50)(38-55)}{(30-5)(30-20)(30-50)(30-55)} = 0.96941$$

$$(6.158) \qquad w_4(38) = \frac{(38-5)(38-20)(38-30)(38-55)}{(50-5)(50-20)(50-30)(5-55)} = 0.59840$$

$$(6.159) \qquad w_5(38) = \frac{(38-5)(38-20)(38-30)(38-50)}{(55-5)(55-20)(55-30)(55-50)} = -0.26068$$

Based on Equation 6.91, the interpolating polynomial is

$$\mu(38) = 0.03482(0.08) - 0.34194(0.015) + 0.96941(0.009) - 0.59840(0.006)$$
$$+ 0.26068(0.0055)$$
$$= 0.00854$$

(6.160)

TABLE 6.13   Viscosity Coefficient Data

T (°C)	μ (N-sec/m^2)
5	0.0800
20	0.0150
30	0.0090
50	0.0060
55	0.0055

Thus the estimated shear stress is

$$\tau = 0.00854 \frac{2}{0.025} = 0.6832\,\text{N/m}^2 \tag{6.161}$$

PROBLEMS

Use the following data set for Problems 6.1 through 6.5:

x	0.4	0.5	0.6	0.7
$f(x)$	0.064	0.125	0.216	0.343

Derive an interpolating polynomial using the method of undetermined coefficients for the function indicated and find the interpolated value of $f(x)$ for the given value of x.

Problem	Function, $f(x)$	x
6.1	$a + bx$	0.47
6.2	$a + bx$	0.62
6.3	$a + bx + cx^2$	0.47
6.4	$a + bx + cx^2$	0.58
6.5	$a + bx + cx^2 + dx^3$	0.58

Use the following data for Problems 6.6 through 6.10:

x	1.6	1.8	2.0	2.2
$f(x)$	3.6	3.1	2.9	2.3

Derive an interpolating polynomial using the method of undetermined coefficients for the function indicated and find the interpolated value of $f(x)$ for the given value of x:

Problem	Function, $f(x)$	x
6.6	$a + bx$	1.62
6.7	$a + bx$	1.70
6.8	$a + bx + cx^2$	2.13
6.9	$a + bx + cx^2$	1.91
6.10	$a + bx + cx^2 + dx^3$	1.74

6.11 When an interpolating polynomial is of the same order as the function used to derive the data, show that the polynomial will provide errorless interpolated values.

6.12 Using the values in the following table, derive second-order interpolating polynomials for estimating the snow-covered area (A) in percent as a function of the accumulated degree-days (D) in °C. Use the method of undetermined coefficients.

Estimate the value of A for D of 50 and 450. If the true values are 100% and 6%, respectively, compute the percentage of error.

A	100	91	44	20	9	4
D	0	100	200	300	400	500

6.13 Using the values in the following table, derive second-order interpolating polynomials for estimating the snow-covered area (A) in percent as a function of the accumulated degree-days (D) in °C. Use the method of undetermined coefficients. Estimate the value of A for D of 25 and 220. If the true values are 99% and 46%, respectively, compute the percentage of error.

A	100	96	81	63	48	39
D	0	50	100	150	200	250

6.14 The following table gives the values of sin(x) for selected angles (x) in degrees:

x (degrees)	10	12	14	16
sin(x)	0.17365	0.20791	0.24192	0.27564

Using the method of undetermined coefficients, derive linear, quadratic, and cubic interpolating polynomials for the interval $12° \leq x \leq 14°$ and estimate the value of sin(x) for angles of 12.5° and 13°. Compare the accuracy achieved given the true values of 0.21644 and 0.22495, respectively.

6.15 The following table gives the values of cos(x) for selected angles (x) in degrees.

x (degrees)	40	42	44	46
cos(x)	0.7660	0.7431	0.7193	0.6947

Using the method of undetermined coefficients, derive linear, quadratic, and cubic interpolating polynomials for the interval $42° \leq x \leq 44°$ and estimate the value of cos(x) for angles of 42.5° and 43°. Compare the accuracy achieved given the true values of 0.7373 and 0.73135, respectively.

6.16 Given the following data set:

x	0.50	0.55	0.60	0.65
f(x)	0.4621	0.5005	0.5370	0.5717

derive an interpolating polynomial using the Gregory–Newton method. Estimate the value of $f(x)$ for $x = 0.475$ and $x = 0.525$. Compare the accuracy of the estimated values if the true values are 0.4422 and 0.4815, respectively.

6.17   Given the following data set:

x	1	2	3	4
$f(x)$	0.26813	0.17973	0.12048	0.08076

derive an interpolating polynomial using the Gregory–Newton method. Estimate the value of $f(x)$ for $x = 0.5$ and $x = 3.5$. Compare the accuracy of the estimated values if the true values are 0.32749 and 0.09864, respectively.

6.18   Use the Gregory–Newton method with the data of Problem 6.12 to derive an interpolation polynomial. Establish the value of A for D of 50 and 450. If the true values are 100% and 6%, respectively, compute the relative error (%).

6.19   Use the Gregory–Newton method with the data of Problem 6.13 to derive an interpolation polynomial. Establish the value of A for D of 25 and 225. If the true values are 99% and 46%, respectively, compute the relative error (%).

6.20   The following table gives the specific enthalpy (h) for superheated steam as a function of temperature (t) and at a constant pressure of 2500 lb/in.^{2}. Using the Gregory–Newton method, derive an interpolating polynomial and estimate the enthalpy at 1100°F.

h (Btu/lb)	1303.6	1458.4	1585.3	1706.1	1826.2
t (°F)	800	1000	1200	1400	1600

6.21   For the data given in Problem 6.12, derive a fifth-order interpolating polynomial using the Gregory–Newton method. Evaluate the error of the estimated value of A for $D = 50$.

6.22   The following table gives the specific weight (γ) of air (lb/ft^3) as a function of temperature T (°F). Using the Gregory–Newton method, derive an interpolating polynomial and estimate the specific weight of air at 50°. If the true value is 7.79, what is the percentage of error?

T	0	20	40	60	80
γ	8.62	8.27	7.94	7.63	7.35

6.23   The following table gives the kinematic viscosity (ν, 1×10^{-5} m^2/s) of air as a function of temperature (T, °C). Using the Gregory–Newton method, derive an interpolations polynomial and estimate the viscosity at 30°C. If the true value is 1.60 m²/s at 30°, what is the relative error?

T	0°	20°	40°	60°	80°
v	1.32	1.50	1.68	1.87	2.09

6.24 The following table gives the sine for angles at an interval of 5°:

Sine	0.000	0.087	0.174	0.259	0.342	0.423	0.500
Angle θ (degrees)	0	5	10	15	20	25	30

a. Using the Gregory–Newton method, derive a fourth-order interpolating polynomial.
b. Estimate the sine of 7°. If the true value is 0.122, compute the error.
c. Derive an interpolating polynomial using an increment of 10° and estimate the sine of 7°. Compare the error with the error for part (b).

6.25 Examine the sensitivity of the interpolating polynomial of Problem 6.24 if there is an error of 5% of the sine of 10°.

6.26 Using the following data, show the error and relative error when using a fourth-order Gregory–Newton interpolating polynomial to estimate the cosine of 42.5° (the true value is 0.7373).

θ	40	45	50	55	60
cosθ	0.7660	0.7071	0.6428	0.5736	0.5000

6.27 The following table includes values of the drag coefficient C_D of a smooth sphere as a function of the Reynolds number R. Develop a Gregory–Newton interpolations polynomial and estimate the drag coefficient at a Reynolds number of 400. If the true C_D is 0.6, estimate the error and relative error.

R	1	10	100	1000	10,000
C_D	28	4	1	0.45	0.38

6.28 Using the data of Problem 6.24, show the change in error for an interpolated value of the sine of 2.5° as the order of the polynomial is increased. Use successively more pairs of values with an increment of 5° when increasing the order of the polynomial.

6.29 Using the data of Problem 6.12, derive both a finite-difference table and an interpolating polynomial based on the table. Evaluate the error of the estimated value of A for $D = 50$.

6.30 Using the data of Problem 6.27, derive both a finite-difference table and an interpolating polynomial based on the table. Evaluate the error of the estimated value of C_D for $R = 4000$ (the true value is 0.40).

6.31 The data in the following table are values of the saturation vapor pressure e_s (mm Hg) in air above a water body as a function of temperature (°C).

T (°C)	10	12	14	16	18
e_s	9.20	10.52	11.98	13.63	15.46

a. Compute the finite-difference table and derive an interpolating polynomial using Newton's method.

b. Estimate e_s at a temperature of 15°C.

6.32 Using the data of Problem 6.24, derive both a finite-difference table and an interpolating polynomial based on the table. Evaluate the error of the estimated value of $\sin(\theta)$ for 7° (the true value is 0.122).

6.33 The data in the following table are the short-wave radiation flux (R) at the outer limit of the atmosphere in gram-calories per square centimeter per day for the month of September:

R	891	856	719	494	219
Latitude (°N)	0°	20°	40°	60°	80°

a. Compute the finite-difference table and derive an interpolating polynomial using Newton's method.

b. Estimate R at a latitude of 35°.

6.34 Using the data of Problem 6.22 compute a finite-difference table and derive an interpolating polynomial using Newton's method. Estimate the specific weight at 30°F.

6.35 Using the data of Problem 6.23, compute a finite-difference table and derive an interpolations polynomial using Newton's method. Estimate the viscosity at a temperature of 50°C.

6.36 Using the data of Problem 6.20, compute a finite-difference table and derive an interpolating polynomial using Newton's method. Estimate the enthalpy at a temperature of 1100°F.

6.37 The following table gives the times of sunrise and sunset in hours and minutes at 40° latitude for 4 days, with an interval of 28 days:

Date	Sunrise	Sunset
May 1	4 hr 59 min	18 hr 53 min
May 29	4 hr 34 min	19 hr 19 min
June 26	4 hr 32 min	19 hr 32 min
July 24	4 hr 50 min	19 hr 21 min

a. Compute finite-difference tables and derive interpolating polynomials using Newton's method for the time of both sunrise and sunset.

b. Estimate the time of sunrise on May 12.

c. Estimate the time of sunset on July 16.

d. Find the date and time of the earliest sunrise.

e. Find the date and time of the latest sunset.

f. Find the date and length of the longest period of daylight.

6.38 The following data give the mean values of the minimum and maximum temperatures (°F) for selected days, each separated by 28 days:

Day	Min	Max
Mar 1	36.4	66.2
Mar 29	44.1	69.4
Apr 26	49.0	72.1
May 24	53.6	75.2
Jun 21	58.3	78.1

a. Compute the finite difference tables and derive interpolations polynomial using Newton's method for the minimum and for the maximum temperatures.

b. Estimate the minimum temperature on May 10.

c. Estimate the maximum temperature on April 12.

d. Find the mean temperature difference on June 8.

6.39 Repeat Problem 6.37 using the following data for 20° latitude:

Date	Sunrise	Sunset
May 1	5 hr 30 min	18 hr 32 min
May 29	5 hr 19 min	18 hr 34 min
June 26	5 hr 22 min	18 hr 42 min
July 24	5 hr 32 min	18 hr 40 min

6.40 The following data give the mean values of the minimum and maximum temperatures (°F) for selected days, each separated by 21 days.

Day	Min	Max
Jun 14	56.4	77.3
Jul 5	61.2	80.1
Jul 26	60.8	80.2
Aug 16	57.0	77.1

a. Compute the finite difference tables and derive interpolating polynomials using Newton's method for the minimum, for the maximum, and for temperature difference.

b. Estimate the minimum temperature on July 15.

c. Estimate the maximum temperature on June 25.

d. Estimate the temperature difference on August 28.

6.41   The following table gives the times of sunrise and sunset in hours and minutes at 60° latitude. The data are given on an unequal time scale:

Date	Sunrise	Sunset
Nov 1	7 hr 21 m	16 hr 04 m
Dec 1	8 hr 34 m	15 hr 02 m
Jan 1	9 hr 02 m	15 hr 04 m
Feb 1	8 hr 13 m	16 hr 13 m
Mar 1	6 hr 58 m	17 hr 26 m

a. Derive Lagrangian interpolation polynomials for the time of sunrise and sunset.

b. Estimate the time of sunrise on February 20.

c. Estimate the time of sunset on November 11.

d. Find the date and time of the latest sunrise.

e. Find the date and time of the earliest sunset.

f. Find the date and length of the shortest period of daylight.

6.42   The following table gives the current density (D) in amperes per square millimeter (A/mm^2) of insulated copper conductors for selected cross-sectional area (A) in square millimeters:

A	0.75	1.0	1.5	2.5	4.0	6.0	10.0	16.0
D	17.4	15.0	13.3	10.8	9.0	7.7	6.8	5.7

Derive a Lagrangian interpolating polynomial for predicting the density D. Estimate D for cross-sectional areas of 1.2 and 2.0 mm^2.

6.43   Using the data of Problem 6.31, derive a Lagrangian polynomial for predicting the saturation vapor pressure at 11°C.

6.44   Using the data of Problem 6.26 derive a Lagrangian polynomial for predicting the $\cos\theta$ at 57.5°.

6.45   The following tables gives the percentage of acetic acid ionized (P) at different concentrations (C) in moles per liter:

C	0.1	0.05	0.02	0.0098	0.0059	0.00103
P	1.35	1.91	2.99	4.22	5.40	12.38

Derive a Lagrangian interpolating polynomial for predicting the percentage ionized. Estimate P for concentrations of 0.015 and 0.003 mole/liter.

6.46   The percentage of light (L) passing through the sea surface that penetrates to specific depths (D) in clean ocean water is as follows:

D (m)	0	1	2	10	50	100
L (%)	100	45	39	22	5	0.5

Find the value of L at a depth of 1.6 m using a Lagrangian, interpolation polynomial.

6.47   Using the data of Problem 6.22 fit linear and quadratic splines. Plot the two curves. Use the two curves to estimate the specific weight at a temperature of 32°.

6.48   Laboratory tests on a flocculant suspension in a settling column give the following values of the percentage of solids removed (S) as a function of the sampling time (T):

T (min)	5	11	23	33	49	60
S (%)	9	31	57	64	68	71

Find the percent removal for detention times of 15 and 30 min using a Lagrangian interpolation polynomial.

6.49   Using the data for the well pumping example (Equation 6.102), estimate the drawdown at a distance of 225 ft from the well.

6.50   Using the data of Problem 6.45, develop linear splines and quadratic splines for the purpose of interpolation. Plot your results.

6.51   Using the data of Problem 6.27, fit linear and quadratic splines. Plot the two curves. Use the two curves to estimate the drag coefficient at a Reynolds number of 550.

6.52   Using the data of Problem 6.46, develop linear splines and quadratic splines for the purpose of interpolation. Plot your results.

6.53   Using the data of Problem 6.48, develop linear splines and quadratic splines for the purpose of interpolation. Plot your results.

6.54   The following data provide the value of a function at any selected pairs of x and y:

	$y = 10$	$y = 20$	$y = 30$
$x = 1$	0.1940	0.4610	0.6781
$x = 2$	0.3125	0.8123	0.9610
$x = 3$	0.8190	2.1670	2.8901
$x = 4$	1.8231	3.1621	4.0070

a. Determine the value of the function using linear interpolation at $x = 2.3$ and $y = 22.3$.

b. Determine the value of the function using linear interpolation at $x = 3.4$ and $y = 10.6$.

Differentiation and Integration

7

7.1 NUMERICAL DIFFERENTIATION

A number of engineering problems require a numerically derived estimate of a derivative of a function $f(x)$, with two general approaches to the problem. First, if the function is known but the derivative cannot be computed analytically, the derivative can be estimated by computing the function for two values of the independent variable(s) separated by distance Δx and dividing the difference by Δx as follows:

$$(7.1) \qquad \frac{df(x)}{dx} \approx \frac{f(x + \Delta x) - f(x)}{\Delta x}$$

in which $f(x)$ and x are the dependent and independent variables, respectively, and $f(x)$ denotes the value of the dependent variable when x is the value of the independent variable.

The second approach to numerical differentiation is to fit a function to a set of points that describes the relationship between the dependent and independent variables and then differentiate the fitted function. Specifically, an interpolation polynomial of order n could be fit to the data and the derivative of the polynomial used as the estimate of the derivative. The selection of one of the two methods depends on the form in which the data are presented and the desired level of accuracy.

7.1.1 FINITE-DIFFERENCE DIFFERENTIATION

Equation 7.1 is just one possible difference method; it is called the *forward difference* equation because it uses the base point x and the value of x that is greater than x by an amount Δx. The following *backward difference* formula could also be used:

$$(7.2) \qquad \frac{df(x)}{dx} \approx \frac{f(x) - f(x - \Delta x)}{\Delta x}$$

For Equation 7.2, the value of x is incremented by $-\Delta x$ and the value of the function is determined for both x and $(x - \Delta x)$; the difference between the values of the two functions is divided by Δx to approximate the derivative. The *two-step method*, which is an alternative to Equations 7.1 and 7.2, computes the value of the function for values of the independent variable of $(x - \Delta x)$ and $(x + \Delta x)$, which provides an estimate of the derivative by

$$(7.3) \qquad \frac{df(x)}{dx} \approx \frac{f(x+\Delta x) - f(x-\Delta x)}{2\Delta x}$$

For many highly variable functions, the two-step method may provide slightly more accurate estimates than either the forward or backward finite-difference approximations.

The finite-difference approach can be used when the data are given as tabular values at some finite increment Δx or when a functional form is given, but the functional form cannot be differentiated analytically. Examples 7.1 and 7.2 are used to illustrate the first and second cases, respectively.

Example 7.1: Evaporation Rates

A design engineer must make estimates of evaporation rates when the amount of water needed to meet irrigation demands is required. One input to a frequently used formula for estimating evaporation rates is the slope of the saturation vapor pressure curve at air temperature T. A small part of the table that is used to make such estimates is given in Table 7.1. If it is necessary to make design estimates at a temperature of 22°C, then the slope of the saturation vapor pressure curve at 22°C could be estimated using, as follows, the forward (Equation 7.1), backward (Equation 7.2), or two-step (Equation 7.3) method:

$$(7.4) \qquad \frac{de_s}{dT} = \frac{e_s(23) - e_s(22)}{23 - 22} = \frac{21.05 - 19.82}{1} = 1.23 \text{ mm Hg/°C}$$

$$(7.5) \qquad \frac{de_s}{dT} = \frac{e_s(22) - e_s(21)}{22 - 21} = \frac{19.82 - 18.65}{1} = 1.17 \text{ mm Hg/°C}$$

$$(7.6) \qquad \frac{de_s}{dT} = \frac{3_s(23) - e_s(21)}{23 - 21} = \frac{21.05 - 18.65}{2} = 1.20 \text{ mm Hg/°C}$$

When applying the two-step method in Equation 7.6, a full increment was actually taken in each direction because this was the smallest increment for which data were available. The true value is 1.20 mm Hg/°C, so the two-step method provided the most accurate estimate.

**TABLE 7.1   Saturation Vapor
Pressure (e_s) in mm Hg as a
Function of Temperature (T) in °C**

T(°C)	e_s (mm Hg)
20	17.53
21	18.65
22	19.82
23	21.05
24	22.37
25	23.75

Example 7.2: Model Fitting

A common problem in engineering analysis is fitting models to engineering data. Simple model forms, such as a straight line or a plane, can be fit using the methods of Chapter 10; these methods require that some objective function be minimized. The minimization of the sum of the squares of the errors is the most frequently used objective function in engineering analysis. A commonly used objective function is

$$(7.7) \qquad F = \min \sum_{i=1}^{n} e_i^2 = \min \sum_{i=1}^{n} (\hat{Y}_i - Y_i)^2$$

in which
    F is the value of the objective function
    n is the number of observations, or sample size
    e_i is the error, or residual, for observation i
    $\hat{Y}_i$ and Y_i are the predicted and measured values of the criterion variable, respectively

If the model used to predict $f(x)$ is relatively simple in structure, Equation 7.7 can be differentiated analytically with respect to the coefficients (i.e., unknowns) of the model. If the model is more complex, then an analytical analysis is not possible, and a numerical solution of the derivatives is necessary to find the value or the values of the coefficients that provide the minimum value of F.

Let us examine the problem using a model employed by the environmental engineer in predicting the dissolved oxygen deficit (ppm) in a stream as a function of time:

$$(7.8) \qquad \hat{Y} = \frac{K_1}{1 + e^{-K_2(t-t_0)}} - K_3$$

in which t is the number of days from the time when a point waste-load is discharged into a stream; K_1, K_2, K_3, and t_0 are unknowns that must be found from measured data; and $\hat{Y}$ is the predicted dissolved oxygen deficit. Substituting Equation 7.8 into Equation 7.7 yields

$$(7.9) \qquad F = \sum_{i=1}^{n} \left(\frac{K_1}{1 + e^{-K_2(t-t_0)}} - K_3 - Y_i \right)^2$$

The subscript i on t_i, which is the time at which the dissolved oxygen deficit y_i is measured, and n refers to the number of the observation in the data set. For purposes of illustration, assume that the data shown in Table 7.2 were measured at a point on a stream approximately 500 ft below the point where an industrial plant discharges waste into the stream.

To find the values of the coefficients, it would be necessary to take the derivative of F with respect to each of the four unknowns using analytical calculus. This would not be a simple task, so the solution can be found using a numerical analysis. The numerical solution requires initial estimates of the unknowns; for example, the following values are assumed: K_1 = 4.5, K_2 = 1.0, K_3 = 0.1, and t_0 = 2. Using these values and the time in days shown in Table 7.2, the following predicted values of the dissolved oxygen deficit are obtained using Equation 7.8: 0.44, 1.11, 2.15, 3.19, and 3.86 ppm. These values can be used to compute the value of F according to Equation 7.7.

$$F = \sum_{i=1}^{5} (\hat{Y}_i - Y_i)^2$$

$$= (0.44 - 0)^2 + (1.11 - 0.62)^2 + (2.15 - 1.82)^2 + (3.19 - 3.03)^2 + (3.86 - 3.64)^2$$

$$= 0.6166$$

(7.10)

To evaluate the derivative of F with respect to K_1, we could increase K_1 by ΔK_1 = 0.1, recompute the values of $\hat{Y}$ using Equation 7.8, and compute a new value of F using Equation 7.7, which is denoted as F_1. Then an estimate of the derivative is given by

$$(7.11) \qquad \frac{\Delta F}{\Delta K_1} = \frac{F_1 - F}{\Delta K_1}$$

TABLE 7.2   Dissolved Oxygen Data

i	1	2	3	4	5
t (days)	0	1	2	3	4
y_i (ppm)	0	0.62	1.82	3.03	3.64

which is the forward difference formula. For K_1 = 4.6, predicted dissolved oxygen concentrations of 0.45, 1.14, 2.20, 3.26, and 3.95 ppm are obtained. These values yield a value for F_1 of

(7.12)
$$
\begin{aligned}
F_1 &= (0.45 - 0)^2 + (1.14 - 0.62)^2 + (2.20 - 1.82)^2 \\
&\quad + (3.26 - 3.03)^2 + (3.95 - 3.64)^2 \\
&= 0.7663
\end{aligned}
$$

Thus the estimate of the derivative is

(7.13)
$$
\frac{\Delta F}{\Delta K_1} = \frac{0.7663 - 0.6166}{0.1} = 1.497
$$

Obviously, if we wish to find the minimum value of F, then the value of K_1 needs to be decreased, because F increased when K_1 was increased from 4.5 to 4.6.

The derivative of F with respect to K_3 is easily computed numerically. If we increase the initial value of K_3 from 0.1 to 0.15, then the predicted dissolved oxygen deficits would be 0.39, 1.06, 2.10, 3.14, and 3.81 ppm, respectively. This would yield a value for F of

$$
F_2 = (0.39 - 0)^2 + (1.06 - 0.62)^2 + (2.10 - 1.82)^2 + (3.14 - 3.03)^2 + (3.81 - 3.64)^2
$$

$$
= 0.4651
$$

(7.14)

Thus the numerical estimate of the derivative would be

(7.15)
$$
\frac{\Delta F}{\Delta K_3} = \frac{F_2 - F}{\Delta K_3} = \frac{0.4651 - 0.6166}{0.05} = -3.03
$$

Since increasing the value of K_3 from 0.1 to 0.15 caused the value of F to decrease, we would want to increase the value of K_3 further to find the minimum value of the objective function.

The objective of this example was to show how the derivative of a complex functional form can be evaluated numerically. The example also illustrates that the numerical evaluation of derivatives can be used to optimize complex engineering models when it is not possible to evaluate the derivatives analytically.

7.1.2 DIFFERENTIATION USING A FINITE-DIFFERENCE TABLE

The finite-difference table introduced in Chapter 6 can also be used to approximate the derivatives of a function that is expressed in incremental form. Consider the problem of estimating the cube of a number. The finite-difference table is given in Table 6.6. The values in the Δf column can be used to make estimates of the first derivative; values in the $\Delta^i f$ column can be used to estimate the ith derivative.

For estimating the first derivative at any value of x, the Δf value in the diagonal row below the row for the value of x can be used to estimate the forward finite difference. The value of Δf in the diagonal row above the row for the value of x can be used to estimate the backward finite difference. The average of these two Δf values can be used to estimate the first derivative with the two-step method. Specifically, for a value of x equal to 11, the estimates of the first derivative by the forward, backward, and two-step methods based on Table 6.6 are, respectively,

$$(7.16) \qquad \frac{df}{dx} = \frac{\Delta f_{i+1}}{\Delta x} = \frac{397}{1} = 397$$

$$(7.17) \qquad \frac{df}{dx} = \frac{\Delta f_i}{\Delta x} = \frac{331}{1} = 331$$

$$(7.18) \qquad \frac{df}{dx} = \frac{\Delta f_i + \Delta f_{i+1}}{2\Delta x} = \frac{331 + 397}{2} = 364$$

The true value can be found by differentiating $f(x) = x^3$ with respect to x, which is $3x^2$. For an x of 11, the true value would be 363. Thus the two-step method provides the most accurate estimate. The large error for the forward and backward difference equations is the result of the relatively large incremental value of the independent variable, that is, Δx.

A value for the second derivative is obtained from the value of $\Delta^2 f$ of Table 6.6. In this case, the estimate equals the value of $\Delta^2 f$ divided by Δx, which for the example would be 66. This agrees with the true value of $6x = 6(11) = 66$.

Example 7.3: Specific Heat Capacity

The specific heat capacity is an important element in thermodynamic processes. For a process in which the pressure is constant, the specific heat capacity (c_p) equals the slope of the relationship between the specific enthalpy (h) and the temperature (T):

$$(7.19) \qquad c_p = \frac{dh}{dT}$$

Table 7.3 gives the finite-difference table for data that defines the relationship between temperature and the enthalpy of a superheated steam. Using the first finite difference (Δh), the specific heat capacity can be estimated. For example, at a temperature of 1200°F, the forward finite-difference equation would yield a specific heat capacity of

$$(7.20) \qquad c_p = \frac{\Delta h}{\Delta T} = \frac{1705 - 1585}{1400 - 1200} = \frac{120}{200} = 0.6 \text{ Btu/lb/°F}$$

TABLE 7.3   **Finite-Difference Table for Specific Enthalpy (*h*) in Btu/lb and Temperature (*T*) in °F**

T	h	Δh	Δ²h	Δ³h	Δ⁴h
800	1305				
		155			
1000	1460		−30		
		125		25	
1200	1585		−5		−20
		120		5	
1400	1705		0		
		120			
1600	1825				

The backward difference formula would yield the following estimate:

$$(7.21) \qquad c_p = \frac{\Delta h}{\Delta T} = \frac{1585 - 1460}{1200 - 1000} = \frac{125}{200} = 0.625 \text{ Btu/lb/°F}$$

Using a full increment on either side of the base point, the two-step method would lead to the following estimate:

$$(7.22) \qquad c_p = \frac{\Delta h}{\Delta T} = \frac{1705 - 1460}{1400 - 1000} = \frac{245}{400} = 0.6125 \text{ Btu/lb/°F}$$

The rate of change of c_p with respect to T can be approximated by the second derivative, which is computed from the second finite difference. For the data of Table 7.3, the rate of change of c_p with respect to T at $T = 1200$°F is

$$(7.23) \qquad \frac{\Delta^2 h}{\Delta T^2} = \frac{-5}{200} = -0.025 \text{ Btu/lb/(°F)}^2$$

At a temperature of 1000°F, the rate of change would be −0.15 Btu/lb/(°F)².

7.1.3 DIFFERENTIATING AN INTERPOLATING POLYNOMIAL

Given data in a tabular form, an interpolating polynomial of the dependent variable as a function of the independent variable can be derived and differentiated. The interpolation methods discussed in Chapter 6 can be used for this purpose. If the interpolating polynomial is expressed as an *n*th-order polynomial, then the derivative is a polynomial of order $(n - 1)$. In general, a polynomial can be expressed as

(7.24) $f(x) = b_n x^n + b_{n-1} x^{n-1} + b_{n-2} x^{n-2} + \ldots + b_1 x + b_0$

The derivative of the polynomial is

(7.25) $\dfrac{df(x)}{dx} = n b_n x^{n-1} + (n-1) b_{n-1} x^{n-2} + \ldots + b_1$

When the interpolating polynomial is expressed in the form of the Gregory–Newton formula (Equation 6.15, Chapter 6), the derivative is somewhat more difficult to evaluate directly. For example, differentiating the first four terms of Equation 6.15 with respect to x yields

(7.26)

$$\frac{df(x)}{dx} = a_2 + a_3[(x - x_2) + (x - x_1)] + a_4[(x - x_2)(x - x_3)$$

$$+ (x - x_1)(x - x_3) + (x - x_1)(x - x_2)]$$

This is obviously more difficult to evaluate than the $(n-1)$th-order derivative of Equation 7.25.

Example 7.4: Specific Heat Capacity

Using the data of Example 6.2 that are given in Table 6.2, the Gregory–Newton method was used to develop the interpolation equation between vapor pressure of water (P) and temperature (T) as given by Equation 6.25. The resulting a_i, where i = 1, 2, 3, 4, and 5, can be used in Equation 7.26 to evaluate the derivative of P with respect to T. Several methods of estimating evaporation rates from open water surfaces require the slope of the vapor pressure curve at some design temperature. The derivative of Equation 6.25 is given by

$$\frac{dP(T)}{dT} = 3.55 + 0.0914063[(T - 48) + (T - 40)]$$

$$+ 0.001172[(T - 48)(T - 56) + (T - 40)(T - 56) + (T - 40)(T - 48)]$$

$$+ 0.00001017[(T - 48)(T - 56)(T - 64) + (T - 40)(T - 56)(T - 64)$$

$$+ (T - 40)(T - 48)(T - 64) + (T - 40)(T - 48)(T - 56)]$$

(7.27)

The slope of the vapor pressure curve at a temperature of 70°C is computed by substituting 70°C for T and yields $dP(T)/dT$ = 10.1084 mm Hg/°C, which is accurate to three significant digits.

7.1.4 DIFFERENTIATION USING TAYLOR SERIES EXPANSION

The basic finite-difference equation for differentiation as given by Equation 7.1 results from the following forward Taylor series expansion:

$$(7.28) \quad f(x+\Delta x)=f(x)+\frac{df(x)}{dx}\Delta x+\frac{df^2(x)}{dx^2}\frac{(\Delta x)^2}{2!}+\frac{df^3(x)}{dx^3}\frac{(\Delta x)^3}{3!}+\dots$$

Truncating terms with the second derivative and higher derivatives results in the following equation:

$$(7.29) \quad f(x+\Delta x)\approx f(x)+\frac{df(x)}{dx}\Delta x$$

Rearranging the terms to solve for $f'(x)=df(x)/dx$ results in the forward finite-difference differentiation equation of Equation 7.1. Therefore, this equation provides a first-order approximation of the first derivative of $f(x)$.

The second-order approximation of the first derivative of $f(x)$ can be obtained by including the second-order term from the Taylor series expansion in Equation 7.29. The resulting second-order approximation, after solving for $f'(x)$, is

$$(7.30) \quad \frac{df(x)}{dx}\approx\frac{f(x+\Delta x)-f(x)}{\Delta x}-\frac{df^2(x)}{dx^2}\frac{\Delta x}{2!}$$

The second-order approximation requires knowledge of the second derivative of $f(x)$. The finite-difference approximation can also be used to compute higher order derivatives. For example, if we let $f'(x)$ be the first derivative of $f(x)$ with respect to x, then the forward difference approximation of the second derivative is given by

$$(7.31) \quad \frac{d^2 f(x)}{dx^2}\approx\frac{f'(x+\Delta x)-f'(x)}{\Delta x}$$

Substituting the first-order approximation of $f'(x)$ in Equation 7.31 produces the following equation for the second derivative:

$$(7.32) \quad \frac{d^2 f(x)}{dx^2}\approx\frac{f(x+2\Delta x)-2f(x+\Delta x)+f(x)}{(\Delta x)^2}$$

Equation 7.32 is a first-order approximation of the second derivative. Now, by substituting Equation 7.32 into Equation 7.30, the second-order approximation of the first derivative is obtained as

$$(7.33) \quad \frac{df(x)}{d(x)}\approx\frac{-f(x+2\Delta x)+4f(x+\Delta x)-3f(x)}{2\Delta x}$$

The resulting Equation 7.33 does not require any higher order derivatives; it can be evaluated using the function $f(x)$. Similarly, the first-order estimate of the second derivative as given by Equations 7.31 and 7.32 can be revised by including a second-order term from the Taylor series expansion to obtain its second-order approximation as

(7.34)
$$\frac{d^2 f(x)}{dx^2} \approx \frac{-f(x+3\Delta x)+4 f(x+2\Delta x)-5 f(x+\Delta x)+2 f(x)}{(\Delta x)^2}$$

Using backward difference in the Taylor series expansion, the following first- and second-order estimates, respectively, for the first derivative of $f(x)$ can be obtained:

(7.35)
$$\frac{df(x)}{dx} \approx \frac{f(x)-f(x-\Delta x)}{\Delta x}$$

(7.36)
$$\frac{df(x)}{dx} \approx \frac{3 f(x)-4 f(x-\Delta x)+f(x-2\Delta x)}{2\Delta x}$$

Using backward difference in the Taylor series expansion, the following first- and second-order estimates, respectively, for the second derivative of $f(x)$ can be obtained:

(7.37)
$$\frac{d^2 f(x)}{dx^2} \approx \frac{f(x)-2 f(x-\Delta x)+f(x-2\Delta x)}{(\Delta x)^2}$$

(7.38)
$$\frac{d^2 f(x)}{dx^2} \approx \frac{2 f(x)-5 f(x-\Delta x)+4 f(x-2\Delta x)-f(x-3\Delta x)}{(\Delta x)^2}$$

Using the two-step method in the Taylor series expansion, the following first- and second-order estimates, respectively, for the first derivative of $f(x)$ can be obtained:

(7.39)
$$\frac{df(x)}{dx} \approx \frac{f(x+\Delta x)-f(x-\Delta x)}{2\Delta x}$$

(7.40)
$$\frac{df(x)}{dx} \approx \frac{-f(x+2\Delta x)+8 f(x+\Delta x)-8 f(x-\Delta x)+f(x-2\Delta x)}{12\Delta x}$$

Using the two-step method in the Taylor series expansion, the following first- and second-order estimates, respectively, for the second derivative of $f(x)$ can be obtained:

$$\frac{d^2 f(x)}{dx^2} \approx \frac{f(x+\Delta x)-2 f(x)+f(x-\Delta x)}{(\Delta x)^2}$$
(7.41)

$$\frac{d^2 f(x)}{dx^2} \approx \frac{-f(x+2\Delta x)+16 f(x+\Delta x)-30 f(x)-16 f(x-\Delta x)-f(x-2\Delta x)}{12(\Delta x)^2}$$
(7.42)

Example 7.5: Evaporation Rates

Example 7.1 was used to illustrate the first-order finite-difference equations for computing the slope of the saturation vapor pressure curve based on the data in Table 7.1. Using the second-order equations (Equations 7.33, 7.36, and 7.40) for the forward, backward, and two-step methods produces, respectively, the following results:

(7.43)
$$\frac{df(x)}{dx} \approx \frac{-e_s(24) + 4e_s(23) - 3e_s(22)}{2(1)}$$

$$\approx \frac{-22.37 + 4(21.05) - 3(19.82)}{2(1)} = 1.185 \text{ mm Hg/}°C$$

(7.44)
$$\frac{df(x)}{dx} \approx \frac{3e_s(22) - 4e_s(21) + e_s(20)}{2(1)}$$

$$\approx \frac{3(19.82) - 4(18.65) + 17.53}{2(1)} = 1.195 \text{ mm Hg/}°C$$

$$\frac{df(x)}{dx} \approx \frac{-e_s(24) + 8e_s(23) - 8e_s(21) + e_s(20)}{12(1)}$$

$$\approx \frac{-(22.37) + 8(21.05) - 8(18.65) + (17.53)}{12(1)} = 1.19667 \text{ mm Hg/}°C$$

(7.45)

The accuracy of the three methods is improved by using the second-order approximation. The two-step method still provides the best estimate, since the true value at $T = 22°C$ is 1.2 mm Hg/°C.

7.2 NUMERICAL INTEGRATION

Integration is introduced in analytical calculus using examples in which the function describes a continuous curve over some interval. For example, we may be interested in the area under the curve $f(x) = x^2$ between the lower and upper limits of a and b, respectively, as shown in Figure 7.1. The integration is expressed as

(7.46)
$$\text{integral} = \int_a^b f(x)\,dx$$

The result of the integral is the area under the $f(x)$ curve between a and b. In engineering, sometimes the function $f(x)$ does not have a known analytical solution. Therefore, a numerical evaluation of the integral is needed.

In many engineering problems, the data are presented as a set of discrete values for some dependent variable. For example, the volume rate of flow (Q) of water in a channel or through a pipe is the integral of the velocity (V) and the incremental area (dA of the channel cross section):

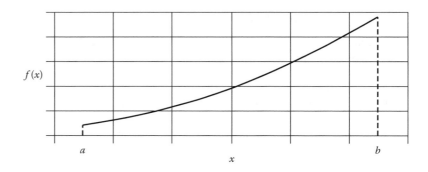

FIGURE 7.1   Analytical integration.

(7.47)
$$Q = \int V\, dA$$

In practical applications for a natural channel that is irregular in shape, velocity measurements are made at a number of elevations and the volume flow rate computed for each subarea, with the total flow in the channel being the sum of the flow rates computed for the subareas. Thus the integral of Equation 7.47 is not computed through analytical integration but by a numerical approximation.

The methods presented in this section for numerical integration can be used for both classes of problems—that is, problems with a known $f(x)$ that cannot be integrated analytically and problems with discrete data. Two general approaches to numerical integration are available. First, a polynomial could be numerically fitted to the discrete points and then the polynomial can be integrated analytically. For example, either the Gregory–Newton or the finite-difference method could be used to derive an interpolating polynomial to a set of points, and then the polynomial could be integrated using analytical calculus. Second, the area can be approximated using simple geometric shapes defined by adjacent points. It can be shown that the two methods are computationally equivalent, although they differ in physical appearance.

Other methods for numerical integration are discussed in Section 8.11 in the next chapter. These methods are based on converting the integration problem of the type given in Equation 7.46 to a differential equation problem; hence, numerical methods for solving differential equations can be used.

7.2.1 INTERPOLATION FORMULA APPROACH

The Gregory–Newton method for deriving an interpolation formula was introduced in Chapter 6. The end result was a polynomial expressed in the following form:

$$f(x) = a_1 + a_2(x - x_1) + a_3(x - x_1)(x - x_2) + \ldots +$$

(7.48)
$$a_{n+1}(x - x_1)(x - x_2) \ldots (x - x_n)$$

The terms could be rearranged to form an nth-order polynomial of the form

(7.49) $$f(x) = b_1 x^n + b_2 x^{n-1} + \ldots + b_{n+1}$$

in which the coefficients b_i, $i = 1, 2, 3, \ldots, n + 1$, are a function of the values of x_i and the a_i. After the coefficients of Equation 7.48 are derived and the polynomial of Equation 7.49 is formed, Equation 7.49 could be integrated analytically as follows:

(7.50) $$\int f(x)\,dx = \frac{b_1}{n+1}x^{n+1} + \frac{b_2}{n}x^n + \ldots + b_{n+1}x$$

The value of Equation 7.50 can be computed for any specified limits of integration.

Example 7.6: Traffic Safety Engineer

A traffic safety engineer is interested in the time gap between successive cars on a freeway during rush hour. In an attempt to develop a relationship between the probability of accidents and the mean gap on a section of a highway, the engineer collects the data given in Table 7.4 for a single location on the highway. The engineer can use the data on the frequency of occurrence to compute the probability for any gap length. For example, if the engineer believes that a gap of 4 seconds is the gap below which accidents are most likely to occur, the following equation can be used to approximate the desired probability:

(7.51) $$P(\text{gap} \le 4\,\text{sec}) = \frac{\displaystyle\int_0^4 f(x)\,dx}{\displaystyle\int_0^{12} f(x)\,dx}$$

TABLE 7.4   Traffic Gap Data

Gap Length (seconds)	Mid-Gap Length, x	Frequency of Occurrence
0–2	1	12
2–4	3	26
4–6	5	18
6–8	7	14
8–10	9	8
10–12	11	6
12 or more	NA	0

The Gregory–Newton formula was used to derive the following interpolating polynomial based on mid-intervals (x) as shown in Table 7.4:

$$f(x) = 12 + 7(x - x_1) - \frac{11}{4}(x - x_1)(x - x_2) + \frac{13}{24}(x - x_1)(x - x_2)(x - x_3)$$

$$- \frac{1}{12}(x - x_1)(x - x_2)(x - x_3)(x - x_4)$$

$$+ \frac{11}{960}(x - x_1)(x - x_2)(x - x_3)(x - x_4)(x - x_5)$$

(7.52)

This interpolating polynomial can be rearranged to form the following fifth-order polynomial:

$$(7.53) \quad f(x) = \frac{11}{960}x^5 - \frac{71}{192}x^4 + \frac{433}{96}x^3 - \frac{2465}{96}x^2 + \frac{20{,}633}{320}x - \frac{1981}{64}$$

The integrals of Equation 7.51 can be computed using the polynomial of Equation 7.53:

$$\int f(x)\,dx = \frac{11}{5760}x^6 - \frac{71}{960}x^5 + \frac{433}{384}x^4 - \frac{2465}{288}x^3 + \frac{20{,}633}{640}x^2 - \frac{1981}{64}x$$

(7.54)

By evaluating Equation 7.54 between the limits for the two integrals of Equation 7.51, the following probability was estimated:

$$(7.55) \qquad P(\text{gap} \le 4\,\text{sec}) = \frac{64.99}{162.19} = 0.40$$

If similar estimates of the probability are computed for other locations on the highway, the traffic safety engineer can develop a relationship between the accident rate and the probability of cars that follow each other with a small time gap.

7.2.2 TRAPEZOIDAL RULE

The trapezoidal rule approximates the area of a function defined by a set of discrete points by fitting a trapezoid to each pair of adjacent points that defines the dependent variable and summing the individual areas, as shown in Figure 7.2a. That is, assuming that the function defining the dependent variable $f(x)$ is quantified by an array of points at selected values of the dependent variable x, the area under the function can be approximated by (see Figure 7.2b)

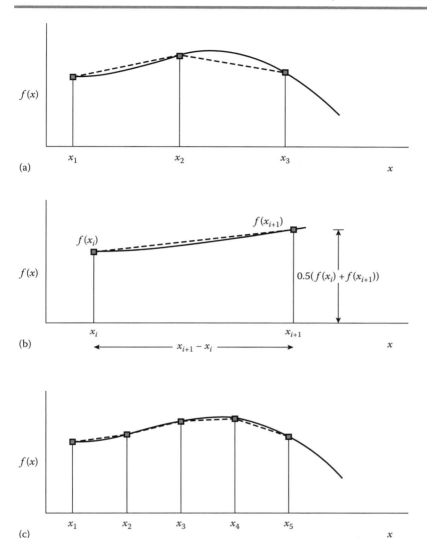

FIGURE 7.2   Trapezoidal rule for integration: (a) wide-interval approximation, (b) terms of Equation 7.56, and (c) narrow-interval approximation.

$$(7.56) \qquad \int_{x_1}^{x_n} f(x)\,dx \approx \sum_{i=1}^{n-1} (x_{i+1} - x_i) \frac{f(x_{i+1}) + f(x_i)}{2}$$

The first term of Equation 7.56 is the base width of the trapezoid, while the second term is the average height of the end points. The accuracy of the estimate can be improved by reducing the width of the interval; however, this increases the computational effort. A comparison of Figure 7.2a and c shows the change in accuracy, where the difference between the curve $f(x)$ and the linear approximations reflects the error. Therefore, Equation 7.56 can be easily applied to a function defined by a set of discrete values.

The trapezoidal rule, Equation 7.56, can be derived by fitting a linear interpolating polynomial to each pair of points. Using the Gregory–Newton formula (Equation 6.10, Chapter 6) with values of the independent variable of x_{i+1} and x_i, the following equation is obtained:

$$(7.57) \qquad f(x) = f(x_i) + \frac{f(x_{i+1}) - f(x_i)}{x_{i+1} - x_i}(x - x_i)$$

Integrating Equation 7.57 between two points, say a and b, produces

$$(7.58) \quad \int_a^b f(x)\,dx \approx \frac{b-a}{x_{i+1} - x_i}\left[\frac{(b+a)}{2}(f(x_{i+1}) - f(x_i)) + x_{i+1}f(x_i) - x_i f(x_{i+1})\right]$$

Letting $b = x_{i+1}$ and $a = x_i$ results in

$$(7.59) \qquad \int_{x_i}^{x_{i+1}} f(x)\,dx \approx \frac{x_{i+1} - x_i}{2}[f(x_{i+1}) + f(x_i)]$$

which is the rule given by Equation 7.56.

Numerical integration can result in errors where a numerical error is defined as the difference between the true value and the estimated value. The true value is unknown; otherwise, it would not be necessary to use the trapezoidal rule. An approximate upper bound on the error can be determined based on the second term of a Taylor series expansion:

$$(7.60) \qquad \frac{x_{i+1} - x_i}{2} f''(x)$$

in which $f''(x)$ is the value of the second derivative evaluated at the value of x that maximizes the value of $f''(x)$ between x_i and x_{i+1}.

Example 7.7: Trapezoidal Rule for Integration

For example, the data shown in Table 7.5 include three intervals; that is, $n = 4$. The discrete $f(x_i)$ values are for the following true (but supposedly unknown) function:

$$(7.61) \qquad f(x) = x^3 - 5x^2 + 4x + 3$$

The points and true function are shown in Figure 7.3, with the true area being the area between the line $f(x) = 0$ and the solid line. Using

TABLE 7.5   Data for Trapezoidal Rule

i	1	2	3	4
x_i	1	2	3	4
$f(x_i)$	3	−1	−3	3

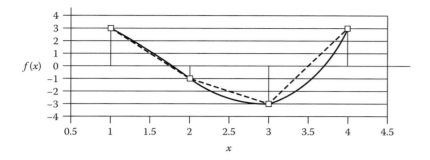

FIGURE 7.3   Trapezoidal rule for integration.

Equation 7.56, the trapezoidal rule provides the following estimate of the integral of Equation 7.61:

$$(7.62) \quad \int f(x)\,dx \approx \frac{2-1}{2}[3+(-1)]+\frac{3-2}{2}[-1+(-3)]+\frac{4-3}{2}[3+(-3)]$$

$$= 1+(-2)+0 = -1$$

The trapezoidal approximation is shown in Figure 7.3 as a dashed line. The true value, which is obtained by integrating Equation 7.61 between the limits of 1 and 4, is −2.25, which indicates that the trapezoidal estimate of Equation 7.62 is in error by 1.25 (absolute value). The error is the algebraic difference between the two curves of Figure 7.3. The error could be reduced by decreasing the grid spacing.

Example 7.8: Flow Rate

The flow rate (Q) of an incompressible fluid is given by the integral:

$$(7.63) \qquad\qquad Q = \int V\,dA$$

in which V is the velocity and A is the area. For a circular pipe of radius r, the incremental area dA is equal to $2\pi r\,dr$, and the quantity must be integrated over the cross section of the pipe. The incremental distance dr is measured from the center of the pipe as shown in Figure 7.4; that is, the expression of Equation 7.63 is integrated from $r = 0$ to $r = r_0$, where r_0 is the radius of the pipe. Thus Equation 7.63 becomes

$$(7.64) \qquad\qquad Q = \int_0^{r_0} V(2\pi r\,dr)$$

If an analytical expression exists for the velocity V as a function of r, Equation 7.64 can be solved analytically. However, if only

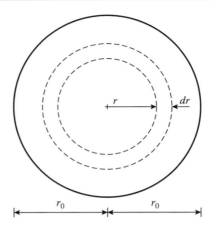

FIGURE 7.4   Flow rate for a pipe.

measurements of V are available at selected distances from the center of the pipe, the integral must be solved numerically.

Table 7.6 gives values of the velocity (V) measured at selected distances from the center of the pipe for a 1 ft diameter pipe. The trapezoidal rule can be used to estimate the value of the integral in Equation 7.64 as follows:

$$(7.65) \qquad Q = \frac{\Delta r}{2} \sum_{i=1}^{6} [V_i(2\pi r_i) + V_{i+1}(2\pi r_{i+1})]$$

in which Δr is the increment of r in which values of V are given in Table 7.6 (i.e., $\Delta r = \frac{1}{12}$ ft), and V_i is the value of V at distance r_i. For a pipe with a diameter of 1 ft, Equation 7.65 becomes

TABLE 7.6   Application of Trapezoidal Rule for Estimating the Flow Rate of a Fluid in a Circular Pipe

i	r_i (ft)	V_i (fps)
1	0	10.000
2	$\frac{1}{12}$	9.722
3	$\frac{1}{6}$	8.889
4	$\frac{1}{4}$	7.500
5	$\frac{1}{3}$	5.556
6	$\frac{5}{12}$	3.056
7	$\frac{1}{2}$	0

$$Q = \frac{2\pi\left(\frac{1}{12}\right)}{2} \sum_{i=1}^{6} [V_i r_i + V_{i+1} r_{i+1}]$$

(7.66)

$$Q = \frac{\pi}{12} \left\{ \left[10(0) + 9.772\left(\frac{1}{12}\right) \right] + \left[9.722\left(\frac{1}{12}\right) + 8.889\left(\frac{1}{6}\right) \right] \right.$$

$$+ \left[8.889\left(\frac{1}{6}\right) + 7.5\left(\frac{1}{4}\right) \right] + \left[7.5\left(\frac{1}{4}\right) + 5.556\left(\frac{1}{3}\right) \right]$$

$$\left. + \left[5.556\left(\frac{1}{3}\right) + 3.056\left(\frac{5}{12}\right) \right] + \left[3.056\left(\frac{5}{12}\right) + 0\left(\frac{1}{2}\right) \right] \right\} = 3.818 \text{ ft}^3/\text{sec}$$

(7.67)

While the true value is usually not known, the data of Table 7.6 were derived from the following relationship:

(7.68) $V = 10(1 - 4r^2)$

Therefore, Equation 7.68 could be solved analytically. The resulting Q is 3.927 ft³/sec, which means that the estimate provided by the trapezoidal rule (Equations 7.66 and 7.67) is in error by 0.109 ft³/sec, or an error of 2.8% of the true value.

7.2.3 SIMPSON'S RULE

The trapezoidal rule is based on a linear interpolating polynomial. The accuracy in an estimate of an integral can usually be improved by using a higher order polynomial as the interpolation formula. Simpson's rule is a numerical method of estimating the value of an integral when a second-order polynomial is used as the interpolating formula. Figure 7.5 shows both a second-order polynomial (Simpson's rule) and the linear equations for the trapezoidal rule in comparison with a higher order function (the solid line). Following the procedure for integrating the linear interpolating polynomial for the trapezoidal rule, it can be shown that Simpson's rule has the following form:

(7.69) $$\int_a^b f(x)\,dx = \frac{\Delta x}{3}[f(a) + 4f(a + \Delta x) + f(b)]$$

in which a and b are the lower and upper bounds on the interval, and the interval width is equal to $2\Delta x$. Thus the interval from $x = a$ to $x = b$ consists of two subintervals, each separated by a distance of $\Delta x = (b - a)/2$. Simpson's rule can only be applied when there are an even number of subintervals—that is, an odd number of data pairs, x and $f(x)$. For $n - 1$ intervals of equal size, Simpson's rule can be expressed as

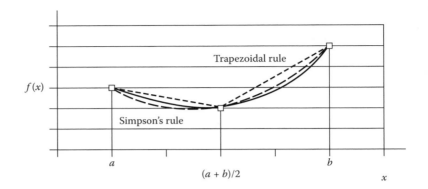

FIGURE 7.5   Simpson's rule for integration.

$$(7.70) \qquad \int_{x_1}^{x_n} f(x)\,dx \approx \sum_{i=1,3,5}^{n-2} \frac{x_{i+1}-x_i}{3}[f(x_i)+4f(x_{i+1})+f(x_{i+2})]$$

Given that Simpson's rule is based on a second-order interpolating polynomial, it can be shown that the error bound is a function of the fourth derivative. Specifically, the absolute value of the upper bound on the error is estimated by

$$(7.71) \qquad \text{error} = \frac{(x_{i+1}-x_i)^5}{90}\frac{d^4 f(x_i)}{dx^4}$$

in which $d^4 f(x_i)/dx^4$ is the fourth derivative of the relationship between the dependent and independent variables, with the derivative being evaluated at the value of the independent variable (x_i) that maximizes the error between x_i and x_{i+1}. The error given by Equation 7.71 is for the specific interval being evaluated, and the bound on the total error equals the sum over all the intervals of the absolute values of the error limits defined by Equation 7.71.

Example 7.9: Flow Rate Problem

Applying Simpson's rule to the data of Table 7.6, the following flow rate (Q) can be obtained using Equation 7.70:

$$Q = \int_0^{r0} V\,2\pi r\,dr = \frac{2\pi\left(\dfrac{1}{12}\right)}{3}\sum_{r=1,3}^{5}[V_i r_i + 4V_{i+1}r_{i+1} + V_{i+2}r_{i+2}]$$

$$= \frac{\pi}{18}\left[10(0)+4(9.722)\left(\frac{1}{12}\right)+8.889\left(\frac{1}{6}\right)\right]+\left[8.889\left(\frac{1}{6}\right)+4(7.5)\left(\frac{1}{4}\right)\right.$$

$$\left.+5.556\left(\frac{1}{3}\right)\right]+\left[5.556\left(\frac{1}{3}\right)+4(3.056)\left(\frac{5}{12}\right)+0\left(\frac{1}{2}\right)\right] = 3.927 \text{ ft}^3/\text{sec}$$

$$(7.72)$$

The estimate from Simpson's rule equals the true value, which is obviously better than the error resulting from the trapezoidal rule. In this case, Simpson's rule gives an exact value because the true function is a second-order polynomial; Simpson's rule will always exactly integrate a quadratic function.

7.2.4 ROMBERG INTEGRATION

It was shown that Simpson's rule provided greater accuracy than the trapezoidal rule, but at the expense of a larger number of evaluations of the function $f(x)$. It then should seem reasonable that the accuracy should improve as the number of evaluations of the function increases. Denoting the trapezoidal estimate as I_{01},

$$(7.73) \qquad I_{01} = \frac{b-a}{2}(f(a)+f(b))$$

where a and b are the start and end of an interval. Then a second estimate of the integral can be made using an average of areas obtained with a center point; this estimate is denoted as I_{11}:

$$(7.74) \qquad I_{11} = \frac{b-a}{2}\left(\frac{1}{2}f(a)+f(m)+\frac{1}{2}f(b)\right)$$

in which $f(m)$ is the value of the function at the midpoint of the interval (a, b). Equation 7.74 can be rewritten as

$$(7.75) \qquad I_{11} = \frac{1}{2}\left[I_{01}+(b-a)f\left(a+\frac{b-a}{2}\right)\right]$$

A third estimate, I_{21}, of the integral can be obtained using three equally spaced intermediate points m_1, m_2, and m_3:

$$(7.76) \quad I_{21} = \frac{b-a}{2}\left[\frac{1}{4}f(a)+\frac{1}{2}f(m_1)+\frac{1}{2}f(m_2)+\frac{1}{2}f(m_3)+\frac{1}{4}f(b)\right]$$

Expressing Equation 7.76 in terms of I_{11} yields

$$(7.77) \qquad I_{21} = \frac{1}{2}\left[I_{11}+\frac{b-a}{2}\sum_{\substack{k=1 \\ k\neq 2}}^{3} f\left(a+\frac{b-a}{4}k\right)\right]$$

Continuing this subdividing of the interval leads to the following recursive relationship:

$$(7.78) \quad I_{i1} = \frac{1}{2}\left[I_{i-1,1}+\frac{b-a}{2^{i-1}}\sum_{k=1,3,5}^{2^i-1} f\left(a+\frac{b-a}{2^i}k\right)\right] \qquad \text{for } i=1,2,\ldots$$

As i increases, the accuracy of the integral I_{i1} is expected to increase. If the integration interval, $b - a$, is halved N times, then the subintervals are of length $h = (b - a)/2^N$ and we have a vector of $N + 1$ values of I_{i1}, $i = 0, 1, 2,..., N$, where I_{01} is the estimate of the integral made with the trapezoidal rule.

Richardson provided a method of combining estimates of integrals to obtain a more accurate estimate. The general extrapolation formula in recursive form is

(7.79)
$$I_{ij} = \frac{4^{j-1} I_{i+1,j-1} - I_{i,j-1}}{4^{j-1} - 1}$$

where $i = 0, 1,..., N - j + 1$ and $j = 1, 2, 3,..., N$. For any value of j, the values I_{ij} form a vector. The values of I_{i2} are computed from the vector I_{i1}; the values of I_{i3} are computed from the vector I_{i2}. When $j = N$, the vector I_{iN} will contain the one value I_{0N}, which is the best estimate of the integral.

The values of I_{ij} are best presented in the following upper triangular matrix form:

(7.80)

$$
\begin{matrix}
I_{01} & I_{02} & I_{03} & I_{04} & \cdots & I_{0,N-1} & I_{0,N} & I_{0,N+1} \\
I_{11} & I_{12} & I_{13} & \cdot & \cdots & I_{1,N-1} & I_{1,N} & \\
I_{21} & I_{22} & \cdot & \cdot & \cdots & I_{2,N-1} & & \\
I_{31} & \cdot & \cdot & \cdot & & & & \\
\cdot & \cdot & \cdot & & & & & \\
\cdot & \cdot & I_{N-2,3} & & & & & \\
\cdot & I_{N-1,2} & & & & & & \\
I_{N1} & & & & & & &
\end{matrix}
$$

To compute the estimate of I_{0N}, Equation 7.78 can be used to obtain the values of I_{i1}. Then the recursive relationship of Equation 7.79 can be used to compute the vector I_{i2} and then successively the values of I_{i3}, I_{i4}, and so on. It is only necessary to store the elements of adjacent vectors I_{ij} and $I_{i,j+1}$. Once the vector $I_{i,j+1}$ has been computed, the vector I_{ij} is not necessary, so the elements can be used to store the values of $I_{i,j+2}$.

Example 7.10: Romberg Method for Integration

To illustrate the Romberg method, a function with a known integral will be used. For the function $f(u) = ue^{au}$, the integral is

(7.81)
$$\int ue^{au}\, du = \frac{e^{au}}{a^2}(au - 1) + c$$

If $a = 2$ and the function is integrated over the interval from 0 to 1, the solution of Equation 7.81 yields an integral of 2.097264 (seven significant digits). The Romberg integration matrix is expressed using six significant digits in Table 7.7. The values in Table 7.7 were computed using Equations 7.78 and 7.79. The table is expressed according to Equation 7.80. For example, the values in Table 7.7 that corresponds to $i = 0, 1, 2, 3, 4,$ and 5 were computed as follows:

For $i = 0$, the trapezoidal rule is used to obtain I_{01}:

$$I_{01} = 0.5(b - a)(f(a) + f(b)) = 0.5(1 - 0)[(0)\exp(2(0)) \\ + (1)\exp(2(1))] = 3.69453$$

For $i = 1$, Equation 7.75 or 7.76 is used to obtain I_{11}:

$$I_{11} = 0.5\left[I_{01} + (b-a)f\left(a + \frac{b-a}{2} \right) \right]$$
$$= 0.5[3.69453 + (1-0)\ 0.5\exp(2(0.5))] = 2.52683$$

For $i = 2$, Equation 7.77 or 7.78 is used to obtain I_{21}:

$$I_{21} = 0.5\left[I_{11} + 0.5(b-a)\left(f\left(a + \frac{(b-a)}{4} \right) + f\left(a + 3\frac{(b-a)}{4} \right) \right) \right]$$
$$= 0.5[2.52683 + 0.5(0.25\exp(2(0.25)) + 0.75\exp(2(0.75))] = 2.20678$$

For $i = 3$, Equation 7.78 is used to obtain I_{31}:

$$I_{31} = 0.5\left[I_{21} + 0.25(b-a)\left(f\left(a + \frac{b-a}{8} \right) + f\left(a + \frac{3(b-a)}{8} \right) \right. \right.$$
$$\left. \left. + f\left(a + \frac{5(b-a)}{8} \right) + f\left(a + \frac{7(b-a)}{8} \right) \right) \right]$$
$$= 0.5[2.20678 + 0.25(0.125\exp(2(0.125)) + 0.375\exp(2(0.375))$$
$$+ 0.625\exp(2(0.625)) + 0.875\exp(2(0.875)))] = 2.12478$$

TABLE 7.7   Computations according to the Romberg Method

i	$j = 1$	2	3	4	5	6
0	3.69453	2.13760	2.09759	2.09727	2.09726	2.09726
1	2.52683	2.10009	2.09727	2.09726	2.09726	
2	2.20678	2.09745	2.09726	2.09726		
3	2.12478	2.09728	2.09726			
4	2.10415	2.09727				
5	2.09899					

For $i = 4$, Equation 7.78 is used to obtain I_{41}:

$$I_{41} = 0.5\left[I_{31} + 0.125(b-a)\left(f\left(a+\frac{b-a}{16}\right) + f\left(a+\frac{3(b-a)}{16}\right) \right. \right.$$

$$+ f\left(a+\frac{5(b-a)}{16}\right) + f\left(a+\frac{7(b-a)}{16}\right) + f\left(a+\frac{9(b-a)}{16}\right)$$

$$+ f\left(a+\frac{11(b-a)}{16}\right) + f\left(a+\frac{13(b-a)}{16}\right)$$

$$\left. \left. + f\left(a+\frac{15(b-a)}{16}\right) \right) \right]$$

$$= 0.5[2.12478 + 0.125(0.0625\exp(2(0.0625)) + 0.1875\exp(2(0.1875))$$

$$+ 0.3125\exp(2(0.3125)) + 0.4375\exp(2(0.4375)) + 0.5625\exp(2(0.5625))$$

$$+ 0.6875\exp(2(0.6875)) + 0.8125\exp(2(0.8125)) + 0.9375\exp(2(0.9375)))]$$

$$= 2.10415$$

Similarly, for $i = 5$, Equation 7.78 is used to obtain $I_{51} = 2.09899$.

The other entries in Table 7.7 were computed using Equation 7.79. For example, I_{02} is computed as follows:

$$I_{02} = \frac{4^{2-1}(2.52683) - 3.69453}{4^{2-1} - 1} = 2.13760$$

The value of I_{03} is computed as follows:

$$I_{03} = \frac{4^{3-1}(2.10009) - 2.13760}{4^{3-1} - 1} = 2.09759$$

Similar computations are performed for the other values in the table. The Romberg method yields an exact value to six significant digits for the integral of 2.09726.

7.3 APPLICATIONS

7.3.1 ESTIMATING AREAS

A land developer must clear a small wooded area on a site to be developed into a commercial shopping center. Figure 7.6 shows the plot. To estimate the time required to clear the wooded area, the developer needs an estimate of the area. This is obtained by assuming that the latitude is the independent variable and the longitude is the dependent variable. The wooded area is the difference between the area below the upper perimeter between latitudes of

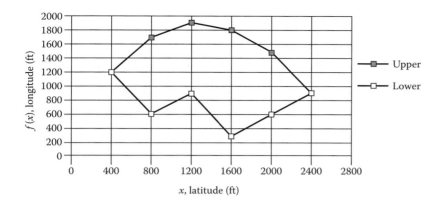

FIGURE 7.6   Plot for commercial shopping center.

400 and 2400 ft and the area below the lower perimeter between the same two points. Six points are used to delineate the two lines. Since the perimeters are defined by an even number of points, the trapezoidal rule is used. The area below the upper perimeter is computed using Equation 7.56 as

$$A_u = \frac{400}{2}[1200 + 2(1700 + 1900 + 1800 + 1500) + 900]$$
(7.82)

$$= 3,180,000\,\text{ft}^2 = 73.00 \text{ acres}$$

The area below the lower perimeter is computed using Equation 7.56 as

$$A_L = \frac{400}{2}[1200 + 2(600 + 900 + 300 + 600) + 900]$$
(7.83)

$$= 1,380,000 \text{ ft}^2 = 31.68 \text{ acres}$$

Therefore, the perimeter of the wooded area encloses 41.32 acres, which is the difference between 73.00 and 31.68 acres.

7.3.2 PIPE FLOW RATE

Using an instrument for measuring flow velocity, velocity measurements were made over the cross section of a 3 ft diameter pipe. Table 7.8 gives the velocity at the center line of the cross section and at distances (r) measured from the center line of the pipe (see Figure 7.7). The flow rate is obtained using the following continuity equation:

$$Q = \int V \, dA = \int V_r \, 2\pi r \, dr = 2\pi \int_0^{1.5} V_r r \, dr$$
(7.84)

If an expression for V_r were available and could be integrated, an analytical solution could be used. In this case, measurements are available at discrete points. An estimate of Q can be made with the trapezoidal rule.

TABLE 7.8   Calculation of Volume Flow Rates Using the Trapezoidal Rule

r (in.)	Velocity (fps)	Incremental area (ft²)	Incremental Q (ft³/sec)
0	5.0		
		0.196	0.970
3	4.9		
		0.589	2.739
6	4.4		
		0.982	4.026
9	3.8		
		1.374	4.534
12	2.8		
		1.767	3.799
15	1.5		
		2.160	1.620
18	0.0		

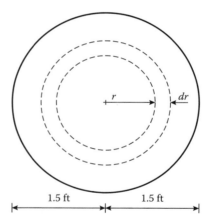

FIGURE 7.7   Flow rate in a circular pipe.

When applying the trapezoidal rule to this case, the difference $(x_{i+1} - x_i)$ in Equation 7.56 must be replaced by the area associated with the points in the form of an increment area (ΔA_i) that is given by $\Delta A_i = \pi\left(r_{i+1}^2 - r_i^2\right)$. Thus the trapezoidal rule becomes

$$(7.85) \qquad Q = \sum_{i=1}^{n-1} \pi\left(r_{i+1}^2 - r_i^2\right)\frac{V_{i+1} + V_i}{2}$$

which uses the product of the average velocity and the incremental area as the incremental flow rate ΔQ_i. Summing the ΔQ_i values gives the total flow rate Q. The calculations are shown in Table 7.8, with a flow rate of 17.7 cfs for the measured data. The flow rate was computed as the summation of the last column in Table 7.8.

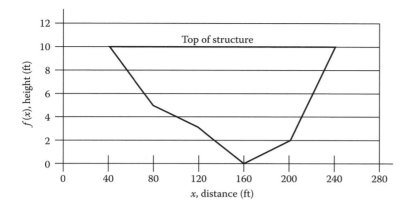

FIGURE 7.8   Cross section of a retaining wall.

7.3.3 VOLUME OF GABION RETAINING WALL

A geotechnical engineer is planning to use a gabion retaining wall to protect a section of highway from debris and mud. A gabion is a wire-mesh container filled with rocks. The engineer must estimate the volume of rock required. The gabion wall will have a width of 6 ft, and the cross-sectional area of the retaining wall is shown in Figure 7.8.

The estimate can be made with the trapezoidal rule using Equation 7.56 as

(7.86)
$$\text{volume} = 6\left[\frac{40}{3}\left(0 + 2(5) + 2(7) + 2(10) + 2(8) + 0\right)\right]$$
$$= 7200 \text{ ft}^3$$

In this case, each value of $f(x)$ is subtracted from 10 ft to compute the ordinates; in essence, the volume of the structure is computed using the top of the structure as the axis. The volume of rock required to fill the wire-mesh gabions would depend on the amount of void space within the baskets holding the rocks.

7.3.4 SEDIMENT LOADS IN RIVERS

Environmental engineers are concerned with the environmental effects of sediment in streams and rivers. Measurements of sediment transport rates are made during flood runoff events—the time when most sediment is transported. Large volumes of sediment that are included in runoff into a reservoir can reduce the storage volume of the reservoir, as well as reduce the quality of the water for fish and wildlife.

At a particular location the following 6 h measurements of the sediment rate (q_s) were made during a storm event: 0, 8, 18, 23, 28, 26, 21, 18, 14, 11, 6, 2, and 2 lb/sec. Each measurement is assumed to reflect the average rate over the 6 h period. A plot of the sediment rate as a function of time (t) is called a sediment graph, or sometimes a pollutograph.

To find the total sediment load (pounds) during the storm (Q_s), Simpson's rule can be used to integrate the function defined by the points:

$$Q_s = \int q_s\, dt \approx \frac{6\,hr}{6}[0 + 4(8) + 2(18) + 4(23) + 2(28) + 4(26)$$

(7.87)

$$+\, 2(21) + 4(18) + 2(14) + 4(11) + 2(6) + 4(2) + 2] = 528\ \text{lb}$$

At a specific weight of 165 lb/ft³, the sediment would have a volume of 11,520 ft³. If the sediment entered a reservoir with a bottom area of 25 acres and settled uniformly over the bottom, it would settle to a depth of 0.0106 ft, or 1/8 in. Following many such storms, the capacity of the reservoir to hold water would be greatly diminished because of sediment accumulation.

7.3.5 PROBABILITY COMPUTATIONS USING THE STANDARD NORMAL PROBABILITY DISTRIBUTION

The standard normal probability distribution is commonly used in engineering. The probability density function $f(x)$ is given by

(7.88)
$$f(x) = \frac{1}{\sqrt{2\pi}}\exp(-0.5x^2)$$

Figure 7.9 shows this symmetric function with respect to $x = 0$. The following integral can be used to compute the probability that x is between a and b:

(7.89)
$$P(a \le x \le b) = \int_a^b \frac{1}{\sqrt{2\pi}}\exp(-0.5x^2)\,dx$$

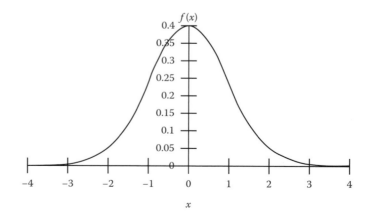

FIGURE 7.9 Standard normal probability density function.

The integral of Equation 7.89 cannot be analytically computed. Therefore, numerical integration can be used. For example, the following probability can be determined using the trapezoidal rule:

$$(7.90) \qquad P(-1 \le x \le 1) = \int_{-1}^{1} \frac{1}{\sqrt{2\pi}} \exp(-0.5x^2) dx$$

Using a constant interval of 0.2, the probability can be computed as

$$(7.91) \quad P(-1 \le x \le 1) = \frac{0.2}{2} \left(f(-1) + f(1) + 2 \sum_{i=1}^{9} f(-1+0.2i) \right) = 0.6810741$$

The exact value can be computed from tables in probability textbooks; in this case, the exact value is 0.68269. The error is 0.0016159 (0.2%). The error level depends on the selected area under the curve and the size of the selected interval.

PROBLEMS

7.1   The following table gives the water vapor capacity of air (V_c) in grains per cubic foot for selected temperatures (T) in degrees Fahrenheit. Determine the slope of the water vapor capacity curve at a temperature of 54°F using the forward, backward, and two-step finite-difference approximations.

T	50	54	58	62	66
V_c	4.108	4.725	5.420	6.203	7.082

7.2   The elevation (E, ft) of the bottom of a stream at various points along the stream is given as a function of the distance (D, ft) from the mouth of the stream. Estimate the slope of the channel, which is important to estimate the velocity of the flowing water, at a distance of 1500 ft using forward, backward, and two-step finite-difference approximations:

D (ft)	0	500	1000	1500	2000	2500
E (ft)	47.2	54.6	59.8	68.2	73.0	77.9

7.3   Given the following values of $f(x)$ and x, determine the first derivative of $f(x)$ with respect to x at $x = 3$ using the forward, backward, and two-step finite-difference approximations. Compare the accuracy if the true value of the first derivative is given by $6x^5 - 25x^4 + 48x^3 - 54x^2 + 70x + 80$.

x	1	2	3	4	5
$f(x)$	5	152	455	1676	6425

7.4   Given the following values of $f(x)$ and (x), determine the first derivative of $f(x)$ with respect to x at $x = 3.8$ using forward, backward, and two-step finite difference approximations. Compare the accuracy if the five value of the first derivation if the true function is $x^5 - 15x^4 + 85x^3 - 225x^2 + 274x - 120$:

x	3.6	3.7	3.8	3.9	4.0
$f(x)$	2352.9	1.2531	−2081.4	0	0

7.5   The following table gives the sine for selected angles (θ). Determine the first derivative of $\sin \theta$ at $\theta = 7°$ using the forward, backward, and two-step finite-difference approximations. Compare the accuracy with the true value. (*Note:* $1° = 0.017453$ radian.)

θ	4	5	6	7	8
$\sin \theta$	0.06976	0.08716	0.10453	0.12187	0.13917

7.6   The following table gives the sine for selected angles (θ). Determine the first derivative of $\sin \theta$ at $\theta = 18°$ using forward, backward, and two-step finite difference approximations. Compare the accuracy with the true value.

θ	17	17.5	18.0	18.5	19.0
$\cos \theta$	0.95630	0.95372	0.95106	0.94832	0.94552

7.7   Using the data in the following table, compute the first derivative at $x = 4$ using the forward, backward, and two-step finite-difference approximations. Compute the second derivative at $x = 4$ using the two-step approximation. Compare these estimates with the true values of 0.25 and −0.0625, respectively.

x	2	3	4	5	6
$f(x)$	0.69315	1.09861	1.38629	1.60944	1.79176

7.8   Using the data of Problem 7.7, estimate the derivative at $x = 6.5$ using forward, backward, and two-step finite difference approximations. Compute the second derivative at $x = 6.5$ using the two-step approximations.

7.9   Using the data in the following table, compute the first derivative at $x = 1.4$ using the forward, backward, and two-step finite-difference approximations. Compute the second derivative at $x = 1.4$ using the two-step approximation.

x	1.0	1.2	1.4	1.6	1.8	2.0
$f(x)$	1.0	0.91817	0.88716	0.89352	0.93138	1.0

7.10   Using the data of Problem 6.16 (Chapter 6), estimate the first derivative at $x = 0.55$ using a finite-difference table.

7.11   Using the data of Problem 6.23 in Chapter 6, estimate the first derivative at a temperature of 40°. Use forward, backward, and two-step approximations.

7.12   The data in the following table are the distances (D) traveled by a car, which was stationary at time $t = 0$, for selected times. Using a finite-difference table, determine (a) the velocity at $t = 6$ sec, (b) the velocity at $t = 8.7$ sec, (c) the acceleration at $t = 1$ sec, and (d) the acceleration at $t = 6.3$ sec.

t (sec)	0	2	4	6	8	10
D (ft)	0	10	50	150	330	610

7.13   A car is traveling at a constant velocity. At time $t = 0$, the driver applies the brakes, with distances (D) traveled at specified times. Using a finite-difference table, estimate (a) the velocity at time $t = 0$ sec, (b) the velocity at time $t = 3.5$ sec, and (c) the rate for deceleration at time $t = 3.5$ sec.

t (sec)	0	1	2	3	4	5
D (ft)	0	90	160	200	220	225

7.14   Fit an interpolating polynomial to the data of Problem 7.1 and estimate the first derivative at (a) $T = 54°$ and (b) $T = 64°$.

7.15   Fit an interpolating polynomial to the data of Problem 6.22 in Chapter 6 and estimate the first derivation of $T = 30°$ and at $T = 40°$.

7.16   Fit an interpolating polynomial to the data of Problem 6.31 (Chapter 6) and estimate the first derivative at $T = 13°$ and at $T = 14°$.

7.17   Fit an interpolating polynomial to the data of Problem 7.3 and estimate the first derivative at (a) $x = 3$ and (b) $x = 5.5$. Compare the estimates with the true values.

7.18   Fit an interpolating polynomial to the data of Problem 7.5 and estimate the first derivative at (a) $\theta = 7°$ and (b) $\theta = 5.5°$. Compare the estimates with the true values.

7.19   Fit an interpolating polynomial to the data of Problem 7.2 and estimate the first and second derivations at $D = 1250$ ft.

7.20   Fit an interpolating polynomial to the data of Problem 7.7 and estimate the first and second derivatives at (a) $x = 4$ and (b) $x = 1.7$. For part (a), compare the estimated values with the true values.

7.21   Fit an interpolating polynomial to the data of Problem 7.9 and estimate the first and second derivatives at $x = 1.4$.

7.22   Fit an interpolating polynomial to the data of Problem 7.6 and estimate the first and second derivatives at $\theta = 17.75°$. Compare the estimates to the true values.

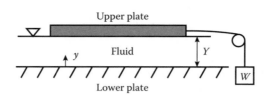

Y (ft)	V (ft/sec)
0	0.000
0.2Y	0.001
0.4Y	0.002
0.6Y	0.005
0.8Y	0.015
1.0Y	0.040

FIGURE P7.23

7.23 Two parallel plates are separated by a fluid having a viscosity (μ) of 10^{-3} lb-sec/ft². While the lower plate is stationary, the upper plate is movable and connected to a weight (W) by a cable, as shown in Figure P7.23. The shear stress across the space is given by $\tau = \mu \, dV/dy$. Velocity measurements were made using stop-action photography when the system was in steady state as shown in the figure. Determine the distribution of shear stress across the space between the plates using an interval of 0.25Y, where $Y = 0.016$ ft.

7.24 The shear stress (τ) between a rotating shaft and a stationary cylinder (see Figure P7.24) is given by $\tau = \mu \, dV/dr$, where μ is the viscosity of the fluid between the cylinder and the shaft, and dV/dr is the velocity gradient within the space (see Figure 7.25). The viscosity of the fluid is 10^{-4} lb-sec/ft², and the radius of the shaft (R_1) and cylinder (R_2) is 3 in. and 3.20 in., respectively. Estimate the angular velocity at the outer surface of the shaft and find the distribution of the shear stress across the fluid.

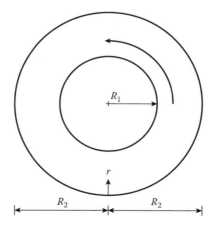

r (inch)	V (ft/sec)
0	0
0.04	1.7
0.08	2.9
0.12	4.8
0.16	7.8

FIGURE P7.24

7.25   For the experimental setup of Problem 7.23 and using $\mu =$ 3.5×10^{-3} lb-sec/ft^2, determine the distribution of the shear stress across the space ($Y = 0.012$ ft) at an interval of 0.2 Y.

Y (ft)	0	0.25 Y	0.5 Y	0.75 Y	Y
V (ft/sec)	0.0000	0.0008	0.0014	0.0031	0.0056

7.26   Using the trapezoidal rule and the following data, find the area under the sine curve from 0 to 90. Compare the result with the true integral.

θ	0	15°	30°	45°	60°	75°	90°
$\sin \theta$	0	0.2588	0.5	0.70701	0.8660	0.9659	1.0

7.27   The system reliability function $R(t)$ is given by $R(t) = 1 - F(t)$, where $F(t)$ is the probability that the component will fail in the interval from 0 to t. $F(t)$ is determined from $F(t) = \int f(t)\, dt$, where $f(t)$ is the probability density of the time to failure of the system. If the following estimates of $f(t)$ are made from sample data, use the trapezoidal rule to estimate (a) the reliability function at different t values and (b) the probability that the system will provide reliable service for 5 days.

t (days)	0	1	2	3	4	5	6	7	8	9
$f(t)$	0.33	0.24	0.17	0.12	0.09	0.06	0.04	0.03	0.02	0.01

7.28   Calculate the area of a quarter of a circle of radius R using the trapezoidal rule and an interval of $R/6$. Compare the estimate to the five value. (*Hint*: $X^2 + Y^2 = R^2$.)

7.29   Calculate the area of a quarter of an ellipse whose semi-axes are A and B. Use the trapezoidal rule. $\left(Hint\colon \dfrac{X^2}{A^2} + \dfrac{Y^2}{B^2} = 1 \right)$. Compare the estimate to the true value.

7.30   Based on the concepts of Problem 7.27, use the following data to estimate (a) the reliability function and (b) the probability that the system will survive for at least 2 h but no more than 6 h. Use the trapezoidal rule.

t (hours)	0	2	4	6	8	10	12	14	16
$f(t)$	0.02	0.05	0.06	0.07	0.08	0.07	0.06	0.05	0.02

7.31   Use the trapezoidal rule to estimate $\int_0^3 x^{1/3}\, dx$ for interval widths of 1, 0.5, and 0.25. Compare these values with the corresponding true values.

7.32   Ground surface elevations $f(x)$ are taken every 25 ft across the stream channel at a bridge opening. Using the trapezoidal

x (ft)	0	25	50	75	100	125	150	175	200
f(x) (ft)	22	15	11	0	3	8	11	16	22

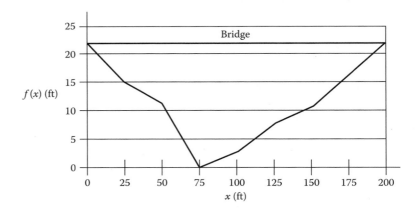

FIGURE P7.32

rule, estimate the flow Q in the river when the water surface is at the bottom of the bridge (22 ft) (see Figure P7.32). The flow Q is given by the integral $Q = \int V\,dA$, where V is the average velocity and equals a constant of 6 ft/sec.

7.33   Use the trapezoidal rule to estimate $\int_0^3 e^{-1/3}\,dx$ using a cell width of 0.2 and a width of 0.25. Compare both estimates to the true value.

7.34   Use the trapezoidal rule to estimate the value of the integral $\int_0^2 (x^4 - 3x^2 + 7)\,dx$, with an interval of 0.25. Compare the estimate with the true value.

7.35   For Problem 7.32, estimate the flow Q if the velocity is not constant but varies with the location according to the following:

x (ft)	0	25	50	75	100	125	150	175	200
V (ft/sec)		2	4	7	8	8	6	6	3

7.36   An engineer is hired to construct an earthen debris dam. The low point of the cross section where the 8.5 ft wide dam will be located is denoted at $h = 0$, with all vertical ground elevations measured above the zero datum. Compacted soil will be placed above the ground surface elevations to a depth 4.6 ft above the low point. If the cost of obtaining the soil and placement at the site is 300 per cubic yard, what is the project cost? x = distance from the west end of the cross section. Apply the trapezoidal rule.

x	0	20	40	60	80	100	120	140	160	180	200
h	4.6	3.1	1.9	0.3	0	0.5	1.3	1.8	2.4	2.6	4.6

7.37 Using the data of Problem 7.26 and Simpson's rule, compute the area under the sine curve and compare the estimated value with both the true value and the value estimated using the trapezoidal rule.

7.38 Using the data of Problem 7.27 and Simpson's rule, compute the probability that the system will provide reliable service for at least 6 days.

7.39 Use Simpson's rule to estimate the project cost for the characteristics described in Problem 7.36.

7.40 For Problem 7.32, estímate the flow of Q using Simpson's rule.

7.41 Use Simpson's rule to find the area under the function of Problem 7.4.

7.42 Using Simpson's rule, estimate $\int_0^3 x^{1/3}\, dx$ for interval widths of 1.5, 0.5, and 0.25. Compare these estimated values with the corresponding true values.

7.43 Using Simpson's rule, estimate $\int_0^2 \dfrac{dx}{(6-x^2)^{3/2}}$ using 2, 4, 6, 8, and 10 intervals.

7.44 Using Simpson's rule, estimate $\int_1^4 \dfrac{dx}{3+2e^x}$ using 4, 8, and 10 intervals.

7.45 Use Simpson's rule to estimate the value of the integral $\int_0^{2.8} 5.3e^{-1.2x}\, dx$ using an increment of 0.4.

7.46 Use Simpson's rule to estimate the value of the integral $\int_{1.6}^{3.2} \dfrac{x^{0.5}\, dx}{2+6x^2}$. Use an interval of 0.4.

Differential Equations

8

8.1 INTRODUCTION

8.1.1 DEFINITIONS

An equation that defines a relationship between an unknown function and one or more of its derivatives is referred to as a *differential equation*. To begin, let us assume that the relationship involves two variables: the independent variable and the dependent variable. The derivatives in the differential equation are of the dependent variable with respect to the independent variable, and the unknown function is a function of the independent variable, the dependent variable, or both. The order of the differential equation is equivalent to the order of the highest derivative in the equation; for example, an equation that includes only first and second derivatives would be a second-order differential equation.

To illustrate these items, a first-order differential equation with y as the dependent variable and x as the independent variable would be

$$(8.1) \qquad \frac{dy}{dx} = f(x, y)$$

in which $f(x, y)$ is a function. A second-order differential equation would have the form

$$(8.2) \qquad \frac{d^2y}{dx^2} = f\left(x, y\frac{dy}{dx}\right)$$

The functions on the right sides of Equations 8.1 and 8.2 do not necessarily have to include all the variables shown. For example, the function of Equation 8.1 could also be either $f(x)$ or $f(y)$, and for Equation 8.2 the function could be $f(x, dy/dx), f(y, dy/dx), f(x, y), f(x)$, $f(y)$, or $f(dy/dx)$. Thus Equations 8.1 and 8.2 are written in general terms.

An ordinary differential equation is one with a single independent variable. Thus, while Equations 8.1 and 8.2 are ordinary differential equations, the following is not:

(8.3)
$$\frac{dy}{dx_1} = f(x_1, x_2, y)$$

in which y is the dependent variable and x_1 and x_2 are independent variables.

Differential equations may be either linear or nonlinear. An nth-order differential equation is linear if the function on the right of the equation is linear in the independent variables, dependent variable, and their derivatives; that is, $f(x, y, dy/dx, d^2y/dx^2, ..., d^{n-1}y/dx^{n-1})$ is linear.

The solution of the differential equation is a function of the independent variable that, when substituted for the dependent variable, reduces the equation to an identity in the independent variable.

Boundary conditions are constraints placed on the solutions; that is, in addition to specifying the differential equation, additional conditions may be placed on the solution. For example, the differential equation might be

(8.4)
$$\frac{dy}{dx} = 5x$$

A particular problem may specify that the desired solution should have $y = 2$ at $x = 1$; this constraint on the solution is called a *boundary condition*. The solution of a differential equation such as Equation 8.4 may be plotted on a graph of y versus x; for the preceding boundary condition, the plotted function passes through the point ($y = 2, x = 1$). For a different boundary condition, the solution would be different.

8.1.2 ORIGIN OF DIFFERENTIAL EQUATIONS

Differential equations can originate from either geometric or physical problems. To illustrate a geometric origin, consider the case of a pair of variables (x and y) in which the slope of the relationship between y and x equals the product of a constant c and the difference between y and x. This geometric problem would define the following differential equation:

(8.5)
$$\frac{dy}{dx} = c(y - x)$$

The solution would be a relationship of the form $y = g(x)$. Equation 8.5 may be subject to one or more boundary conditions.

Physical problems can also be defined by differential equations. Simple problems in electrical circuits and heat transfer are commonly used to introduce differential equations. Simple motion problems can also be governed by differential equations. For example, consider a mass of m slugs that falls from rest in a medium in which the resistance is proportional to the square of the velocity (V) of the mass at any time t. The net force (F) on the mass equals the difference between the force exerted by gravity and the resistance of the medium. For a medium resistance defined as R (in mass units per second), the governing equation is

$$(8.6) \qquad F = ma = m\frac{dV}{dt}$$

The net force is also given by

$$(8.7) \qquad F = mg_r - RV$$

where a = acceleration and g_r = gravitational acceleration. The fact that the mass begins at rest defines a boundary condition that V (the dependent variable) equals zero at time (the independent variable) zero ($V = 0$ at $t = 0$). Dividing through by m yields

$$(8.8) \qquad \frac{dV}{dt} = g_r - \frac{R}{m}V \quad \text{such that } V = 0 \text{ at } t = 0$$

A solution that shows V as a function of t depends on the values of R and m.

8.2 TAYLOR SERIES EXPANSION

8.2.1 FUNDAMENTAL CASE

Assume that the problem is a first-order ordinary differential equation of the form

$$(8.9) \qquad \frac{dy}{dx} = f(x) \quad \text{subject to } y = y_0 \text{ at } x = x_0$$

If we separate the variables and integrate both sides, we have

$$(8.10) \qquad \int_{y_0}^{y} dy = \int_{x_0}^{x} f(x)\, dx$$

or

$$(8.11) \qquad y = y_0 + \int_{x_0}^{x} f(x)\, dx$$

For a single independent variable x, a Taylor series expansion has the form

$$(8.12) \qquad g(x) = g(x_0) + (x - x_0)g'(x_0) + \frac{(x - x_0)^2}{2!}g''(x_0) + \ldots$$

where x_0 is a value where the function $g(x_0)$ is known, and the derivatives $g'(x_0)$, $g''(x_0)$, ..., and so on, with respect to x, can be computed; x is some other value of the independent variable for which the value of the dependent variable $y = g(x)$ is not known and needs to be determined. If we can compute the derivatives, the values of the function can be computed for any value of the independent variable.

Equations 8.11 and 8.12 can be used to solve a differential equation. The terms on the right side of Equation 8.12 beginning with $(x - x_0)g'(x_0)$ can be used to represent the integral on the right side of Equation 8.11. Since $y_0 = g(x_0)$ and $g'(x_0) = f(x_0)$, the solution to the differential equation of Equation 8.9 can be obtained using the Taylor series expansion of Equation 8.12.

Example 8.1: First-Order Differential Equation

Given the following differential equation,

$$\frac{dy}{dx} = 3x^2 \quad \text{such that } y = 1 \text{ at } x = 1 \tag{8.13}$$

the higher order derivatives are easily computed:

$$\frac{d^2 y}{dx^2} = 6x \tag{8.14}$$

$$\frac{d^3 y}{dx^3} = 6 \tag{8.15}$$

$$\frac{d^n y}{dx^n} = 0 \quad \text{for } n \geq 4 \tag{8.16}$$

Based on Equations 8.13 and 8.14, the Taylor series of Equation 8.12 yields a function $y = g(x)$ that describes the dependent variable for any value of x:

$$g(x) = 1 + (x-1)\frac{dy}{dx} + \frac{(x-1)^2}{2!}\frac{d^2 y}{dx^2} + \frac{(x-1)^3}{3!}\frac{d^3 y}{dx^3} \tag{8.17}$$

$$= 1 + (x-1)\left(3x_0^2\right) + \frac{(x-1)^2}{2!}(6x_0) + \frac{(x-1)^3}{3!}(6) \tag{8.18}$$

$$= 1 + 3x_0^2(x-1) + 3x_0(x-1)^2 + (x-1)^3 \tag{8.19}$$

where $x_0 = 1$. Equation 8.19 can be used to compute the value of y [i.e., $g(x)$] for any value of x. The solution is given in Table 8.1 and shown graphically in Figure 8.1 for values of x from 1 to 2. Note that the higher order terms are small near the boundary condition of $x = 1$, but they become larger as x deviates farther from the boundary condition. For the case of Figure 8.1, the derivatives beyond the third equal zero. Thus there was no error in the solution when all of the derivatives were used, as in Equations 8.17 through 8.19. If only the first and second derivatives were used, the solution would be in error by an amount equal to the term containing the third derivative. Thus the error would be the difference between the upper two lines of Figure 8.1. At x equal to 2, the exact solution is $y = 8$. The error would be 12.5% of the true value of y if the term with the third derivative were omitted. If the terms with the second and third derivatives were omitted, then, at x equal to 2, the error would be 50% of the true value of y.

TABLE 8.1   Taylor Series Solution

x	One Term	Two Terms	Three Terms	Four Terms
1	1	1	1	1
1.1	1	1.3	1.33	1.331
1.2	1	1.6	1.72	1.728
1.3	1	1.9	2.17	2.197
1.4	1	2.2	2.68	2.744
1.5	1	2.5	3.25	3.375
1.6	1	2.8	3.88	4.096
1.7	1	3.1	4.57	4.913
1.8	1	3.4	5.32	5.832
1.9	1	3.7	6.13	6.859
2	1	4	7	8

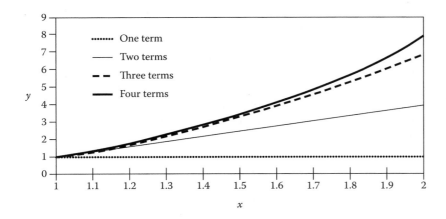

FIGURE 8.1   Graphical representation of the terms of a Taylor series expansion.

8.2.2 GENERAL CASE

Assume that the problem is a first-order ordinary differential equation of the following general form:

$$(8.20) \qquad \frac{dy}{dx} = f(x,y) \quad \text{subject to } y = y_0 \text{ at } x = x_0$$

For this general case, a Taylor series expansion has the form

$$(8.21) \quad g(x) = g(x_0, y_0) + (x - x_0)g'(x_0, y_0) + \frac{(x - x_0)^2}{2!} g''(x_0, y_0) + \ldots$$

where x_0 and y_0 are the values when the function $g(x_0, y_0)$ is known, and the derivatives $g'(x_0, y_0)$, $g''(x_0, y_0)$, ..., and so on, with respect to x, can be computed; x is some other value of the independent variable for which the value of the dependent variable $y = g(x)$ is not known and needs to be determined. If we can compute the derivatives, the values of the function can be computed for any value of the independent variable. The derivatives in this case are more difficult to evaluate. Implicit differentiation using the chain rule for differentiation might be necessary.

8.3 EULER'S METHOD

In some cases the derivatives are not easily computed. Therefore, the Taylor series of Equations 8.12 and 8.21 can be truncated so that only the term with the first derivative is used. Thus the value of the dependent variable $g(x)$ can be computed using

$$(8.22) \qquad g(x) = g(x_0) + (x - x_0)\frac{dy}{dx} + e$$

The derivative dy/dx in Equation 8.22 is evaluated at the beginning of the interval (i.e., x_0). It should be evident from Figure 8.1 that, as the distance or interval $(x - x_0)$ increases, the error increases because the nonlinear terms (i.e., second order and higher) of the Taylor series expansion become more important. Thus it is important to keep the distance $(x - x_0)$ small. In practice, Equation 8.22 is applied iteratively over small distances in order to reduce the error. The computations according to this method are performed one step at a time. Therefore, the method is sometimes referred to as the one-step Euler's method. This is best illustrated using an example.

Example 8.2: First-Order Differential Equation

Consider the differential equation

$$(8.23) \qquad \frac{dy}{dx} = 4x^2, \quad \text{such that } y = 1 \text{ at } x = 1$$

This equation is easily solved analytically, and the analytical solution is useful here as a basis for assessing the accuracy of estimates obtained numerically. For example, for x equal to 1.1, the true value of y is

(8.24)
$$\int_1^y dy = \int_1^{1.1} 4x^2 \, dx$$

(8.25)
$$y - 1 = \frac{4}{3} x^3 \Big|_1^{1.1} = 0.44133^+$$

Therefore, at $x = 1.1$, $y = 1.44133^+$.

Using Equation 8.22 with a step size of $\Delta x = (x - x_0) = 0.1$, we get

(8.26)
$$g(1.1) = 1 + 0.1[4(1)^2] = 1.4$$

Thus, when a step size of 0.1 is used, Euler's equation results in an error (in absolute value) of 0.04133^+. If, instead, we use a step size of 0.05 and apply Euler's equation twice (at $x = 1$ and at $x = 1.05$), we get a value of y at $x = 1.1$ of

(8.27) $g(1.05) = g(1) + (1.05 - 1.00)[4(1)^2] = 1 + 0.2 = 1.2$

(8.28) $g(1.10) = g(1.05) + (1.10 - 1.05)[4(1.05)^2] = 1.4205$

Thus, by decreasing the step size from 0.1 to 0.05, the error was decreased from 0.04133^+ to 0.020833^+, a reduction of 50%. For a step size of 0.02, the estimated value, after five steps, is 1.43296, which is in error by 0.008373^+.

8.3.1 ERRORS WITH EULER'S METHOD

Errors in numerical solutions of differential equations can be classified as global or local errors. The local errors occur over one step size, whereas the global errors are cumulative over the range of solution. The global errors can be computed as the difference between the numerical and exact solutions. The local errors can be computed as the difference between a numerical solution at the end of the step using the exact value at the beginning of the step and the exact solution at the end of the step.

The error ($\in$) using Euler's method can be approximated using the second term of the Taylor series expansion as

(8.29)
$$\in = \frac{(x - x_0)^2}{2!} \frac{d^2 y}{dx^2}$$

where the second derivative is evaluated at the point where it is maximum over the interval $(x - x_0)$. If the range is divided into n increments, then the error at the end of the range of interest for x would be $n\epsilon$.

Example 8.3: Analysis of Errors

Considering the differential equation in Example 8.2, the local and global errors can be computed as shown in Table 8.2 for step sizes of 0.1, 0.05, and 0.01, respectively. It can be observed from these tables that by decreasing the step size both the local and global errors can decrease.

In this example, d^2y/dx^2 is $8x$. Thus an approximate assessment of the error is given by

$$(8.30) \qquad \epsilon = \frac{(x - x_0)^2}{2}(8x) = 4x(x - x_0)^2$$

For increments of 0.1, 0.05, and 0.02, the upper limits on the error would be, respectively,

$$(8.31) \qquad \epsilon_{0.1} = 4(1.1)(0.1)^2 = 0.044$$

$$(8.32) \qquad \epsilon_{0.05} = 2(4)(1.1)(0.05)^2 = 0.022$$

$$(8.33) \qquad \epsilon_{0.02} = 5(4)(1.1)(0.02)^2 = 0.0088$$

The estimated errors given by Equations 8.31 through 8.33 are greater than the actual errors.

TABLE 8.2   Local and Global Errors for Example 8.3

x	Exact Solution	Numerical Solution	Local Error (%)	Global Error (%)
		Using a Step Size of 0.1		
1	1	1	0	0
1.1	1.4413333	1.4	−2.8677151	−2.8677151
1.2	1.9706667	1.884	−2.300406	−4.3978349
1.3	2.596	2.46	−1.9003595	−5.238829
1.4	3.3253333	3.136	−1.6038492	−5.6936648
1.5	4.1666667	3.92	−1.376	−5.92
1.6	5.128	4.82	−1.1960478	−6.0062402
1.7	6.2173333	5.844	−1.0508256	−6.004718
1.8	7.4426667	7	−0.9315657	−5.947689
1.9	8.812	8.296	−0.8321985	−5.8556514
2	10.333333	9.74	−0.7483871	−5.7419355

(Continued)

TABLE 8.2 (CONTINUED)   Local and Global Errors for Example 8.3

x	Exact Solution	Numerical Solution	Local Error (%)	Global Error (%)
		Using a Step Size of 0.05		
1	1	1	0	0
1.05	1.2101667	1.2	−0.8401047	−0.8401047
1.1	1.4413333	1.4205	−0.7400555	−1.4454209
1.15	1.6945	1.6625	−0.6589948	−1.8884627
1.2	1.9706667	1.927	−0.5920162	−2.2158322
1.25	2.2708333	2.215	−0.5357798	−2.4587156
1.3	2.596	2.5275	−0.4879301	−2.6386749
1.35	2.9471667	2.8655	−0.4467568	−2.771023
1.4	3.3253333	3.23	−0.4109864	−2.8668805
1.45	3.7315	3.622	−0.3796507	−2.9344768
1.5	4.1666667	4.0425	−0.352	−2.98
1.55	4.6318333	4.4925	−0.3274441	−3.0081681
1.6	5.128	4.973	−0.3055122	−3.0226209
1.65	5.6561667	5.485	−0.2858237	−3.0261956
1.7	6.2173333	6.0295	−0.2680678	−3.0211237
1.75	6.8125	6.6075	−0.2519878	−3.0091743
1.8	7.4426667	7.22	−0.2373701	−2.9917592
1.85	8.1088333	7.868	−0.2240355	−2.9700121
1.9	8.812	8.5525	−0.2118323	−2.9448479
1.95	9.5531667	9.2745	−0.2006316	−2.9170083
2	10.333333	10.035	−0.1903226	−2.8870968
		Using a Step Size of 0.02		
1	1	1	0	0
1.02	1.0816107	1.08	−0.1489137	−0.1489137
1.04	1.1664853	1.163232	−0.1408219	−0.2789005
1.06	1.254688	1.24976	−0.1334728	−0.392767
1.08	1.3462827	1.339648	−0.1267688	−0.4928138
1.1	1.4413333	1.43296	−0.120629	−0.5809436
1.2	1.9706667	1.95312	−0.0963464	−0.8903924
1.3	2.596	2.56848	−0.0793015	−1.0600924
1.4	3.3253333	3.28704	−0.0667201	−1.1515638
1.5	4.1666667	4.1168	−0.057088	−1.1968
1.6	5.128	5.06576	−0.049506	−1.2137285
1.7	6.2173333	6.14192	−0.0434055	−1.212953
1.8	7.4426667	7.35328	−0.0384092	−1.2010032
1.9	8.812	8.70784	−0.0342563	−1.1820245
2	10.333333	10.2136	−0.0307613	−1.1587097

8.4 MODIFIED EULER'S METHOD

To this point, two methods for solving differential equations were introduced. The use of a Taylor series provided accurate estimates, but it required estimates of the higher order derivatives. Euler's method did not require higher order derivatives, but it required a small step size and an iterative solution. A third method, the modified Euler's method, attempts to improve the accuracy of the estimated solution while using only the first derivative. It is structurally similar to the Euler's method of Section 8.3, but it uses an average slope, rather than the slope at the start of the interval (or step).

The general procedure for the most basic form is as follows:

1. Evaluate the slope (i.e., dy/dx) at the start of the interval.
2. Estimate the value of the dependent variable y at the end of the interval Δx (i.e., y at $x + \Delta x$) using the Euler's method.
3. Evaluate the slope at the end of the interval.
4. Find the average slope using the slopes of steps 1 and 3.
5. Compute a revised value of the dependent variable at the end of the interval using the average slope of step 4 with Euler's method.

Example 8.4: Modified Euler's Method

The procedure is best illustrated using an example. Consider the differential equation

(8.34) $$\frac{dy}{dx} = x\sqrt{y} \quad \text{such that } y = 1 \text{ at } x = 1$$

The five steps of the first iteration for $\Delta x = 0.1$ are as follows:

1. $\left.\dfrac{dy}{dx}\right|_1 = 1\sqrt{1} = 1$

2. $g(1.1) = g(1.0) + (1.1 - 1.0)\left.\dfrac{dy}{dx}\right|_1 = 1 + 0.1(1) = 1.1$

3. $\left.\dfrac{dy}{dx}\right|_{1.1} = 1.1\sqrt{1.1} = 1.15369$

4. $\left.\dfrac{dy}{dx}\right|_a = \dfrac{1}{2}(1 + 1.15369) = 1.07684, \quad \text{where } a = \text{average}$

5. $g(1.1) = g(1.0) + (1.1 - 1.0)\left.\dfrac{dy}{dx}\right|_a = 1 + 0.1(1.07684) = 1.10768$

where, for example, dy/dx_1 means the first derivative of y with respect to x evaluated at $x = 1$ and the corresponding y value, and dy/dx_a means the average of the first derivative of y with respect to x evaluated at the start and end of an interval. It is especially important to note the following: (1) estimates of the dependent variable were made at both steps 2 and 5, and (2) dy/dx_1 was used in step 2, whereas dy/dx_a was used in step 5. An estimate could be obtained at $x = 1.2$ by performing a second execution with the ends of the interval at $x = 1.1$ and $x = 1.2$. The steps for the second interval are as follows:

1. $\dfrac{dy}{dx}\bigg|_{1.1} = x\sqrt{y} = 1.1\sqrt{1.10768} = 1.15771$

2. $g(1.2) = g(1.1) + (1.2 - 1.1)\, dy/dx_{1.1} = 1.10768 + 0.1(1.15771)$
   $= 1.22345$

3. $\dfrac{dy}{dx}\bigg|_{1.2} = 1.2\sqrt{1.22345} = 1.32732$

4. $\dfrac{dy}{dx}\bigg|_{a} = \dfrac{1}{2}\left(\dfrac{dy}{dx}\bigg|_{1.1} + \dfrac{dy}{dx}\bigg|_{1.2}\right) = 1.24251$

5. $g(1.2) = g(1.1) + (1.2 - 1.1)\,dy/dx_a = 1.23193$

The procedure could be continued for additional values of x.

8.5 RUNGE–KUTTA METHODS

The Euler's and modified Euler's methods are considered to be one-step methods since they use information from one interval in estimating the value of y at the end of the interval. A class of methods called Runge–Kutta methods, which include the Euler's and modified Euler's methods as special cases, is widely used. In these methods, the y value at the end of an interval is determined based on the value at the beginning of the interval, the step size, and some representative slope over the interval. For example, Euler's method uses the slope at the beginning of the interval as the representative slope over the interval.

The Runge–Kutta methods are classified according to the order that they represent in their solutions. However, it should be noted that these methods do not require higher order derivatives for their solutions. In this section, example cases of the second- and third-order Runge–Kutta methods are provided. Only the fourth-order Runge–Kutta methods are described in detail. The use of second- and third-order Runge–Kutta methods is similar to that of fourth-order Runge–Kutta methods. Also the second- and third-order methods are used in the applications section of this chapter (see Section 8.12).

8.5.1 SECOND-ORDER RUNGE–KUTTA METHODS

The modified Euler's method is considered a case of the second-order Runge–Kutta methods. It can be expressed as

(8.35) $y_{i+1} = y_i + 0.5[f(x_i, y_i) + f(x_i + h, y_i + hf(x_i, y_i))]h$

where $y_i = g(x_i)$, $y_{i+1} = g(x_i + \Delta x)$, $x_i = x$ value at the start of an interval, and $x_{i+1} = x_i + \Delta x$; $h = \Delta x$. The computations according to this method can be summarized in the following steps:

1. Evaluate the slope at the start of an interval—that is, at (x_i, y_i). It is called slope S_1—that is,

(8.36) $$S_1 = f(x_i, y_i)$$

2. Evaluate the slope at the end of the interval (x_{i+1}, y_{i+1}), called slope S_2, as follows:

(8.37) $$S_2 = f(x_i + h, y_i + hS_1)$$

3. Evaluate y_{i+1} using the average slope of S_1 and S_2 as

(8.38) $$y_{i+1} = y_i + 0.5(S_1 + S_2)h$$

A general representation of the second-order Runge–Kutta methods is given by the following:

(8.39) $$y_{i+1} = y_i + [f(x_i + 0.5h, y_i + 0.5hf(x_i, y_i))]h$$

The computational steps for this method are summarized as follows:

1. Evaluate the slope at (x_i, y_i). It is called slope S_1 and is given by Equation 8.36.
2. Evaluate the slope at mid-interval, S_2, as

(8.40) $$S_2 = f(x_i + 0.5h, y, + 0.5hS_1)$$

3. Evaluate the quantity of interest y_{i+1} as

(8.41) $$y_{i+1} = y_i + S_2h$$

A third example of the second-order Runge–Kutta methods is given by the following predictor:

(8.42) $$y_{i+1} = y_i + \left[\frac{f(x_i, y_i)}{3} + \frac{2f(x_i + 0.75h, y_i + 0.75hf(x_i, y_i))}{3} \right]h$$

The computational steps in this case are summarized as follows:

1. Evaluate the slope at (x_i, y_i), called S_1, using Equation 8.36.
2. Evaluate a second slope estimate at an intermediate point in the interval, called S_2, as

(8.43) $$S_2 = f(x_i + 0.75h, y_i + 0.75hS_1)$$

3. Evaluate the quantity of interest y_{i+1} as

(8.44)
$$y_{i+1} = y_i + \left(\frac{1}{3}S_1 + \frac{2}{3}S_2\right)h$$

8.5.2 THIRD-ORDER RUNGE–KUTTA METHODS

The following is an example of the third-order Runge–Kutta method:

(8.45)
$$y_{i+1} = y_i + \frac{1}{6}[f(x_i, y_i) + 4f(x_i + 0.5h, y_i + 0.5hf(x_i, y_i))$$
$$+ f(x_i + h, y_i - hf(x_i, y_i) + 2hf(x_i + 0.5h, y_i + 0.5hf(x_i, y_i)))]h$$

The computational steps for this method are as follows:

1. Evaluate the slope at (x_i, y_i), called S_1, using Equation 8.35.
2. Evaluate a second slope estimate at the midpoint of the step, called S_2, as

(8.46)
$$S_2 = f(x_i + 0.5h, y_i + 0.5hS_1)$$

3. Evaluate a third slope S_3 as

(8.47)
$$S_3 = f(x_i + h, y_i - hS_1 + 2hS_2)$$

4. Estimate the quantity of interest y_{i+1} as

(8.48)
$$y_{i+1} = y_i + \frac{1}{6}[S_1 + 4S_2 + S_3]h$$

8.5.3 FOURTH-ORDER RUNGE–KUTTA METHODS

The fourth-order Runge–Kutta methods are a commonly used level of these methods and are introduced here in greater detail than the previous Runge–Kutta methods. For the differential equation

(8.49)
$$\frac{dy}{dx} = f(x, y) \quad \text{such that } y = y_0 \text{ at } x = x_0$$

the following procedure can be used to estimate the value of the dependent variable y at the end of the interval, denoted as y_{i+1}, based on values y_i and x_i at the beginning of the interval and the width of the interval $\Delta x = h$:

1. Compute the slope S_1 at (x_i, y_i) using $f(x, y)$ according to Equation 8.36:

$$S_1 = f(x_i, y_i)$$

2. Estimate y at the midpoint of the interval using y_i, $f(x_i, y_i)$, and $h/2$:

(8.50)
$$y_{i+1/2} = y_i + \frac{h}{2} f(x_i, y_i)$$

where the subscript $\left(i + \frac{1}{2} \right)$ is used to denote the center of the interval.

3. Estimate the slope at mid-interval S_2 as

(8.51)
$$S_2 = f(x_i + 0.5h, y_i + 0.5hS_1)$$

4. Revise the estimate of y at mid-interval $\left(i + \frac{1}{2} \right)$ using y_i, S_2, and $h/2$:

(8.52)
$$y_{i+1/2} = y_i + \frac{h}{2} S_2$$

5. Compute a revised estimate of the slope at mid-interval S_3 using $y_{i+1/2}$ of step 4 by substituting in $f(x, y)$. The slope S_3 can be computed as

(8.53)
$$S_3 = f(x_i + 0.5h, y_i + 0.5hS_2)$$

6. Estimate y at the end of the interval (y_{i+1}) using y_i, S_3, and h:

(8.54)
$$y_{i+1} = y_i + hS_3$$

7. Estimate the slope S_4 at the end of the interval using

(8.55)
$$S_4 = f(x_i + h, y_i + hS_3)$$

8. Using S_1 of step 1, S_2 of step 3, S_3 of step 5, and S_4 of step 7, estimate y_{i+1} using

(8.56)
$$y_{i+1} = y_i + \frac{h}{6}(S_1 + 2S_2 + 2S_3 + S_4)$$

Equation 8.56 uses four estimates of the slope to estimate the value of the y_{i+1}.

Example 8.5: Fourth-Order Runge–Kutta Method

The fourth-order Runge–Kutta method is illustrated using the following differential equation:

(8.57) $$\frac{dy}{dx} = \frac{xy}{1+x^2} \quad \text{such that } y = 1 \text{ at } x = 1$$

Using an interval of $h = 0.1$, the following steps can be used to estimate y at $x = 1.1$—that is, $g(1.1)$:

1. Compute the slope at $x = 1$ by entering the values of the boundary condition into Equation 8.57.

$$S_1 = \frac{1(1)}{1+(1)^2} = 0.5$$

2. Compute y at $x = 1.05$ using Euler's method according to Equation 8.50:

$$g(1.05) = g(1.0) + \frac{h}{2}S_1 = 1 + \frac{0.10}{2}(0.5) = 1.025$$

3. Use Equation 8.51 to compute the slope at $x = 1.05$ as

$$S_2 = \frac{1.05(1.025)}{1+(1.05)^2} = 0.51189$$

4. Revise y at $x = 1.05$ as

$$g(1.05) = g(1.0) + \frac{h}{2}S_2 = 1 + \frac{0.10}{2}(0.51189) = 1.02559$$

5. Revise the estimate of the slope at the midpoint of the interval:

$$S_3 = \frac{1.05(1.02559)}{1+(1.05)^2} = 0.51219$$

6. Make an initial estimate of y at the end of the interval:

$$g(1.1) = g(1.0) + hS_3 = 1 + 0.1(0.51219) = 1.05122$$

7. Estimate the slope at $x = 1.1$:

$$S_4 = \frac{1.1(1.05122)}{1+(1.1)^2} = 0.52323$$

8. Revise the estimate of y at $x = 1.1$ using Equation 8.56:

$$g(1.1) = g(1.0) + \frac{h}{6}(S_1 + 2S_2 + 2S_3 + S_4)$$

$$= 1 + \frac{0.1}{6}[0.5 + 2(0.51189) + 2(0.51219) + 0.52323]$$

$$= 1.05119$$

The following is the exact solution to Equation 8.57:

(8.58)
$$y = \left(\frac{1+x^2}{2}\right)^{0.5}$$

Thus the exact value of y at $x = 1.1$ is 1.05119, which is the same as the fourth-order Runge–Kutta solution to five significant digits. Euler's method would provide the estimate of $y = 1.05000$, which is in error by 0.113%.

8.6 PREDICTOR–CORRECTOR METHODS

Unless the step sizes are small, procedures such as Euler's method and the Runge–Kutta methods may not yield precise solutions. A class of algorithms called predictor–corrector methods can be used to improve the accuracy of the solution. These methods make use of solutions from previous intervals to project to the end of the next interval. The disadvantage of predictor–corrector algorithms is that, for the initial intervals, values for previous intervals may not be available; in such cases, one-step methods, such as Euler's method, may have to be used to generate values for the first few steps. In addition to the multistep aspect of the solution, predictor–corrector algorithms iterate several times over the same interval until the solution converges to within an acceptable tolerance; this is the corrector aspect of the algorithm.

The method has two parts: the predictor and the corrector parts. The predictor part is used to compute an initial estimate of the dependent variable at the end of the interval; this is used with the corrector algorithm to improve the estimate. The solution iterates with the corrector algorithm to improve the accuracy of the estimate.

8.6.1 EULER–TRAPEZOIDAL METHOD

The Euler–trapezoidal method uses Euler's method as the predictor algorithm and the trapezoidal rule as the corrector equation. Using the first subscript to indicate the interval and the second subscript to indicate the trial number, Euler's formula is

(8.59)
$$y_{i+1,0} = y_{i,*} + h\frac{dy}{dx}\bigg|_{i,*}$$

where 0 and * as the second subscripts indicate the initial and final estimate, respectively. For the corrector equation, the trapezoidal rule is given in the form

(8.60)
$$y_{i+1,j} = y_{i,*} + \frac{h}{2}\left[\frac{dy}{dx}\bigg|_{i,*} + \frac{dy}{dx}\bigg|_{i+1,j-1}\right]$$

where j is an iteration counter for applying the corrector equation. The corrector equation can be applied as many times as necessary to get convergence.

Example 8.6: Euler–Trapezoidal Method

The problem of Equation 8.34,

$$\frac{dy}{dx} = x\sqrt{y} \quad \text{such that } y = 1 \text{ at } x = 1$$

is used to illustrate the Euler–trapezoidal method. The initial (predictor) estimate for y at $x = 1.1$ is

$$\frac{dy}{dx}\bigg|_{0,0} = 1\sqrt{1} = 1$$

$$y_{1,0} = y_{0,*} + 0.1\left[\frac{dy}{dx}\bigg|_{0,0}\right]$$

$$y_{1,0} = 1 + 0.1(1) = 1.1$$

The corrector equation of Equation 8.60 can be used to improve the estimate:

$$\frac{dy}{dx}\bigg|_{1,0} = 1.1\sqrt{1.1} = 1.15369$$

$$y_{1,1} = y_{0,*} + \frac{h}{2}\left[\frac{dy}{dx}\bigg|_{0,0} + \frac{dy}{dx}\bigg|_{1,0}\right] = 1 + \frac{0.1}{2}[1 + 1.15369] = 1.10768$$

$$\frac{dy}{dx}\bigg|_{1,2} = 1.1\sqrt{1.10768} = 1.15771$$

$$y_{1,2} = y_{0,\cdot} + \frac{h}{2}\left[\left.\frac{dy}{dx}\right|_{0,1} + \left.\frac{dy}{dx}\right|_{1,1}\right] = 1 + \frac{0.1}{2}[1 + 1.15771] = 1.10789$$

$$\left.\frac{dy}{dx}\right|_{1,3} = 1.1\sqrt{1.10789} = 1.15782$$

$$y_{1,3} = y_{0,\cdot} + \frac{h}{2}\left[\left.\frac{dy}{dx}\right|_{0,1} + \left.\frac{dy}{dx}\right|_{1,2}\right] = 1 + \frac{0.1}{2}[1 + 1.15782] = 1.10789$$

Since $y_{1,3} = y_{1,2}$ further iteration will not change the estimate of y at $x = 1.1$. Thus an estimate of y at $x = 1.2$ can be made using the predictor equation (Equation 8.59) as

$$y_{2,0} = y_{1,3} + h\left.\frac{dy}{dx}\right|_{1,\cdot} = 1.10789 + 0.1(1.15782) = 1.22367$$

The corrector estimate using Equation 8.60 is

$$\left.\frac{dy}{dx}\right|_{2,1} = 1.2\sqrt{1.22367} = 1.32744$$

$$y_{2,1} = y_{1,3} + \frac{h}{2}\left[\left.\frac{dy}{dx}\right|_{1,\cdot} + \left.\frac{dy}{dx}\right|_{2,1}\right] = 1.10789 + \frac{0.1}{2}[1.15782 + 1.32744] = 1.23215$$

$$\left.\frac{dy}{dx}\right|_{2,2} = 1.2\sqrt{1.23215} = 1.33203$$

$$y_{2,2} = y_{1,3} + \frac{h}{2}\left[\left.\frac{dy}{dx}\right|_{1,\cdot} + \left.\frac{dy}{dx}\right|_{2,2}\right] = 1.10789 + \frac{0.1}{2}[1.15782 + 1.33203] = 1.23238$$

$$\left.\frac{dy}{dx}\right|_{2,3} = 1.2\sqrt{1.23238} = 1.33215$$

$$y_{2,3} = y_{1,3} + \frac{h}{2}\left[\left.\frac{dy}{dx}\right|_{1,\cdot} + \left.\frac{dy}{dx}\right|_{2,3}\right] = 1.10789 + \frac{0.1}{2}[1.15782 + 1.33215] = 1.23239$$

Again, the corrector algorithm converged in three iterations.

The Euler–trapezoidal solution can be compared with the modified Euler's solution of Example 8.4. At $x = 1.1$, the modified Euler's method yielded an estimate of 1.10768; the predictor–corrector analysis yields 1.10789. At $x = 1.2$, the modified Euler's and Euler–trapezoidal methods yielded estimates of 1.23193 and 1.23239, respectively.

8.6.2 MILNE–SIMPSON METHOD

The Milne–Simpson method uses Milne's equation as the predictor equation and Simpson's rule for integration as the corrector formula. The Milne equation uses values for three prior ordinates y and has the form

(8.61) $$y_{i+1,0} = y_{i-3,\bullet} + \frac{4h}{3}\left[2\frac{dy}{dx}\bigg|_{i,\bullet} - \frac{dy}{dx}\bigg|_{i-1,\bullet} + 2\frac{dy}{dx}\bigg|_{i-2,\bullet}\right]$$

The corrector equation is

(8.62) $$y_{i+1,j} = y_{i-1,\bullet} + \frac{h}{3}\left[\frac{dy}{dx}\bigg|_{i+1,j-1} + 4\frac{dy}{dx}\bigg|_{i,\bullet} + \frac{dy}{dx}\bigg|_{i-1,\bullet}\right]$$

The predictor equation requires estimates at two previous sampling points. For the two initial sampling points, a one-step method such as Euler's equation (Equation 8.22) can be used.

Example 8.7: Milne–Simpson Method

The problem of Example 8.6 can be used to illustrate the Milne–Simpson method. To use the predictor method, a value would be needed for $y_{i-3,\bullet}$, which is y at $x = 0.7$. Since this value is not available, estimates for y at $x = 1.1$, 1.2, and 1.3 are needed before the predictor equation can be used. Therefore, the following values were obtained from the Euler–trapezoidal method in Example 8.6:

x	y	$\dfrac{dy}{dx}$
1	1	1
1.1	1.10789	1.15782
1.2	1.23239	1.33215

To compute the initial estimate of y at $x = 1.3$, Euler's method can be used:

$$y_{3,0} = y_{2,\bullet} + h\frac{dy}{dx}\bigg|_{2,\bullet} = 1.23239 + 0.1(1.33215) = 1.36560$$

$$\frac{dy}{dx}\bigg|_{3,0} = 1.3\sqrt{1.36560} = 1.51917$$

The corrector formula of Equation 8.62 can now be used:

$$y_{3,1} = y_{1,\bullet} + \frac{0.1}{3}\left[\frac{dy}{dx}\Big|_{3,0} + 4\frac{dy}{dx}\Big|_{2,\bullet} + \frac{dy}{dx}\Big|_{1,\bullet}\right]$$

$$= 1.10789 + \frac{0.1}{3}[1.51917 + 4(1.33215) + 1.15782]$$

$$= 1.37474$$

$$\frac{dy}{dx}\Big|_{3,1} = 1.3\sqrt{1.37474} = 1.52424$$

$$y_{3,2} = 1.10789 + \frac{0.1}{3}[1.52424 + 4(1.33215) + 1.15782]$$

$$= 1.37491$$

$$\frac{dy}{dx}\Big|_{3,2} = 1.3\sqrt{1.37491} = 1.52434$$

$$y_{3,3} = 1.10789 + \frac{0.1}{3}[1.52434 + 4(1.33215) + 1.15782]$$

$$= 1.37492$$

Thus, computations for $x = 1.3$ are complete. Now the Milne predictor equation (Equation 8.61) can be used to estimate y at $x = 1.4$:

$$y_{4,0} = y_{0,\bullet} + \frac{4h}{3}\left[2\frac{dy}{dx}\Big|_{3,\bullet} - \frac{dy}{dx}\Big|_{2,\bullet} + 2\frac{dy}{dx}\Big|_{1,\bullet}\right]$$

$$= 1 + \frac{4(0.1)}{3}[2(1.52434) - 1.33215 + 2(1.15782)] = 1.53762$$

$$\frac{dy}{dx}\Big|_{4,1} = 1.4\sqrt{1.53762} = 1.73601$$

$$y_{4,1} = y + \frac{h}{3}\left[\frac{dy}{dx}\Big|_{4,0} + 4\frac{dy}{dx}\Big|_{3,\bullet} + \frac{dy}{dx}\Big|_{2,\bullet}\right]$$

$$= 1.23239 + \frac{0.1}{3}[1.73601 + 4(1.52434) + 1.33215] = 1.53791$$

$$\frac{dy}{dx}\Big|_{4,1} = 1.4\sqrt{1.53791} = 1.73617$$

$$y_{4,2} = 1.23239 + \frac{0.1}{3}[1.73617 + 4(1.52434) + 1.33215] = 1.53791$$

Additional intervals could be similarly evaluated.

8.7 LEAST-SQUARES METHOD

The methods detailed previously present the solution as a series of points using the step size or interval h. The least-squares method provides a function that can be used to approximate the value of the dependent variable $y = g(x)$ for any value of the independent variable x. The procedure for deriving the least-squares function is as follows:

1. For a given ordinary differential equation (Equation 8.1), assume that the solution is a differentiable function. For example, an nth-order polynomial could be used:

(8.63) $$\hat{y} = b_0 + b_1 x + b_2 x^2 + \ldots + b_n x^n$$

2. Use the boundary condition of the ordinary differential equation to evaluate one of the coefficients of the assumed function or to provide a condition toward the solution of the coefficients.
3. Define the least-squares objective function F, which is the integral of the errors (e) squared:

(8.64) $$F = \int_x e^2 \, dx$$

where the error is defined as the difference between the derivatives for the assumed function (Equation 8.63) and the actual function (Equation 8.1):

(8.65) $$e = \frac{d\hat{y}}{dx} - \frac{dy}{dx}$$

4. Find the minimum of F with respect to the unknowns (b_1, b_2, b_3, ..., b_n) of the assumed function of Equation 8.63 by taking the derivatives of Equation 8.64 with respect to the unknowns and setting the derivatives equal to zero:

(8.66) $$\frac{\partial F}{\partial b_i} = \int_{\text{all } x} 2e \frac{\partial e}{\partial b_i} \, dx = 0$$

5. The integrals of Equation 8.66 are called the normal equations; the solution of the normal equations yields values of the unknown coefficients.

Using the coefficients with Equation 8.63, an estimate of y for any x can be made. The accuracy of the estimate depends on how well the function of Equation 8.63 can approximate the actual solution and how far from the boundary condition the estimate is made.

The least-squares method has the advantage that the solution is represented by a function. It has the following disadvantages: (1) The computational effort is greater than for the step methods, and (2) it is more difficult to provide a generalized computer program that provides a solution.

Note that the objective function of Equation 8.64 corresponds to the least-squares solution for a set of discrete points (see Equations 10.17 through 10.25 in Chapter 10). For solution of an ordinary differential equation, the error is defined in terms of the derivative (Equation 8.65), rather than values of the dependent variable (y of Equation 10.17).

Example 8.8: Least-Squares Method

The least-squares method can be used to solve the following ordinary differential equation:

$$(8.67) \qquad \frac{dy}{dx} = xy \quad \text{such that } y = 1 \text{ at } x = 0$$

As a basis for comparison, Equation 8.67 can be solved analytically:

$$(8.68) \qquad y = e^{x^2/2}$$

To illustrate the method, a solution can be obtained for the interval $0 \leq x \leq 1$.

As a first attempt, a linear model is used for the approximating function:

$$(8.69) \qquad \hat{y} = b_0 + b_1 x$$

Using the boundary condition

$$\hat{y} = 1 = b_0 + b_1(0)$$

yields $b_0 = 1$. Thus the linear model is

$$(8.70) \qquad \hat{y} = 1 + b_1 x$$

and the derivative is

$$\frac{d\hat{y}}{dx} = b_1$$

The error function of Equation 8.65 is

$$(8.71) \qquad e = b_1 - xy = b_1 - x(1 + b_1 x)$$

Since the analysis has only one unknown, b_1, only one derivative is needed for Equation 8.60; the derivative is computed from Equation 8.71 as

(8.72)
$$\frac{de}{db_1} = 1 - x^2$$

With Equation 8.72, the derivative of Equation 8.66 is

(8.73)
$$\int_0^x 2e \frac{de}{db_1} dx = \int_0^x 2[b_1 - x(1 + b_1 x)](1 - x^2) dx = 0$$

where the lower limit is established from the boundary condition. Dividing through by 2 yields

(8.74)
$$\left(b_1 x - \frac{x^2}{2} - \frac{2b_1 x^3}{3} + \frac{x^4}{4} + \frac{b_1 x^5}{5} \right)\Bigg|_0^x = 0$$

If we are interested in the value of y for the range $0 \le x \le 1$, then the integral of Equation 8.74 with an upper limit of integration $x = 1$ yields $b_1 = \frac{15}{32}$. Thus the approximating function of Equation 8.70 is

(8.75)
$$\hat{y} = 1 + \frac{15}{32} x$$

A comparison of the approximating function of Equation 8.75 and the true value given by Equation 8.68 is given in Table 8.3. The linear model produces very poor estimates.

To improve the accuracy of estimates, a quadratic model is used instead of the linear model of Equation 8.69:

(8.76)
$$\hat{y} = b_0 + b_1 x + b_2 x^2$$

TABLE 8.3   A Solution Using a Linear Model for the Least-Squares Method

x	True y Value (Equation 8.68)	Numerical y Value (Equation 8.75)	Error (%)
0	1.0	1.0	–
0.2	1.0202	1.0938	7.2
0.4	1.0833	1.1875	9.6
0.6	1.1972	1.2812	7.0
0.8	1.3771	1.3750	0.0
1.0	1.6487	1.4688	−10.9

The boundary condition again yields $b_0 = 1$. The derivative is

$$\frac{d\hat{y}}{dx} = b_1 + 2b_2 x$$

and the error function is

(8.77)
$$e = b_1 + 2b_2 x - xy = b_1 + 2b_2 x - x(1 + b_1 x + b_2 x^2)$$
$$= b_1(1 - x^2) + b_2(2x - x^3) - x$$

Thus, derivatives are computed for both b_1 and b_2 as follows:

(8.78)
$$\frac{\partial e}{\partial b_1} = 1 - x^2$$

(8.79)
$$\frac{\partial e}{\partial b_2} = 2x - x^3$$

The integral of Equation 8.66 using the derivative of Equation 8.78 is

$$\int_0^x [b_1(1 - x^2) + b_2 + (2x - x^3) - x](1 - x^2)\, dx = 0$$

$$\left[b_1 x - \frac{2b_1 x^3}{3} + b_2 x^2 - \frac{3b_2 x^4}{4} - \frac{x^2}{2} + \frac{b_1 x^5}{5} + \frac{b_2 x^6}{6} + \frac{x^4}{4} \right]_0^x = 0$$

Using $x = 1$ as the upper limit yields the following first of two normal equations:

(8.80)
$$\frac{8b_1}{15} + \frac{5b_2}{12} = \frac{1}{4}$$

Using the derivative of Equation 8.79 with the integral of Equation 8.66 gives

$$2\int_0^x [b_1(1 - x^2) + b_2(2x - x^3) - x](2x - x^3)\, dx = 0$$

$$\left[b_1 x^2 - \frac{2b_1 x^4}{4} + \frac{4b_2 x^3}{3} - \frac{4b_2 x^5}{5} - \frac{2x^2}{3} + \frac{b_1 x^5}{5} + \frac{b_2 x^7}{7} + \frac{x^5}{5} \right]_0^x = 0$$

**TABLE 8.4   A Solution Using a Quadratic Model
for the Least-Squares Method**

x	True y Value (Equation 8.68)	Numerical y Value (Equation 8.82)	Error (%)
0	1.0	1.0	–
0.2	1.0202	1.0022	–1.8
0.4	1.0833	1.0674	0.0
0.6	1.1972	1.1956	0.0
0.8	1.3771	1.3868	0.0
1.0	1.6487	1.6411	0.0

For the upper limit $x = 1$, the second normal equation is

(8.81)
$$\frac{9b_1}{20} + \frac{71b_2}{105} = \frac{7}{15}$$

Solving the two normal equations, Equations 8.80 and 8.81, yields the following quadratic approximating equation:

(8.82)
$$\hat{y} = 1 - 0.14669x + 0.78776x^2$$

The values of y estimated from Equation 8.82 are given in Table 8.4. This quadratic equation (that is, Equation 8.82) gives more accurate estimates than the linear equation (Equation 8.75). Unfortunately, it is not possible to know in advance the order of the polynomial necessary to provide sufficient accuracy for any specific problem.

8.8 GALERKIN METHOD

The Galerkin method is similar to the least-squares method in that an approximating equation is necessary. The method uses the error function of Equation 8.65 and an approximating equation such as Equation 8.63. The normal equations are computed from the following:

(8.83)
$$\int_x w_i e \, dx = 0 \quad i = 1, 2, \ldots, n$$

where n is the number of unknowns to be determined in Equation 8.63, and w_i is a weighting factor. For the polynomial of Equation 8.63, the weight w_i is $d\hat{y}/db_i$.

Example 8.9: Galerkin Method

The ordinary differential equation of Equation 8.67 can be used to illustrate the Galerkin method. Using the quadratic approximating

---

TABLE 8.5   Solution Using a Quadratic Model for the Galerkin Method

x	True y Value (Equation 8.68)	Numerical y Value (Equation 8.88)	Error (%)
0	1.0	1.0	–
0.2	1.0202	0.9816	0.0
0.4	1.0833	1.0316	0.0
0.6	1.1972	1.1500	0.0
0.8	1.3771	1.3368	0.0
1.0	1.6487	1.5921	0.0

---

equation of Equation 8.76 with $b_0 = 1$ and the error function of Equation 8.77 in Equation 8.83 yields the following system of integral equations:

$$\text{(8.84)} \qquad \int_0^1 [b_1(1-x^2)+b_2(2x-x^3)-x]x \, dx = 0$$

$$\text{(8.85)} \qquad \int_0^1 [b_1(1-x^2)+b_2(2x-x^3)-x]x^2 \, dx = 0$$

which yields the following set of normal equations:

$$\text{(8.86)} \qquad \frac{1}{12}b_1 + \frac{7}{15}b_2 = \frac{1}{3}$$

$$\text{(8.87)} \qquad \frac{2}{12}b_1 + \frac{1}{3}b_2 = \frac{1}{4}$$

The solution of the normal equations yields the following approximating equation:

$$\text{(8.88)} \qquad \hat{y} = 1 - 0.26316x + 0.85526x^2$$

For the range $0 \le x \le 1$, the predicted values from Equation 8.88 are given in Table 8.5.

8.9 HIGHER ORDER DIFFERENTIAL EQUATIONS

Many engineering problems require the solution of higher order differential equations. For example, Equation 8.2 is a second-order differential equation and is given by

$$\frac{d^2y}{dx^2} = f\left(x, y, \frac{dy}{dx}\right)$$

The function $f(\cdot)$ does not need to include all the parameters x, y, and dy/dx. The methods previously discussed can be used to solve higher order differential equations after transforming them into systems of first-order differential equations. For example, Equation 8.2 can be transformed into the following system of first-order differential equations:

$$(8.89) \qquad \frac{dy_2}{dx} = f(x, y_1, y_2)$$

$$(8.90) \qquad \frac{dy_1}{dx} = y_2$$

where

$$(8.91) \qquad y_1 = y$$

and

$$(8.92) \qquad y_2 = \frac{dy_1}{dx} = \frac{dy}{dx}$$

The solution of this system of equations requires the boundary conditions of both y_1 and y_2 at the starting x value.

In general, any system of n equations of the following type can be solved using any of the previously discussed methods:

$$(8.93) \qquad \frac{dy_1}{dx} = f_1(x, y_1, y_2, \ldots, y_n)$$

$$(8.94) \qquad \frac{dy_2}{dx} = f_2(x, y_1, y_2, \ldots, y_n)$$

$$(8.95) \qquad \frac{dy_3}{dx} = f_3(x, y_1, y_2, \ldots, y_n)$$

$$\vdots$$

$$(8.96) \qquad \frac{dy_n}{dx} = f_n(x, y_1, y_2, \ldots, y_n)$$

For the single-step methods, the solution procedure requires the knowledge of n boundary conditions for $y_1, y_2, ..., y_n$ at the starting x value. Multistep methods require additional boundary conditions, as was discussed in previous sections. A system of equations is solved by applying any of the previously discussed methods for each equation in the system at each step. All of the equations in the system should be solved at each step before proceeding to the next step. The discussion in the section is limited to solving higher order differential equations and systems of equations using step-based methods. Other methods can be similarly used.

Example 8.10: System of Equations for a Beam

A simply supported beam AB supports a uniformly distributed load of intensity 2 kips/ft as shown in Figure 8.2. The bending moment at a distance X along the span of the beam is given by

$$(8.97) \qquad M(X) = 10X - X^2$$

The deflection and rotation at any point along the span of the beam can be determined by solving the following differential equation:

$$(8.98) \qquad \frac{d^2Y}{dX^2} = \frac{M}{EI} = \frac{10X - X^2}{EI}$$

This equation can be expressed as

$$(8.99) \qquad \frac{dZ}{dX} = \frac{M}{EI} = \frac{10X - X^2}{EI}$$

$$(8.100) \qquad \frac{dY}{dX} = Z$$

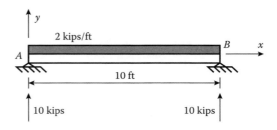

FIGURE 8.2   Simply supported beam.

where Y = deflection in feet, Z = rotation (or slope), E = modulus of elasticity of the material, and I = moment of inertia of the cross section of the beam. Assume that EI = 3600 kip/ft^2, and at X = 0, Y = 0 and Z = −0.02314. Using Euler's method, the following equations result:

(8.101) $$Z_{i+1} = Z_i + f_2(X_i, Y_i, Z_i)h$$

(8.102) $$Y_{i+1} = Y_i + f_1(X_i, Y_i, Z_i)h$$

where $f_1(X_i, Y_i, Z_i) = Z$ evaluated at Z_i, and $f_2(X_i, Y_i, Z_i) = M/EI$ evaluated at $M(X_i)$. Applying the preceding equations sequentially for a step size of 0.1 ft results in Table 8.6. The results of the example are also shown in Figures 8.3 and 8.4. In this case, Euler's method results in accurate prediction of the deflection and rotation along the beam. The relatively high accuracy of the method can be attributed to the small step size.

TABLE 8.6   Deflection and Rotation along a Beam Using a Step Size of 0.1 Foot

X (ft)	$\dfrac{dY}{dX}$	$Z = \dfrac{dY}{dX}$	Y (ft)	Exact Z	Exact Y (ft)
0	0	−0.0231481	0	−0.0231481	0
0.1	0.000275	−0.0231481	−0.0023148	−0.0231344	−0.0023144
0.2	0.0005444	−0.0231206	−0.0046296	−0.0230933	−0.004626
0.3	0.0008083	−0.0230662	−0.0069417	−0.0230256	−0.0069321
0.4	0.0010667	−0.0229854	−0.0092483	−0.0229319	−0.0092302
0.5	0.0013194	−0.0228787	−0.0115469	−0.0228125	−0.0115177
0.6	0.0015667	−0.0227468	−0.0138347	−0.0226681	−0.0137919
0.7	0.0018083	−0.0225901	−0.0161094	−0.0224994	−0.0160505
0.8	0.0020444	−0.0224093	−0.0183684	−0.0223067	−0.018291
0.9	0.002275	−0.0222048	−0.0206093	−0.0220906	−0.020511
1	0.0025	−0.0219773	−0.0228298	−0.0218519	−0.0227083
2	0.0044444	−0.0185565	−0.0434305	−0.0183333	−0.042963
3	0.0058333	−0.0134412	−0.0598019	−0.0131481	−0.0588194
4	0.0066667	−0.007187	−0.0704998	−0.0068519	−0.0688889
5	0.0069444	−0.0003495	−0.0746352	0.00000000	−0.072338
6	0.0066667	0.0065157	−0.0718747	0.0068519	−0.0688889
7	0.0058333	0.0128532	−0.0624406	0.0131481	−0.0588194
8	0.0044444	0.0181074	−0.0471107	0.0183333	−0.042963
9	0.0025	0.0217227	−0.0272183	0.0218519	−0.0227083
10	0.000000	0.0231435	−0.0046523	0.0231481	0.000000

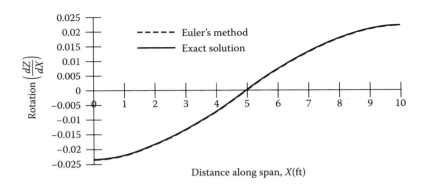

FIGURE 8.3   Rotation along span.

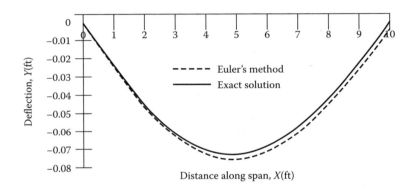

FIGURE 8.4   Deflection along span.

8.10 BOUNDARY-VALUE PROBLEMS

The previously discussed methods for solving ordinary differential equations can be used to solve high-order differential equations as discussed in Section 8.9. These methods require the initial conditions to be at the same x value. For example, the bending moment at a distance x along the span of the beam in Example 8.10 is given by Equation 8.97. The initial conditions were assumed to be at $x = 0$, $y = 0$ and $Z = -0.02314$. The specified initial condition for Z in Example 8.10 is not commonly known. However, in this problem it is known that at $x = 0$, $y = 0$ and at $x = 10$, $y = 0$. These two conditions have different x values. Therefore, they cannot be used in solving the differential equation according to the previously discussed methods. In this case, we are dealing with a boundary-value problem.

Boundary-value problems can be solved using other methods, such as the shooting method and finite-difference methods. In this section these methods are described.

8.10.1 SHOOTING METHOD

The shooting method is a trial-and-error method that utilizes any of the previously discussed methods for solving differential equations. According to this method, the boundary conditions are revised to make all of them the same x value. Then the differential equation is solved, and the resulting solution at the original (known) boundary conditions is compared with the original boundary conditions. The difference between the solution at the boundary conditions and the original boundary conditions can be computed and used to revise the boundary conditions again for another trial. Linear interpolation between two successive differences can be used to revise the conditions. This method becomes inefficient, especially for higher order differential equations where several boundary conditions are needed.

Example 8.11: System of Equations for a Beam

The simply supported beam AB with a uniformly distributed load of intensity 2 kips/ft as shown in Figure 8.2 was used in Example 8.10. The deflection and rotation at any point along the span of the beam can be determined by solving Equations 8.99 and 8.100. Assume that at $x = 0$, $y = 0$, and at $x = 10$, $y = 0$. This boundary-value problem can be solved using the shooting method in conjunction with Euler's method. The boundary conditions are revised to at $x = 0$, $y = 0$ and at $x = 0$, $Z = -0.02314$. Now the problem can be solved as was described in Example 8.10. The results are shown in Table 8.6 and Figures 8.3 and 8.4. The solution produced $y = -0.0046523$ at $x = 10$, which is in error by 0.0046523 as compared to the specified boundary condition at $x = 10$, $y = 0$. The assumed boundary conditions can be revised to (second trial) $x = 0$, $y = 0$, and at $x = 0$, $Z = -0.0225$. The calculations are not shown herein, but the new y value at $x = 10$ is 0.0018292, which is better than -0.0046523. Using linear interpolation, the next trial is at $x = 0$, $Z = -0.0226806$. The resulting y value at $x = 10$ is 0.00002317, which is zero to four significant digits.

8.10.2 FINITE-DIFFERENCE METHODS

In these methods, the derivatives in the differential equation are replaced by finite-difference equations, which are described in Section 7.1 (Chapter 7). Then the resulting differential equation, which is now in a finite-difference form, is used at some interior points using a selected step size at the specified boundary conditions. Each use of the finite-difference equation results in a linear equation in terms of the unknown solutions at the selected interior points. The resulting system of equations needs to be solved simultaneously to obtain the solution at the interior points. The system of linear equations can be solved using the numerical methods discussed in Chapter 5.

8.11 INTEGRAL EQUATIONS

The previously discussed methods for solving differential equations can be used to solve integration problems of the following type:

$$(8.103) \qquad y = \int_{x_l}^{x_u} f(x,y)\, dx$$

where x_l = the lower limit of x, x_u = the upper limit of x, and (x, y) = any function of x and y. Equation 8.103 can be rewritten as

$$(8.104) \qquad \frac{dy}{dx} = f(x,y)$$

The solution methods of Equation 8.104 require the knowledge of the boundary condition, y at $x = x_l$. Any of the previously discussed methods can be used to solve the resulting differential equation. Several steps should be used between x_l and x_u to obtain an accurate result for the integral.

8.12 APPLICATIONS

8.12.1 MOTION OF A FALLING BODY

The mass system defined by Equations 8.6 through 8.8 can be solved using a Taylor series expansion. For g_r = 32.2 ft/sec^2, a resistance R equal to 1 slug/sec, and a mass of 750 slugs, Equation 8.8 can be written as

$$(8.105) \qquad \frac{dV}{dt} = 32.2 - \frac{1}{750}V = 32.2 - 0.00133V \quad \text{such that } V = 0 \text{ at } t = 0$$

The solution to Equation 8.105 would be a relationship between V and t. To provide an accurate estimate of this relationship, higher order derivatives must be computed. The second and third derivatives are

$$(8.106) \qquad \begin{aligned} \frac{d^2V}{dt^2} &= -\frac{R}{m}\left(\frac{dV}{dt}\right) \\[2mm] &= -\frac{R}{m}\left(g_r - \frac{R}{m}V\right) \end{aligned}$$

$$(8.107) \qquad \frac{d^3V}{dt^3} = \left(\frac{R}{m}\right)^2\left(g_r - \frac{R}{m}V\right)$$

Using only the terms including the first three derivatives of Equation 8.12 yields

$$(8.108) \qquad V(t) = V(t_0) + (t - t_0)\frac{dV}{dt} + \frac{(t - t_0)^2}{2!}\frac{d^2V}{dt^2} + \frac{(t - t_0)^3}{3!}\frac{d^3V}{dt^3} + \in$$

in which $\in$ is the error that results from not including the higher order terms, $V(t)$ is the velocity at time t, and t_0 is the value of the independent variable for the boundary condition. For the boundary condition $V = 0$ at $t = 0$ and the derivatives of Equations 8.8, 8.106, and 8.107, Equation 8.108 can be rewritten as

$$V(t) = 0 + (t-0)\left[g_r - \frac{R}{m}V \right] + \frac{(t-0)^2}{2!}\left[-\frac{R}{m}\left(g_r - \frac{R}{m}V \right) \right]$$
$$+ \left(\frac{R}{m} \right)^2 \left(g_r - \frac{R}{m}V \right)\frac{(t-0)^3}{6} + \in$$

(8.109)

Therefore, the velocity V is given by

(8.110) $$V(t) = g_r t - \frac{R}{m}Vt - \frac{R}{2m}t^2\left(g_r - \frac{R}{m}V \right) + \frac{R^2}{m^2}\frac{t^3}{6}\left(g_r - \frac{R}{m}V \right) + \in$$

Since V equals zero at the boundary condition, we get

(8.111) $$V(t) = g_r t - \frac{R}{2m}t^2 g_r + \frac{R^2}{m^2}\frac{t^3}{6}g_r + \in$$

For t equal to 3 sec, the velocity would be 96.407 ft/sec, with an error $\in$. Equation 8.111 could be used to plot a graph of V versus t. The true solution to Equation 8.105 is

(8.112) $$V(t) = \frac{g_r m}{R}\left[1 - e^{-(R/m)t} \right]$$

According to Equation 8.112, the true velocity at $t = 3$ sec is 96.407 ft/sec, which is the same as the solution according to Equation 8.111.

8.12.2 ELECTRICAL CIRCUIT

For the electrical circuit shown in Figure 8.5, the relationship between the dependent variable i (the current) and the independent variable t (time) is described by the following differential equation:

(8.113) $$L\frac{di}{dt} + Ri = E, \quad \text{such that } i = 0 \text{ at } t = 0$$

The differential equation can be rearranged as

(8.114) $$\frac{di}{dt} = \frac{E - Ri}{L}$$

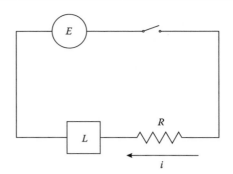

FIGURE 8.5   Circuit with constant voltage source (*E*), a resistance (*R*), and inductance (*L*) in series.

Using Euler's equation, the current at any time *t*, denoted as *i(t)*, is given by

(8.115)
$$i(t) = i(t = t_0) + (t - t_0) \frac{di}{dt}$$

(8.116)
$$i(t) = i(t = t_0) + \left[\frac{E - Ri}{L} \right](t - t_0)$$

The bound on the error can be approximated by

(8.117)
$$\epsilon \cong \frac{(\Delta t)}{2} \frac{d^2 i}{dt^2} = \frac{(\Delta t)}{2} \left[\left(\frac{R}{L} \right)^2 i - \frac{RE}{L^2} \right]$$

To illustrate the solution, assume that *L* = 10 henries, *E* = 6 volts, *R* = 3 ohms, and Δ*t* = 1 second. Thus the current for times of 5 sec and less is computed by Equations 8.115 and 8.116 as follows:

$$i(1) = 0 + (1 - 0) \left[\frac{6 - 3(0)}{10} \right] = 0.6 \, \text{A}$$

$$i(2) = 0.6 + (1 - 0) \left[\frac{6 - 3(0.6)}{10} \right] = 1.02 \, \text{A}$$

$$i(3) = 1.02 + (1 - 0) \left[\frac{6 - 3(1.02)}{10} \right] = 1.314 \, \text{A}$$

$$i(4) = 1.314 + (1 - 0) \left[\frac{6 - 3(1.314)}{10} \right] = 1.5198 \, \text{A}$$

$$i(5) = 1.5198 + (1 - 0) \left[\frac{6 - 3(1.5198)}{10} \right] = 1.66386 \, \text{A}$$

The error bound at time 5 sec can be computed using Equation 8.117:

$$|\epsilon| \cong \left| \frac{5(1)}{2} \left[\left(\frac{3}{10} \right)^2 (1.66386) - \frac{3(6)}{(10)^2} \right] \right| = 0.0756315\,\text{A}$$

An analytical solution for $i(5)$ yields 1.55374 A, which differs from the computed value by 0.11012 A. The true error in this case is larger than the error bound because the derivative evaluation in Equation 8.117 was not performed at a point with an interval that maximizes the derivative (see Equation 8.29).

8.12.3 ONE-DIMENSIONAL HEAT FLOW

One-dimensional heat flow is governed by the equation

(8.118)
$$H = KA\frac{dT}{dr}$$

in which

K is the coefficient of thermal conductivity (calories per second-cm-°C)

H is the quantity of heat (calories per second) flowing through a material

A is the area (cm²) perpendicular to the flow of heat

T is the temperature (°C)

r is a distance measurement (cm)

For a pipe with a circular cross section carrying steam, the flow of heat would be perpendicular to the flow of steam (see Figure 8.6). In such a case, the surface area A at a radius r from the center of the pipe would be $2\pi r$. Equation 8.118 can be used to calculate the temperature distribution across the cross section of the pipe. For example, if steam at 250°C is flowing through a pipe having internal (r_0) and external radii of 5 and

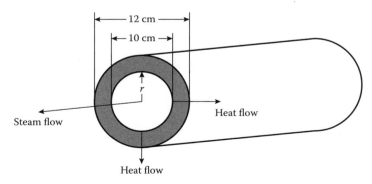

FIGURE 8.6 Calculation of the temperature distribution within the wall of a steam pipe.

6 cm, respectively, and a coefficient of thermal conductivity of 0.0005, then Equation 8.118 can be used to estimate the temperature distribution within the walls of the pipe. Placing Equation 8.118 in standard form yields

(8.119) $$\frac{dT}{dr} = \frac{H}{2\pi rK} \quad \text{such that } T = T_0 \text{ at } r = r_0$$

in which T_0 is the temperature of the steam on the internal wall of the pipe. Thus, assuming a flow rate of 0.8 cal/sec yields

(8.120) $$\frac{dT}{dr} = \frac{0.8}{2\pi(0.0005)r} \quad \text{such that } T = 250°C \text{ at } r = 5\text{ cm}$$

Using a step size of $\Delta r = 0.25$ cm, we can use the modified Euler's method to determine the temperature distribution within the wall of the pipe. For the wall thickness of 1 cm, four steps will be required. The calculations are as follows:

$$T_1 = T_0 + \frac{\Delta r}{2}\left[\frac{dT}{dr}\bigg|_0 + \frac{dT}{dr}\bigg|_1\right]$$

$$= 250 + \frac{0.25}{2}\left[\frac{-0.8}{2\pi(5)(0.0005)} + \frac{-0.8}{2\pi(5.25)(0.0005)}\right]$$

$$= 237.571°C$$

$$T_2 = T_1 + \frac{\Delta r}{2}\left[\frac{dT}{dr}\bigg|_1 + \frac{dT}{dr}\bigg|_2\right]$$

$$= 237.571 + \frac{0.25}{2}\left[\frac{-0.8}{2\pi(5.25)(0.0005)} + \frac{-0.8}{2\pi(5.5)(0.0005)}\right]$$

$$= 225.7205°C$$

$$T_3 = T_2 + \frac{\Delta r}{2}\left[\frac{dT}{dr}\bigg|_2 + \frac{dT}{dr}\bigg|_3\right]$$

$$= 225.7205 + \frac{0.25}{2}\left[\frac{-0.8}{2\pi(5.5)(0.0005)} + \frac{-0.8}{2\pi(5.75)(0.0005)}\right]$$

$$= 214.397°C$$

$$T_4 = T_3 + \frac{\Delta r}{2}\left[\frac{dT}{dr}\bigg|_3 + \frac{dT}{dr}\bigg|_4\right]$$

$$= 214.397 + \frac{0.25}{2}\left[\frac{-0.8}{2\pi(5.75)(0.0005)} + \frac{-0.8}{2\pi(6)(0.0005)}\right]$$

$$= 203.556°C$$

Thus the temperature at the outer wall of the pipe would be 203.556°C. The use of Euler's method would provide values of 237.268°C, 225.1415°C, 213.5666°C, and 202.495°C for values of r of 5.25, 5.50, 5.75, and 6.00 cm, respectively.

8.12.4 ESTIMATING BIOCHEMICAL OXYGEN DEMAND

Rivers and streams contain microorganisms that utilize organic materials, such as sewage, for growth. As part of the process, oxygen is utilized in the biological decomposition of the organic matter. The demand for oxygen is referred to as the biochemical oxygen demand (BOD). Environmental engineers use estimates of the oxygen demanded by the microorganisms as a measure of the health of the river or stream. For 1 or 2 days after organic pollution is discharged into the water body, the temporal relationship between the BOD (y) and time (t) is defined by the differential equation

$$\frac{dy}{dt} = -ky \tag{8.121}$$

where k is called the deoxygenation constant. At the time the organic matter is discharged into the water body, the initial BOD is y_0; letting this time be $t = 0$, the boundary condition is $y = y_0$ at $t = 0$.

The solution is easily computed analytically as

$$y_t = y_0 e^{-kt} \tag{8.122}$$

where y_t is the value of y at time t.

The least-squares method can be used to approximate the time variation of BOD. A quadratic model will be used:

$$y = b_0 + b_1 t + b_2 t^2 \tag{8.123}$$

If y_0 is 20 mg/L, then the boundary condition yields $b_0 = 20$. The time derivative of Equation 8.123 is

$$\frac{dy}{dt} = b_1 + 2b_2 t \tag{8.124}$$

The error function is

$$
e = \frac{d\hat{y}}{dt} - \frac{dy}{dt} = b_1 + 2b_2 t - (-ky) \tag{8.125}
$$
$$
= b_1(1 + kt) + b_2(2t + kt^2) + 20k
$$

Thus, the derivatives of e with respect to b_1 and b_2 are, respectively,

(8.126)
$$\frac{\partial e}{\partial b_1} = 1 + kt$$

and

(8.127)
$$\frac{\partial e}{\partial b_2} = 2t + kt^2$$

Thus, the least-squares integrals are

(8.128)
$$\int_0^T [b_1(1+kt) + b_2(2t+kt^2) + 20k](1+kt)\,dt = 0$$

and

(8.129)
$$\int_0^T [b_1(1+kt) + b_2(2t+kt^2) + 20k](2t+kt^2)\,dt = 0$$

where T is the longest time for which the solution is needed.

If the problem assumes $K = 0.25$ day^{-1} and $T = 2$ days, then the integrals of Equations 8.128 and 8.129 yield the following normal equations:

(8.130)
$$\frac{19}{6}b_1 + \frac{25}{4}b_2 = -12.5$$

and

(8.131)
$$\frac{25}{4}b_1 + \frac{226}{15}b_2 = -\frac{70}{3}$$

Solution of the normal equations yields the following quadratic approximating equation:

(8.132)
$$\hat{y} = 20 - 4.9141t + 0.4898t^2$$

Estimates of the BOD can be made with Equation 8.132 for times in the range from 0 to 2 days and compared with the true values from Equation 8.122. The comparison is given in Table 8.7. The small errors indicate that the numerical approximation provides accurate estimates.

**TABLE 8.7   Comparison of BOD Estimates Made with the
Least-Squares Method and the True Solution**

t (days)	True BOD (mg/L)	Estimated BOD (mg/L)	Error (mg/L)
0	20	20	0
0.25	18.788	18.802	0.014
0.50	17.650	17.665	0.015
0.75	16.581	16.590	0.009
1.00	15.576	15.576	0.000
1.25	14.632	14.623	−0.009
1.50	13.746	13.731	−0.015
1.75	12.913	12.900	−0.013
2.00	12.131	12.131	0.000

8.12.5 ACCUMULATION OF ERODED SOIL IN A SEDIMENT TRAP

Flood runoff from a construction site enters a sediment trap at a rate (q) of 25 ft³/sec. The runoff contains a sediment load of 3.3 lb/ft³. The tank initially contains 4000 ft³ of water that is essentially sediment free. The turbulence from the runoff into the trap keeps the concentration of sediment fairly consistent with location and depth. The outlet permits stored water to drain at a rate of 5 ft³/sec.

Let S be the weight of the sediment in the trap at the end of time t. In the interval from t to $t + \Delta t$, $25\Delta t$ ft³ of runoff enters the trap and $5\Delta t$ ft³ of runoff flows out of the trap. Thus the volume of water in the trap at time t is $4000 + (25-5)t$, and the concentration of sediment in the tank at time t is $S/(4000 + 20t)$. The change in sediment, inflow minus outflow, is ΔS:

$$\Delta S = \left(25\frac{ft^3}{sec}\right)\left(3.3\frac{1b}{ft^3}\right)(\Delta t\,sec) - \frac{(5ft^3/sec)(S\,lb)}{(4000+20t)ft^3}(\Delta t\,sec)$$

or

(8.133)
$$\frac{\Delta S}{\Delta t} = 82.5 - \frac{5S}{4000+20t}$$

As $\Delta t \to 0$, Equation 8.133 becomes the following differential equation:

(8.134)
$$\frac{dS}{dt} = 82.5 - \frac{5S}{4000+20t}$$

which can be solved to find the sediment in the sediment trap as a function of time.

The modified Euler's method is used to compute the accumulation of sediment in the trap for a 30 min storm; a time increment of 5 min (Δt = 300 sec) is used. The results are given in Table 8.8. The total accumulated sediment in the 30 min storm is 124,903 lb, which would have a volume of 757 ft³ (γ = 165 lb/ft³).

TABLE 8.8   Accumulation of Soil in a Trap by Modified Euler's Estimation

t (sec)	S_i (lb)	$\left.\dfrac{dS}{dt}\right|_i$	S_{i+1} (lb)	$\left.\dfrac{dS}{dt}\right|_{i+1}$	$\left.\dfrac{dS}{dt}\right|_{average}$
0	0	82.50	24,750	70.13	76.31
300	22,894	71.05	44,210	68.68	69.87
600	43,854	68.80	64,493	67.84	68.32
900	64,350	67.87	84,713	67.37	67.62
1200	84,637	67.39	104,853	67.08	67.23
1500	104,807	67.09	124,933	66.88	66.985
1800	124,903				

8.12.6 GROWTH RATE OF BACTERIA

A culture of bacteria increases at a rate that is proportional to the number of bacteria present at that instance. Assuming that the number doubles every 5 h, a biomedical engineer can estimate the number of bacteria present at a future time using the differential equation

(8.135)
$$\frac{dy}{dt} = ky$$

where y = the number of bacteria present at time t. Since the number will double in 5 h, it can be shown that $k = (\ln 2)/5$. Assuming that the culture consists of 1 unit at time $t = 0$, Equation 8.135 becomes

(8.136)
$$\frac{dy}{dt} = \frac{(\ln 2)y}{5} \quad \text{such that } y = 1 \text{ at } t = 0$$

To estimate the number at the end of 1 h, the fourth-order Runge–Kutta method was used with a step size $h = 1.0$. The following are the computational steps:

1. At $t = 0$, the slope is given by Equation 8.57 as

$$S_1 = \frac{(\ln 2)(1)}{5} = 0.138629$$

2. The value of y at 0.5 h according to Equation 8.50 is

$$y(\text{at } t = 0.5\,\text{h}) = y(0) + \frac{h}{2}S_1 = 1 + \frac{1.0}{2}(0.138629) = 1.0693147$$

3. The estimated slope at 0.5 h (Equation 8.51) is

$$S_2 = \frac{(\ln 2)(1.0693147)}{5} = 0.1482385$$

4. The revised estimate of y at $t = 0.5$ h is

$$y(0.5) = y(0) + \frac{h}{2}S_2 = 1 + \frac{1.0}{2}(0.1482385) = 1.074119$$

5. The revised estimate of the derivative is

$$S_3 = \frac{(\ln 2)(1.074119)}{5} = 0.1489045$$

6. The initial estimate of y at $t = 0.1$ is

$$y(1.0) = y(0) + hS_3 = 1 + 1.0(0.1489045) = 1.1489045$$

7. The slope estimated at $t = 1$ is

$$S_4 = \frac{(\ln 2)(1.1489045)}{5} = 0.159272$$

8. The final estimate of y at $t = 1$ is (Equation 8.56)

$$y(1.0) = y(0) + \frac{h}{6}(S_1 + 2S_2 + 2S_3 + S_4)$$

$$= 1 + \frac{1}{6}[0.138629 + 2(0.148238) + 2(0.1489045) + 0.159272]$$

$$= 1.1486977$$

8.12.7 BENDING MOMENT AND SHEAR FORCE FOR A BEAM

A simply supported beam AB supports a uniformly distributed load of intensity 2 kips/ft as shown in Figure 8.7. The shear force V at a distance along the span of the beam is given by

(8.137) $$V = 10 - 2X$$

The bending moment at any point along the span of the beam can be determined by solving the following differential equation:

(8.138) $$\frac{dM}{dX} = V = 10 - 2X$$

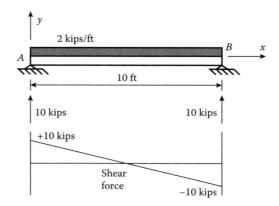

FIGURE 8.7   Simply supported beam and its shear force diagram.

where M = bending moment in kip-ft and V = shear force in kips. The true solution is

(8.139) $$M = 10X - X^2$$

Assume that, at $X = 0$, $M = 0$ and $V = -10$ kips. Using Euler's method, the following equation results:

(8.140) $$M_{i+1} = M_i + f(X_i, V_i)h$$

where $f(X_i, V_i) = V(X_i)$. Applying the preceding equation for step sizes of 0.1 and 0.5 ft results in Table 8.9. The results of the example are also shown in Figure 8.8.

Using the second-order (Equation 8.35), third-order (Equation 8.56), and fourth-order (Equation 8.45) Runge–Kutta methods, the following equations result:

(8.141)   Second order: $$M_{i+1} = M_i + \frac{1}{2}[V(X_i) + V(X_i + h)]h$$

(8.142) Third order: $$M_{i+1} = M_i + \frac{1}{6}[V(X_i) + 4V(X_i + 0.5h) + V(X_i + h)]h$$

(8.143) Fourth order: $$M_{i+1} = M_i + \frac{1}{6}[V(X_i) + 4V(X_i + 0.5h) + V(X_i + h)]h$$

Applying the preceding equations for step sizes of 0.1 and 0.5 ft results in Table 8.10. The results of the example are also shown in Figure 8.9. Since the true solution is to the second order, the second-, third-, and fourth-order Runge–Kutta methods result in solutions without error.

TABLE 8.9   Bending Moment for a Beam Using Euler's Method with Step Sizes of 0.1 and 0.5 Foot

x	$V = \dfrac{dM}{dX}$	M (h = 0.1)	M (h = 0.5)	M exact
0	10	0	0	0
0.1	9.8	1		−0.99
0.2	9.6	1.98		−1.96
0.3	9.4	2.94		−2.91
0.4	9.2	3.88		−3.84
0.5	9	4.8	5	4.75
1	8	9.1	9.5	9
2	6	16.2	17	16
3	4	21.3	22.5	21
4	2	24.4	26	24
5	0	25.5	27.5	25
6	−2	24.6	27	24
7	−4	21.7	24.5	21
8	−6	16.8	20	16
9	−8	9.9	13.5	9
10	−10	1	5	0

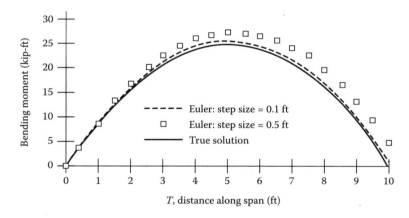

FIGURE 8.8   Bending moment for a beam.

8.12.8 DYNAMIC RESPONSE OF A STRUCTURE

The one-story frame shown in Figure 8.10 is displaced a distance of 0.7 in. and then released from rest (velocity = 0) to vibrate freely. The frame has the following properties: mass m = 2 kip·sec²/in., and stiffness k = 40 kips/in. This system can be modeled as a spring–mass system, in which the stiffness of the columns represents the stiffness of the spring, and the mass of the girders and floor loads represents the vibrating mass in the spring–mass system. Assuming the system to be undamped—that

TABLE 8.10 Bending Moment for a Beam Using the Runge–Kutta Method with Step Size of 0.5 Foot

x	Second-Order Runge–Kutta Method	Third-Order Runge–Kutta Method	Fourth-Order Runge–Kutta Method	True Solution
0	0	0	0	0
0.1	0.99	0.99	0.99	0.99
0.2	1.96	1.96	1.96	1.96
0.3	2.91	2.91	2.91	2.91
0.4	3.84	3.84	3.84	3.84
0.5	4.75	4.75	4.75	4.75
1	9	9	9	9
2	16	16	16	16
3	21	21	21	21
4	24	24	24	24
5	25	25	25	25
6	24	24	24	24
7	21	21	21	21
8	16	16	16	16
9	9	9	9	9
10	0	0	0	0

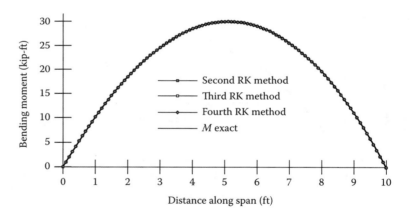

FIGURE 8.9 Bending moment for a beam using the Runge–Kutta methods.

is, without internal friction that opposes the motion and without any external forces—the following equation of motion can be written:

$$(8.144) \qquad m\frac{d^2Y}{dT^2} + kY = 0$$

In this application, the response of the system in the form of the horizontal displacement Y as a function of time T is of interest. Also, the velocity V as a function of time is of interest. The velocity is given by

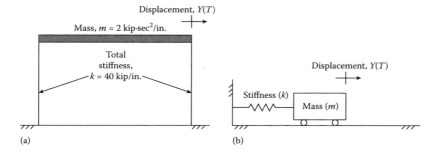

FIGURE 8.10 (a) One-story frame and (b) its spring–mass model.

(8.145)
$$V = \frac{dY}{dT}$$

The second-order differential equation (Equation 8.144) can be written as the following system of two equations:

(8.146)
$$\frac{dY}{dT} = V$$

(8.147)
$$\frac{dV}{dT} = \frac{kY}{m}$$

Using Euler's method with three step sizes—0.1, 0.05, and 0.005 sec—the response of the structure can be determined as shown in Table 8.11 and Figures 8.11a through f. The true solution is given by

(8.148)
$$Y = Y_0 \cos\left(\sqrt{\frac{k}{m}}\,t\right) + \frac{V_0}{k/m}\sin\left(\sqrt{\frac{k}{m}}\,t\right)$$

(8.149)
$$V = -Y_0\sqrt{\frac{k}{m}}\sin\left(\sqrt{\frac{k}{m}}\,t\right) + V_0\cos\left(\sqrt{\frac{k}{m}}\,t\right)$$

It can be observed from Table 8.11 and Figures 8.11 through 8.16 that, to produce accurate numerical results, a small step size is required.

8.12.9 DEFLECTION OF A BEAM

A cantilevered beam AB supports a uniformly distributed load of intensity 1.5 kips/ft as shown in Figure 8.12. The deflection of any point along the span of the beam can be determined by solving the following differential equation:

(8.150)
$$\frac{d^4Y}{dX^4} = \frac{-w}{EI}$$

TABLE 8.11 Velocity and Displacement for a One-Story Frame Using Euler's Method

T (sec)	$\dfrac{dV}{dT}$	$V = \dfrac{dY}{dT}$ (in./sec)	V Exact (in./sec)	Y (in.)	Y Exact (in.)
(a) with Step Size of 0.1 Second					
0	−14	0	0	0.7	0.7
0.1	−14	−1.4	−1.3537978	0.7	0.6311589
0.2	−11.2	−2.8	−2.4413187	0.56	0.4381759
0.3	−5.6	−3.92	−3.0486595	0.28	0.1590087
0.4	2.24	−4.48	−3.0563632	−0.112	−0.1514339
0.5	11.2	−4.256	−2.4629144	−0.56	−0.432091
1	16.91648	6.8096	3.0405805	−0.845824	−0.1665639
1.5	−54.935552	−0.3050701	−1.2908213	2.7467776	0.6377217
2	45.803268	−16.456352	−1.4470025	−2.2901634	−0.6207328
2.5	63.412004	27.089275	3.0772121	−3.1706002	0.1286013
(b) with Step Size of 0.05 Second					
0	−14	0	0	0.7	0.7
0.05	−14	−0.7	−0.6941812	0.7	0.6825728
0.1	−13.3	−1.4	−1.3537978	0.665	0.6311589
0.15	−11.9	−2.065	−1.946006	0.595	0.5483184
0.2	−9.835	−2.66	−2.4413187	0.49175	0.4381759
0.25	−7.175	−3.15175	−2.8150732	0.35875	0.3062158
0.5	10.513567	−3.2305438	−2.4629144	−0.5256783	−0.432091
1	7.0137979	4.8520768	3.0405805	−0.3506899	−0.1665639
2	30.118545	−4.8616409	−1.4470025	−1.5059272	−0.6207328
2.5	−0.1813191	10.60089	3.0772121	0.009066	0.1286013
(c) with Step Size of 0.005 Second					
0	−14	0	0	0.7	0.7
0.005	−14	−0.07	−0.0699942	0.7	0.699825
0.01	−13.993	−0.14	−0.1399533	0.69965	0.6993001
0.015	−13.979	−0.209965	−0.2098425	0.69895	0.6984256
0.02	−13.958004	−0.27986	−0.2796268	0.6979002	0.6972019
0.025	−13.930018	−0.34965	−0.3492713	0.6965009	0.6956296
0.5	8.8563255	−2.5259855	−2.4629144	−0.4428163	−0.432091
1	3.5126826	3.19585	3.0405805	−0.1756341	−0.1665639
2	13.709301	−1.6037152	−1.4470025	−0.685465	−0.6207328
2.5	−2.8853428	3.4880371	3.0772121	0.1442671	0.1286013

where Y = deflection in feet, E = modules of elasticity of the material, and I = moment of inertia of the cross section of the beam. Equation 8.150 is a fourth-order differential equation. The following system of four differential equations can be developed for the purpose of solving for the deflection:

(8.151)
$$\frac{dY}{dX} = S$$

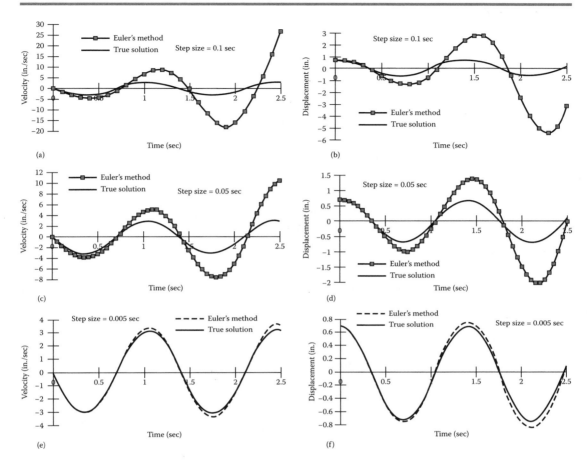

FIGURE 8.11   (a) Velocity for a one-story frame using Euler's method with step size of 0.1 sec. (b) Displacement for a one-story frame using Euler's method with step size of 0.1 sec. (c) Velocity for a one-story frame using Euler's method with step size of 0.05 sec. (d) Displacement for a one-story frame using Euler's method with step size of 0.05 sec. (e) Velocity for a one-story frame using Euler's method with step size of 0.005 sec. (f) Displacement for a one-story frame using Euler's method with step size of 0.005 sec.

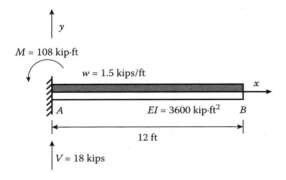

FIGURE 8.12   Cantilever beam.

(8.152)
$$\frac{dS}{dX} = \frac{M}{EI}$$

(8.153)
$$\frac{dM}{dX} = V$$

(8.154)
$$\frac{dV}{dX} = -w$$

where

    S = slope of the tangent to the deflected shape of the beam
    M = bending moment
    V = shear force
    w = intensity of the uniform load

Assume that EI = 3600 kip-ft² and that, at X = 0, Y = 0, S = 0, V = 18 kips, and that M = −108 kip-ft. The boundary conditions were determined based on mechanics of materials and statics. The preceding four first-order equations can then be solved using any of the methods discussed in this chapter. Using Euler's equation, the solution can be expressed in the following form:

(8.155)
$$y_{i+1} = y_i + f(x_i, y_i)h$$

Applying Equation 8.155 sequentially for the four differential equations with a step size of 0.5 ft results in Table 8.12. The deflections for the example are shown in Figure 8.13. The true solutions for shear, moment, rotation, and deflection are given, respectively, by

(8.156)
$$V = -wx + 18$$

(8.157)
$$M = -wx^2/2 + 18x - 108$$

(8.158)
$$S = \frac{1}{EI}\left(\frac{-w}{6}x^3 + 9x^2 - 108x\right)$$

and

(8.159)
$$Y = \frac{1}{EI}\left(\frac{-w}{24}x^4 + 3x^3 - 54x^2\right)$$

It can be observed from Figure 8.17 that there is a good agreement between the numerical and analytical solutions.

TABLE 8.12 Shear, Moment, Rotation, and Deflection for a Cantilever Beam with a Step Size of 0.1 Second

X (ft)	$\dfrac{dV}{dX}=-w$(kips/ft)	$V=\dfrac{dM}{dX}$(kips)	V Exact (kips)	Mr. (kips/ft)	M Exact (kip-ft)	$\dfrac{dS}{dX}=\dfrac{M}{EI}$	$S=\dfrac{dY}{dX}$	S Exact	Y (ft)	Y, Exact (ft)
0	-1.5	18	18	-108	-108	-0.03	0	0	0	0
0.5	-1.5	17.25	17.25	-99	-99.1875	-0.0275	-0.015	-0.0143837	0	-0.0036469
1	-1.5	16.5	16.5	-90.375	-90.75	-0.0251042	-0.02875	-0.0275694	-0.0075	-0.014184
1.5	-1.5	15.75	15.75	-82.125	-82.6875	-0.0228125	-0.0413021	-0.0396094	-0.021875	-0.0310254
2	-1.5	15	15	-74.25	-75	-0.020625	-0.0527083	-0.0505556	-0.042526	-0.0536111
2.5	-1.5	14.25	14.25	-66.75	-67.6875	-0.0185417	-0.0630208	-0.0604601	-0.0688802	-0.0814073
3	-1.5	13.5	13.5	-59.625	-60.75	-0.0165625	-0.0722917	-0.069375	-0.1003906	-0.1139063
4	-1.5	12	12	-46.5	-48	-0.0129167	-0.0879167	-0.0844444	-0.1768229	-0.1911111
5	-1.5	10.5	10.5	-34.875	-36.75	-0.0096875	-0.1	-0.0961806	-0.2679688	-0.281684
6	-1.5	9	9	-24.75	-27	-0.006875	-0.1089583	-0.105	-0.3703906	-0.3825
7	-1.5	7.5	7.5	-16.125	-18.75	-0.0044792	-0.1152083	-0.1113194	-0.4810677	-0.4908507
8	-1.5	6	6	-9	-12	-0.0025	-0.1191667	-0.1155556	-0.5973958	-0.6044444
9	-1.5	4.5	4.5	-3.375	-6.75	-0.0009375	-0.12125	-0.118125	-0.7171875	-0.7214063
10	-1.5	3	3	0.75	-3	0.0002083	-0.121875	-0.1194444	-0.8386719	-0.8402778
11	-1.5	1.5	1.5	3.375	-0.75	0.0009375	-0.1214583	-0.1199306	-0.9604948	-0.9600174
12	-1.5	0	0	4.5	0	0.00125	-0.1204167	-0.12	-1.0817188	-1.08

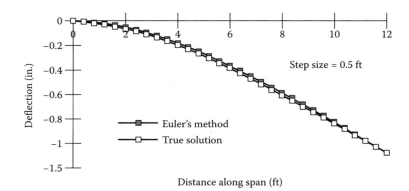

FIGURE 8.13 Deflection for a cantilever beam with a step size of 0.1 sec.

PROBLEMS

8.1 Formulate a differential equation for each of the following geometric problems:

a. The slope of the relationship between x and y equals 2.5 times the sum of x and y, with $y = 2$ at $x = 1$.

b. The slope of the relationship between x and y equals a second-order polynomial $2x^2 + 3x - 7$, with $y = 3$ at $x = 0$.

c. The slope of the slope of the relationship between y and x equals $2x^2 + 3y^3$, with $y = 1$ and $dy/dx = 0$ at $x = 1$.

8.2 Formulate a differential equation for each of the following geometric problems:

a. The slope of the relation between x and y equals the difference between x and y divided by 1.4, with $y = 3$ at $x = 1$.

b. The slope of the relation between x and y equals the second-order polynomial $2x^2 - x + 3$, with $y = 2$ at $x = 1$.

For Problems 8.3 through 8.8, estimate values of the function $y(x)$ using a Taylor series expansion for the range and increment (Δx) shown. Also, compare the accuracy of the solution by comparing the computed values with the true values computed from the corresponding true solutions.

Problems 8.3 through 8.8 are contained in the table.

Problem Number	Differential Equation	Boundary Condition	Δx	Range	True Solution
8.3	$\dfrac{dy}{dx} = 5x^4 - 12x^2 + 6$	$y = 3$ at $x = 1$	0.05	$1.0 \leq x \leq 1.3$	$y = x^5 - 4x^3 + 6x$
8.4	$\dfrac{dy}{dx} = 1 - y$	$y = 2$ at $x = 0$	0.1	$0 \leq x \leq 0.5$	$y = 1 + e^{-x}$
8.5	$\dfrac{dy}{dx} = 3x^2 y$	$y = 1$ at $x = 0$	0.1	$0 \leq x \leq 0.6$	$y = e^{x^3}$

(Continued)

Problem Number	Differential Equation	Boundary Condition	Δx	Range	True Solution
8.6	$\dfrac{d^2y}{dx^2}=2x$	$y=0$ and $dy/dx=0$ at $x=0$	0.1	$0 \leq x \leq 0.5$	$y=x^3/3$
8.7	$\dfrac{dy}{dx}=e^x+xe^x+1$	$y=0$ at $x=0$	0.2	$0 \leq x \leq 1$	$y=xe^x+x$
8.8	$\dfrac{dy}{dx}=x+y-z$	$y=2$ at $x=1$	0.08	$1 \leq x \leq 1.4$	

8.9   Formulate a differential equation for the set of curves that intersects orthogonally the set of parabolas $y^2 = 9bx$, where b is a constant. Then evaluate the curve of the set that passes through the point $y = 3$ at $x = 0$ using the Taylor series expansion for the interval from 0 to 0.6 using an increment of 0.1.

For Problems 8.10 through 8.18, determine the function $y(x)$ using Euler's one-step method for the range and increment (Δx) shown. Also, compare the accuracy of the solution by comparing the computed values with the true values computed from the corresponding true solutions.

Problem Number	Differential Equation	Boundary Condition	Δx	Range	True Solution
8.10	$\dfrac{dy}{dx}+y-1=0$	$y=2$ at $x=0$	0.2	$0 \leq x \leq 1$	$y=1+e^{-x}$
8.11	$\dfrac{dy}{dx}+y-1=0$	$y=2$ at $x=0$	0.1	$0 \leq x \leq 1$	$y=1+e^{-x}$
8.12	$\dfrac{dy}{dx}-5x+2=0$	$y=1$ at $x=1$	0.1	$1 \leq x \leq 1.5$	$y=2.5x^2-2x+0.5$
8.13	$(1+x^2)\dfrac{dy}{dx}-xy=0$	$y=5$ at $x=2$	0.1	$2 \leq x \leq 3$	$y=\sqrt{5(1+x^2)}$
8.14	$(x^2-x)\dfrac{dy}{dx}+y^2-y=0$	$y=2$ at $x=2$	0.05	$2 \leq x \leq 2.25$	$(x-1)(y-1)=0.25xy$
8.15	$x+y\dfrac{dy}{dx}-2x^3-2xy^2=0$	$y=1$ at $x=0$	0.1	$0 \leq x \leq 0.5$	$\ln(x^2+y^2)-2x^2=0$
8.16	$2xy\dfrac{dy}{dx}+x^2-3y^2=0$	$y=2$ at $x=1$	0.2	$1 \leq x \leq 2$	$y=\sqrt{3x^3+x^2}$
8.17	$(x^2+y^2)\dfrac{dy}{dx}-xy=0$	$y=0.584$ at $x=1$	0.05	$1 \leq x \leq 1.5$	$\ln y-\dfrac{x^2}{2y^2}=-2$
8.18	$(x^2-y^2)\dfrac{dy}{dx}+xy=0$	$y=2$ at $x=1$	0.1	$1 \leq x \leq 1.5$	

8.19   If interest is added continuously to the principal P at an annual rate r, estimate the total account at the end of each year for 10 years. Use an interest rate of 12% on a principal of $100,000. Use a modified Euler's method.

8.20 The rate at which a radioactive isotope decomposes is at any instant proportional to the remaining mass of the isotope. If a piece initially has a mass of 10 milligrams and the mass decreases to half its original mass in 1000 years, estimate the decay of the piece for a 100-year period using a time increment of 10 years. Use a modified Euler's method.

For Problems 8.21 through 8.28, estimate values of the function $y(x)$ using the fourth-order Runge–Kutta method for the range and increment (Δx) shown. Also, compare the accuracy of the solution by comparing the computed values with the values computed from the corresponding true solutions.

Problem Number	Differential Equation	Boundary Condition	Δx	Range	True Solution
8.21	$\dfrac{dy}{dx}+y-1=0$	$y=2$ at $x=0$	0.2	$0\le x\le 1$	$y=1+e^{-x}$
8.22	$\dfrac{dy}{dx}+x^2-2x+7$	$y=6$ at $x=1$	0.2	$1\le x\le 2$	$y=\dfrac{x^3}{3}-x^2+7x-\dfrac{1}{3}$
8.23	$\dfrac{dy}{dx}+x^2-2x+7$	$y=6$ at $x=1$	0.1	$1\le x\le 2$	$y=\dfrac{x^3}{3}-x^2+7x-\dfrac{1}{3}$
8.24	$(1+x^2)\dfrac{dy}{dx}-xy=0$	$y=2$ at $x=1$	0.25	$1\le x\le 2$	$y=\sqrt{2(1+x^2)}$
8.25	$x+y\dfrac{dy}{dx}-2x^3-2xy^2=0$	$y=1$ at $x=0$	0.1	$0\le x\le 1$	$\ln(x^2+y^2)-2x^2=0$
8.26	$2xy\dfrac{dy}{dx}+x^2-3y^2=0$	$y=2$ at $x=1$	0.1	$1\le x\le 2$	$3x^3+x^2-y^2=0$
8.27	$(x^2-x)\dfrac{dy}{dx}+y^2-y=0$	$y=3$ at $x=3$	0.05	$3\le x\le 3.3$	$(x-1)(y-1)=\dfrac{4xy}{9}$
8.28	$(x^2+y^2)\dfrac{dy}{dx}^2-xy=0$	$y=0.584$ at $x=1$	0.05	$1\le x\le 1.5$	$\ln y-\dfrac{x^2}{2y^2}=-2$

8.29 If the angular velocity w of a cylinder rotating in a viscous fluid decreases at a rate proportional to w and if w is initially 100 revolutions per minute, estimate w versus time (t) for 10 min, at an increment of 1 min if the angular velocity decreases to 90 rpm after 1 min; that is, the proportionality constant can be shown to be ln(0.9). Use the fourth-order Runge–Kutta method.

8.30 Find the total accumulated soil in the sediment trap discussed in Section 8.12.5 using a second-order Runge-Kutta method.

8.31 Find the total accumulated soil in the sediment trap discussed in Section 8.12.5 using (a) Euler's method and (b) the fourth-order Runge–Kutta method.

8.32 Solve Problem 8.10 using the Euler–trapezoidal (predictor–corrector) method.

8.33 Solve Problem 8.12 using the Euler–trapezoidal (predictor–corrector) method.

8.34 Solve Problem 8.14 using the Euler–trapezoidal (predictor–corrector) method.

8.35 Solve Problem 8.16 using the Euler–trapezoidal (predictor–corrector) method.

8.36 Solve Problem 8.17 using the Milne–Simpson (predictor–corrector) method. Use the Euler–trapezoidal (predictor–corrector) method to obtain the needed boundary conditions to start the Milne–Simpson method.

8.37 Solve Problem 8.10 using the least-squares method with a quadratic model (i.e., $\hat{y} = b_0 + b_1 x + b_2 x^2$). Assess the accuracy of the model as compared to the true solution.

8.38 Solve Problem 8.12 using the least-squares method with a quadratic model (i.e., $\hat{y} = b_0 + b_{1x} + b_2 x^2$). Assess the accuracy of the model as compared to the true solution.

8.39 Solve Problem 8.10 using the Galerkin method with a quadratic model (i.e., $\hat{y} = b_0 + b_1 x + b_2 x^2$). Assess the accuracy of the model as compared to the true solution.

8.40 Solve Problem 8.12 using the Galerkin method with a quadratic model (i.e., $\hat{y} = b_0 + b_1 x + b_2 x^2$). Assess the accuracy of the model as compared to the true solution.

8.41 Using the Euler method with $h = 0.5$, find the values of y and y' at $x = 1.0$ for the following differential equation:

$$\frac{d^2 y}{dx^2} - 0.005\frac{dy}{dx} + 0.15 y = 0$$

The boundary conditions are at $x = 0$, $y = 1$, and $y' = 0$.

8.42 Solve the following differential equation:

$$\left(\frac{dy}{dx}\right)^2 - xy^2 - 4 = 0$$

Use Euler's method with a step size $(h) = 0.1$, $y = 5$ at $x = 1$, and from $x = 1$ to $x = 1.5$.

8.43 The beam example of Section 8.12.7 involves a boundary-value problem. Define the boundary-value problem for the example and then solve it using Euler's method in conjunction with the shooting method.

8.44 The beam example of Section 8.12.7 involves a boundary-value problem. Define the boundary-value problem for the example and then solve it using the second Runge–Kutta method in conjunction with the shooting method.

8.45 The beam example of Section 8.12.7 involves a boundary-value problem. Define the boundary-value problem for the example and then solve it using the finite-difference method.

8.46   Evaluate the following integral using a step size of 0.1 for $x_0 = 1$:

$$y = 0.5 + \int_0^{x0} \frac{1}{\sqrt{2\pi}} \exp(-0.5x^2)\,dx$$

where $\pi = 3.1416$ and $y = 0.5$ at $x = 0$.

Data Description and Treatment

9

9.1 INTRODUCTION

It is common in engineering to deal with certain types of dispersion and uncertainty by collecting data and information about some variables that are needed to solve a problem of interest. The data can then be used to establish some understanding about the relations among the different variables. After collecting the data, it is necessary to utilize techniques for describing, treating, and analyzing them. The objective of this chapter is to introduce some of the techniques that are commonly used in engineering to describe and summarize data. Before quantitative statistics are performed, graphical analyses are very beneficial.

9.2 CLASSIFICATION OF DATA

Graphical analyses are widely used in engineering for both making initial assessments of data characteristics and presenting results to be used by engineers. A number of ways of graphically presenting data, each having advantages for specific types of data, are available. Therefore, before discussing the graphical methods, it is necessary to review methods for classifying data.

Data can be measured on one of four scales: nominal, ordinal, interval, and ratio. The four scales are given in order of numerical value. Variables defined on one scale can be reduced to a lower scale of measurement, but cannot be described on a higher scale of measurement. However, when a variable is transformed to a lower scale of measurement, an accompanying loss of information occurs.

In addition to the scale of measurement, data can be classified based on their dimensionality. The dimensionality is a measure of the number of axes needed to graphically present the data. Graphical analyses are usually limited to use with one-, two-, and three-dimensional data.

9.2.1 NOMINAL SCALE

The nominal scale of measurement is at the lowest level because the data cannot be ordered (i.e., placed in an order). Measurements consist of simply identifying the sample as belonging to one of several categories. Nominal measurement scales are both discrete and qualitative. However, numbers may be assigned to the categories for the purpose of coding.

Frequently used examples of variables measured on a nominal scale include (1) gender: female or male; (2) political affiliation: Republican, Democrat, Independent, or other; and (3) college major: engineering, sciences, physical education, or other. Engineering data are sometimes provided using a nominal scale—for example: (1) project failed or did not fail, (2) fatal and nonfatal accidents, and (3) land use: urban, rural, forest, institutional, commercial, other. If we assigned numbers to the six land uses identified, the numbers would not indicate a level of importance.

9.2.2 ORDINAL SCALE

The ordinal scale of measurement is considered to be a higher scale than the nominal scale because it has the added property that order exists among the groups. However, the magnitude of the differences between groups is not meaningful. For example, military ranks are measured on an ordinal scale. The major is above the sergeant and the sergeant is above the private, but we cannot say that a major is two or three times higher than a sergeant.

Variables of interest in engineering that are measured on an ordinal scale include the infiltration potential of soil texture classes and hazard classifications for dam design (high hazard, moderate hazard, low hazard). Soils are classified into one of several categories, such as sand, sandy loam, clay loam, and clay. In this respect, soil texture is measured on a nominal scale. However, if we consider the infiltration potential of the soil, then we can put the soil textures in numerical order according to the infiltration potential, high to low.

9.2.3 INTERVAL SCALE

The interval scale of measurement has the characteristics of the ordinal scale, in addition to having a meaningfulness in the separation between any two numbers on the scale. Temperature is defined on the interval scale. We recognize that a difference in temperature of 5°C is less than a difference of 10°C. Values on an interval scale may be treated with arithmetic operators. For example, the mean value of a set of test grades requires addition and division.

Engineering data are frequently recorded on an interval scale. The yield strength of steel, the compression strength of concrete, and the shear stress of soil are variables measured on an interval scale. The annual number of traffic fatalities and the number of lost worker-hours on construction sites due to accidents are also engineering variables recorded on an interval scale.

9.2.4 RATIO SCALE

The ratio scale represents the highest level of measurement. In addition to the characteristics of the interval scale, the ratio scale has a true zero point as its origin—unlike the interval scale, for which the zero point is set by some standard. For example, in the interval scale the zero point for temperature (°C) is set at the point where water freezes. However, it could have been set at the point where water boils or based on some other substance.

The standard deviation, which is discussed in Section 9.5.2, is measured on a ratio scale. The zero point is that for which there is no variation. The coefficient of a variation and many dimensionless ratios such as the Mach and Reynolds numbers are measured on a ratio scale.

9.2.5 DIMENSIONALITY OF DATA

The dimensionality of data was defined as the number of axes needed to represent the data. Tabular data with one value per classification are an example of one-dimensional data. For example, if a transportation engineer gives the number of fatal traffic accidents for each state in 1991, the variable is described using one-dimensional, interval scale data. It can be represented in tabular form as a function of the state. It could also be represented pictorially with the number of fatalities as the ordinate (vertical axis) and the state as the abscissa (horizontal axis). In this case, the ordinate is on an interval scale, while the abscissa is on a nominal scale; when presented this way, it appears as a two-dimensional graph.

Two-dimensional plots are very common in engineering. In Chapter 10, the solution of a dependent variable y could be plotted as a function of the independent variable x. In this case, both variables are expressed on an interval scale. As another example, the corrosion rate of structural steel as a function of the length of time that the steel has been exposed to the corrosive environment is a two-dimensional plot. If we have data for different types of steel (carbon steel, copper steel, and weathered steel), the three relationships can be presented on the same graph, with the steel types identified. In this case, steel type is a nominal variable, so the two-dimensional plot includes two variables on interval scales and one variable on a nominal scale.

9.3 GRAPHICAL DESCRIPTION OF DATA

The first step in data analysis is often a graphical study of the characteristics of the data sample. Depending on the objectives of the analysis and the nature of the problem under consideration, one or more of the following graphical descriptors are commonly used: area charts, pie charts, bar charts, column charts, scatter diagrams, line charts, combinations, and three-dimensional charts. The selection of a graphical descriptor type should be based on (1) its intended readers, (2) the types of data, (3) the dimensionality of the problem, and (4) the ability of the graphical descriptor to emphasize certain characteristics or relations of the parameters of the problem. In this section, examples are used to illustrate the different types.

9.3.1 AREA CHARTS

Area charts are useful for three-dimensional data that include both nominal and interval independent variables, with the value of the dependent variable measured on an interval scale and cumulated over all values of the nominal variable. The independent variable measured on an interval scale is shown on the abscissa. At any value along the abscissa, the values of the dependent variable are cumulated over the independent variable measured on the nominal scale.

Example 9.1: Area Chart for Traffic Analysis

A traffic engineer is interested in analyzing the traffic at an intersection. An incoming vehicle to the intersection can proceed in one of the following directions: straight, left turn, or right turn (U-turns are assumed to be illegal). The vehicles were counted at the intersection and classified according to direction. The counts were established for 24 hours in a typical business day. The results are shown in Figure 9.1 in the form of an area chart. The area chart shows that at 0800 approximately 50 vehicles made a right turn, approximately 30 vehicles made a left turn, and approximately 135 vehicles went through the intersection. Therefore, the total is about 215 vehicles.

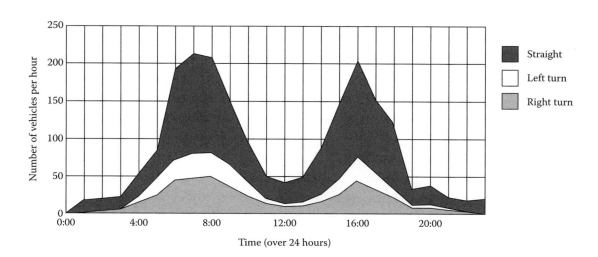

FIGURE 9.1   Traffic at an intersection.

9.3.2 PIE CHARTS

Pie charts are commonly used to graphically present data recorded as fractions, percentages, or proportions. The 100% of the circular pie chart is separated into pie slices based on fractions or percentages. The variable presented in a pie chart is measured on an interval scale. If it were necessary to represent subdata for measures of a nominal variable, the diameter of the pie chart could be used as a measure of the nominal variable.

Example 9.2: Pie Chart for Shipment Breakdown

A shipping company facilitates for customers the transfer of any items from any location to another within the United States. The company transfers 25%, 30%, and 45% of the items by air, ground, and sea transportation, respectively. In this case, the variable is the form of transportation, with values recorded as a percentage of all transportation. The breakdown of these items by shipping method is shown in Figure 9.2 in the form of a pie chart. If the breakdown was needed for the four seasons, the four pie charts could be provided, with the diameter of the circle used to reflect the relative amounts of items shipped.

9.3.3 BAR CHARTS

Bar charts are also useful for data recorded on an interval scale with one or more independent variables recorded on nominal or ordinal scales. The dependent variable can be a magnitude or a fraction. Both one- and two-dimensional bar charts can be used, with a dimension that is based on the number of independent variables.

Example 9.3: Bar Chart for Reinforcing Steel Production

A reinforcing-steel manufacturer provides steel of three different yield strengths: 40, 50, and 60 ksi. The production manager would like to keep track of production for both reinforcing-steel type and

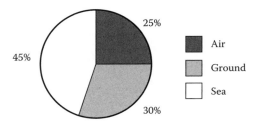

FIGURE 9.2 Shipping methods.

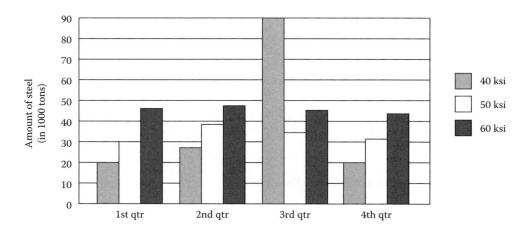

FIGURE 9.3 Reinforcing steel production by yield strength and quarter.

the four quarters of a production year. Thus the amount of steel produced is the dependent variable. There are two independent variables: the steel type and the quarters. The dependent variable is on an interval scale, while the independent variables are on ordinal and nominal scales, respectively. A bar chart, in this case, can meet the requirements of the manager as shown in Figure 9.3. In this case, the bars are shown vertically because the descriptors can be easily included on the abscissa and as a side note.

Example 9.4: Bar Chart for Capacity of Desalination Plants

In this case, the dependent variable is the capacity (in million gallons per day, mgd) of water from desalination plants worldwide. The independent variable is the process used in the desalination. Figure 9.4 shows the distribution. The bar chart in this example is

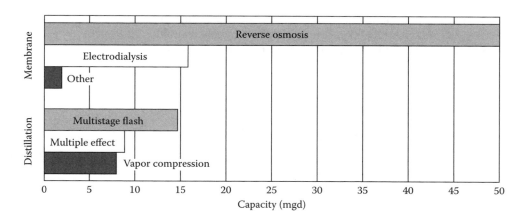

FIGURE 9.4 Worldwide desalination capacity (mgd) for available processes.

shown sideways to facilitate presenting the descriptors of the six processes.

9.3.4 COLUMN CHARTS

Column charts are very similar to bar charts, but with the added constraint that the dependent variable is expressed as a percentage (or fraction) of a total. In this case, one of the independent variables is used for the abscissa, while the dependent variable is shown as percentages (or fractions) of the second independent variable. The dependent variable is shown as the ordinate.

Example 9.5: Column Chart for Reinforcing Steel Production

An alternative method for displaying the reinforcing steel production of Example 9.3 is by column charts. Figures 9.5a and b show example charts. Note that, in Figure 9.5a, steel production is displayed as a percentage of the total steel produced for each quarter. In Figure 9.5b, the production is shown as an amount.

9.3.5 SCATTER DIAGRAMS

When both the independent and dependent variables are measured on interval or ratio scales, the data are best presented with scatter plots. Usually, the variable to be predicted (dependent variable) is shown on the ordinate and the independent variable on the abscissa. The ranges for the variables on axes are based on the minimum and maximum values of the measured data, possible values for the variables, or values that may be expected to occur in the future.

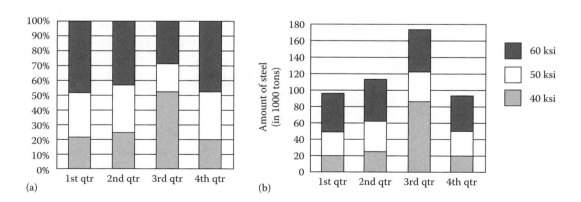

FIGURE 9.5 Reinforcing steel production by yield strength and quarter. (a) Amounts as percentages. (b) Amounts in tons.

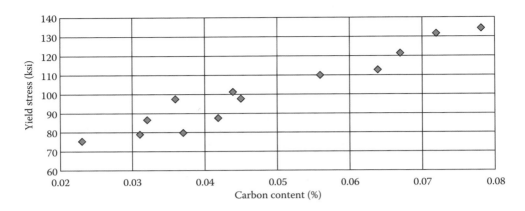

FIGURE 9.6 Yield strength and carbon content.

Example 9.6: Scatter Diagram for Yield Strength and Carbon Content

The yield strength of steel depends on several factors, such as the carbon content. An experimental program was designed to understand this relationship. Figure 9.6 shows the measured values. Since we are interested in the yield strength as a function of the carbon content, the yield strength is placed on the ordinate. The limits of the plot were set by the values of the data. An alternative plot with a range of 0 to 200 ksi and 0% to 0.1% carbon content could also be used to display the data. It is evident from this figure that under similar test conditions, the yield strength increases as the carbon content increases. While the graph shows the direction of the trend, statistical analyses (see Chapter 10) would be necessary to quantitatively assess the trend.

9.3.6 LINE GRAPHS

Line graphs are commonly used to illustrate mathematical equations. With both variables measured on interval or ratio scales, the variable to be predicted is usually shown as the ordinate. Line graphs are frequently used for design work when the design method is relatively simple.

Example 9.7: Line Chart for Peak Discharge Rates

For one section of the state of Maryland, peak discharge rates (Q in ft³/sec, cfs) can be estimated as a function of drainage area (A in mile²) by the following equations:

(9.1) $$Q_2 = 55.1A^{0.672}$$

(9.2) $$Q_{10} = 172A^{0.667}$$

(9.3) $$Q_{100} = 548A^{0.662}$$

in which the subscripts on Q reflect the return frequency (T in years) of the storm. Thus Q_{100} is the 100-year peak discharge. Designs with

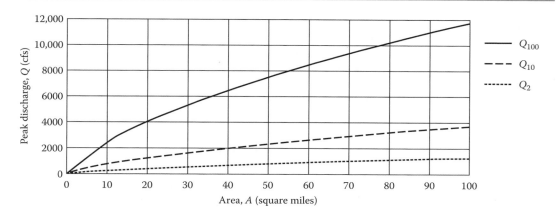

FIGURE 9.7   Peak discharge rates versus drainage area and return frequency.

these equations are usually limited to drainage areas of less than 100 square miles. Figure 9.7 shows the peak discharge for drainage areas up to 100 square miles. The two-dimensional line chart shows the dependent variable on the ordinate, one independent variable as the abscissa, and the second independent variable in the legend taking three values. All three variables are measured on interval scales.

Example 9.8: Line Chart for Yield Strength and Carbon Content

To establish the relationship between yield strength and carbon content, two independent laboratories were requested to perform similar tests on similar sets of specimens. The results from the two laboratories are shown in Figure 9.8. The lengths of the lines connecting the points reflect the magnitude of the variations that are reported based on measurements made at different laboratories. The apparent discontinuity between carbon content of 0.036 and 0.037 could be due to sampling variation or some other factor that was not controlled (e.g., the percentage of silicon).

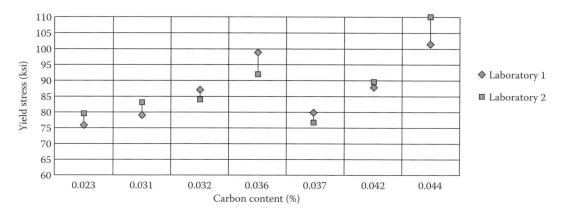

FIGURE 9.8   Line chart for yield strength and carbon content.

9.3.7 COMBINATION CHARTS

In combination charts, two or more of the previously discussed graphical methods are used to present data. For example, a line graph and bar chart can be combined in the same plot. A combination chart that includes both a scatter plot and a line graph is also commonly used to present experimental data and theoretical (or fitted) prediction equations.

Example 9.9: Operation of a Marine Vessel

The annual number of operation-hours of a high-speed Coast Guard patrol boat is of interest. The annual number of hours was recorded for 10 years. The measured data are shown in Figure 9.9 as vertical bars; they represent sample information. An analytical model is used to fit the data; this is shown in the figure as a solid line representing an assumed frequency model for the population. The combination chart in this case provides an effective presentation tool.

Example 9.10: Sample Distribution and Probability Function

A combination graph is useful in comparing sample measurements and an assumed probability function. Measurements of daily evaporation (in./day) were collected for each day in a month. A histogram of 30 values is shown in Figure 9.10. An increment of 0.02 in./day is used to develop this histogram. The measured evaporation ranges from 0.046 to 0.157 in./day; thus lower and upper bounds of 0.04 and 0.16 in./day are used. Based on the figure, it is reasonable to assume that each increment on the abscissa of the graph is equally likely. Therefore, a uniform frequency of 5, which results from 30 measurements divided by 6 intervals, can be used as the model.

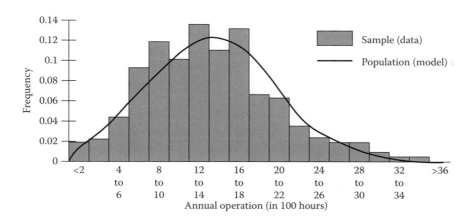

FIGURE 9.9   Combination chart for operational profile.

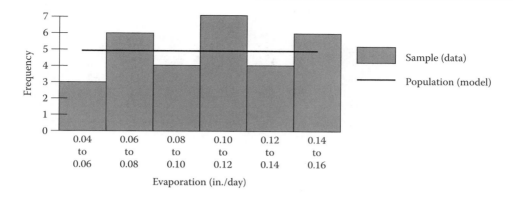

FIGURE 9.10   Combination chart for evaporation data.

Example 9.11: Corrosion Penetration versus Exposure Time

Measurements of corrosion were made yearly on a set of 10 steel specimens. The measured data are plotted in the combination graph of Figure 9.11. A power model was fitted to the data (see Section 10.9 in Chapter 10) and is also shown in Figure 9.11. The power model is

$$(9.4) \qquad\qquad \text{Penetration} = 28.075 t^{0.39943}$$

where t = time exposure.

The combination chart of Figure 9.11 is useful because it shows both the measured data and a graph that can be used to make predictions. The scatter of the measured points about the fitted line indicates the accuracy of the model.

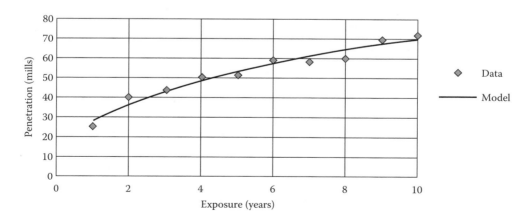

FIGURE 9.11   Combination chart for corrosion prediction.

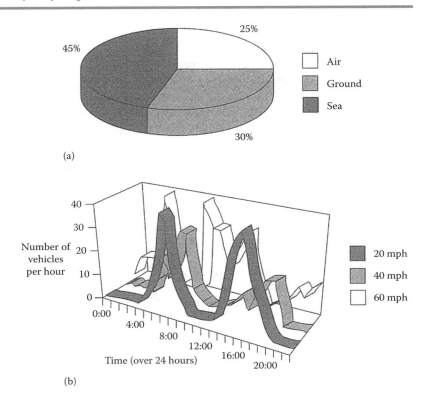

FIGURE 9.12   (a) Three-dimensional pie chart and (b) three-dimensional surface chart.

9.3.8 THREE-DIMENSIONAL CHARTS

Any of the charts described in the previous examples can be displayed in three dimensions. However, three-dimensional charts are commonly used to describe the relationships among three variables. For example, Figure 9.12a shows a three-dimensional pie chart for the data of Example 9.2. Another example is shown in Figure 9.12b, in which the speed (in miles per hour) and number of vehicles per hour passing through an intersection are shown at different times over a 24-hour period.

9.4 HISTOGRAMS AND FREQUENCY DIAGRAMS

Histograms and frequency diagrams are special types of charts that are commonly used to display and describe data. They are developed from the data for some variables of interest. A histogram is a plot (or tabulation) of the number of data points versus selected intervals or values for a parameter. A frequency diagram (or frequency histogram) is a plot (or tabulation) of the frequency of occurrence versus selected intervals or values of the parameter. The number of intervals (k) can be subjectively selected depending on the sample size (n). The number of intervals can be approximately determined as

(9.5) $k = 1 + 3.3 \log_{10}(n)$

Also, the number of intervals can depend on the level of dispersion in the data. Relative frequency diagrams can be derived from histograms by dividing the number of data points that correspond to each interval by the sample size. The meaning of these diagrams and their usefulness and development are illustrated in Example 9.12.

Example 9.12: Grades of Students

Students are always interested in the frequency histogram of test scores. This can be developed by determining the number of test scores in various groups, such as the letter grades or intervals of 10 points. If a test is given to 50 students and the numbers of scores are tabulated for each grade level (i.e., A, B, C, D, and F), a histogram can be plotted. If the number of students receiving grades of A, B, C, D, and F were 5, 11, 18, 10, and 6, respectively, a graph of the number of students versus the grade level indicates that the grades have a bell-shaped plot. The resulting histogram and frequency diagrams are shown in Figures 9.13a and b, respectively. Alternatively, the frequency of grades could be computed for intervals, such as 0–10,

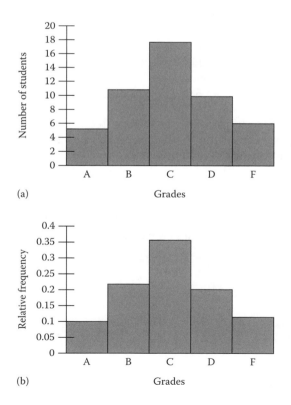

(a)

(b)

FIGURE 9.13   (a) Histogram of grades and (b) frequency histogram of grades.

11–20, 21–30, ..., 91–100. The frequency could be plotted against these intervals.

The effectiveness of a graphical analysis in identifying the true shape of the frequency diagram depends on the sample size and the intervals selected to plot the abscissa. For small samples, it is difficult to separate the data into groups that provide a reliable indication of the frequency of occurrence. With small samples, the configuration of the histogram may be very different for a small change in the intervals selected for the abscissa; for example, the grade distribution might appear very different if the intervals 0–15, 16–25, 26–35, ..., 76–85, 86–100 were used instead of the 10-point intervals described previously.

Example 9.13: Thickness Measurement of Corroded Steel Plates

Exposing steel to a corrosive environment, such as in the case of a steel bridge spanning a waterway or a cargo ship making voyages regularly, leads to loss of thickness of structural components. A corroded steel plate was measured at 20 locations and produced the following measurements in millimeters: 7.807, 8.886, 8.694, 8.185, 9.235, 8.526, 6.890, 8.953, 6.284, 6.533, 8.953, 8.112, 7.372, 9.640, 7.344, 8.837, 8.900, 9.048, 7.253, and 8.588. The minimum and maximum values are 6.284 and 9.640, respectively. These extreme values and examination of the data can be used to select a suitable constant-interval size. Equation 9.5 would suggest the use of five intervals. As an example, a size of 0.500 mm was selected, starting with a thickness of 6.000 mm and incrementally increasing the thickness to 10.000 mm. Table 9.1 shows these intervals (or bins), the counts in each bin, and the relative frequency. The relative frequency equals the count in the table divided by 20. The results are shown in Figures 9.14a and b.

TABLE 9.1   Frequency and Fraction Histogram of Thickness Measurements

Interval or Bin for x	Frequency	Relative Frequency
$x \leq 6.0$	0	0
$6.0 < x \leq 6.5$	1	0.05
$6.5 < x \leq 7.0$	2	0.10
$7.0 < x \leq 7.5$	3	0.15
$7.5 < x \leq 8.0$	1	0.05
$8.0 < x \leq 8.5$	2	0.10
$8.5 < x \leq 9.0$	8	0.40
$9.0 < x \leq 9.5$	2	0.10
$9.5 < x \leq 10.0$	1	0.05

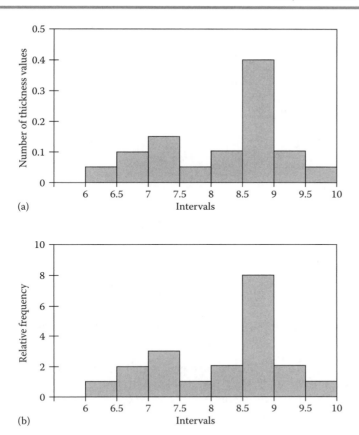

FIGURE 9.14   (a) Frequency histogram of plate thickness. (b) Relative frequency histogram of plate thickness.

Example 9.14: Combustion of Gases

Thirty-eight laboratory measurements are made of the work developed by combustion of gases within an enclosed piston and cylinder during a single stroke. The initial condition of a volume of 2.5 in.^{3} is maintained for each test. The resulting work values (lb-ft) are 76, 78, 81, 82, 84, 84, 86, 86, 87, 88, 88, 88, 89, 91, 91, 92, 92, 92, 94, 94, 98, 101, 103, 103, 103, 104, 104, 106, 108, 109, 112, 113, 114, 116, 116, 118, 118, 119.

To examine the uncertainty in the work done by combustion, histograms are developed using different cell boundaries as shown in Figure 9.15. Figure 9.15a presents the data using a cell interval of 5 lb-ft, while Figures 9.15b and c use an interval of 10 lb-ft. Figure 9.15a does not suggest a shape, while Figures 9.15b and c reflect a uniform and gamma distribution, respectively. Figure 9.15c is obviously skewed.

This example illustrates that histograms are subject to variation based on the cell width selected and cell boundaries used. Obviously,

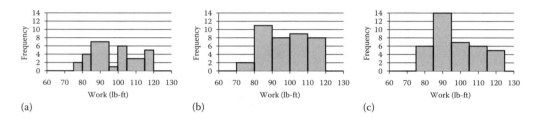

FIGURE 9.15   Frequency histogram of work by gas combustion: (a) cell width 5 lb-ft; (b) cell width 10 lb-ft; (c) cell width 10 lb-ft.

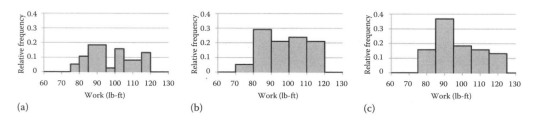

FIGURE 9.16   Relative frequency histogram of work by gas combustion: (a) cell width 5 lb-ft, (b) cell width 10 lb-ft, and (c) cell width 10 lb-ft.

small samples can lead to histograms that portray quite different images of the underlying population. Therefore caution should be used when interpreting histograms based on small samples.

Figure 9.16 shows the ordinate in relative frequency instead of frequency as shown in Figure 9.15. The relative frequencies were computed as the frequency divided by the sample size of 38, and sum to one for each case. The relative frequency can be interpreted as sample probability. For example, based on the relative frequency of Figure 9.16a, the probability of the work being less than 90 lb-ft is (0.053 + 0.105 + 0.184 = 0.342). In Figure 9.16c, the probability of the work being less than 95 lb-ft is (0.184 + 0.158 + 0.132 = 0.474).

9.5 DESCRIPTIVE MEASURES

In engineering it is sometimes desirable to characterize some data by certain descriptive measures (or statistics). These measures, which take numerical values, can be easily communicated to others and quantify the main characteristics of the data.

Most data analyses include the following three descriptive measures at the fundamental level:

1. Central tendency measures
2. Dispersion measures
3. Percentile measures

In this section an introductory description of these measures is provided. A formal discussion of these measures is provided in probability and statistics textbooks. Additionally, box-and-whisker plots are introduced as a graphical means of presenting these measures.

9.5.1 CENTRAL TENDENCY MEASURES

A very important descriptor of data is the central tendency measure. The following three descriptors can be used:

1. Average or mean value
2. Median value
3. Mode value

The average value is the most commonly used central tendency descriptor. For n observations, if all observations are given equal weights, the average value is given by

(9.6)
$$\bar{X} = \frac{1}{n}\sum_{i=1}^{n} x_i$$

where x_i = a sample point, and i = 1, 2, ..., n. An average value is often referred to as the mean.

The median value x_m is defined as the point that divides the data into two equal parts; that is, 50% of the data are above x_m and 50% are below x_m. The median value can be determined by ranking the n values in the sample in decreasing order, 1 to n. If n is an odd number, the median is the value with a rank of $(n + 1)/2$. If n is an even number, the median equals the average of the two middle values—that is, those with ranks $n/2$ and $(n/2) + 1$.

The mode value x_d is defined as the point of the highest percent of frequency of occurrence. This point can be determined with the aid of the frequency histogram.

Although these measures convey certain information about the underlying sample, they do not completely characterize the underlying variables. Two variables can have the same mean, but different histograms. Thus measures of central tendency cannot fully characterize the data. Other characteristics are also important and necessary.

Example 9.15: Mean Value of Grades

Consider the grades of the students discussed in Example 9.9. Assume the following grade points (levels) that correspond to the letter grades:

A = 4
B = 3
C = 2
D = 1
F = 0

Therefore, the average grade of the class can be computed as follows:

$$\bar{X} = \frac{4+4+4+4+4+3+3+\ldots+3+2+2+\ldots+2+1+1+\ldots+1+0+0+\ldots+0}{50} = 1.98$$

(9.7)

This simple equation includes 5 of the 4 value, 11 of the 3 value, 18 of the 2 value, 10 of the 1 value, and 6 of the 0 value. Therefore, this equation can be rewritten as

(9.8) $$\bar{X} = \frac{5(4)+11(3)+18(2)+10(1)+6(0)}{50} = 1.98$$

The average value indicates that, on the average, the class is at the C level. By inspecting the frequency histogram in Figure 9.13b, the median value is also C, since 32% of the grades are in the A and B levels and 32% of the grades in the D and F levels. Therefore, the C grade divides the grades into two equal percentages. This method of finding the median is proper for values on an ordinal scale, but not on a continuous scale. Again, the C grade is the mode, since it has the highest frequency of 36%. In this example, the average, median, and mode values have about the same value. However, in general, these values can be different.

9.5.2 DISPERSION MEASURES

The dispersion measures describe the level of scatter in the data about the central tendency location. The most commonly used measures are the variance and other quantities that are derived from it. For n observations in a sample that are given equal weight, the variance (S^2) is given by

(9.9) $$S^2 = \frac{1}{n-1}\sum_{i=1}^{n}(x_i - \bar{X})^2$$

The units of the variance are the square of the units of the variable x; for example, if the variable is measured in pounds per square inch (psi), the variance has units of (psi)^2. Computationally, the variance of a sample can be determined using the following alternative equation:

(9.10) $$S^2 = \frac{1}{n-1}\left[\sum_{i=1}^{n}x_i^2 - \frac{1}{n}\left(\sum_{i=1}^{n}x_i\right)^2\right]$$

Equation 9.10 provides the same answer as Equation 9.9 when computations are made using an appropriate number of significant digits.

Equations 9.9 and 9.10 provide an estimate of the variance that is an average of the squared difference between the x values and their average value. In these equations, $(n - 1)$ is used to compute the average, instead of (n), in order to obtain an unbiased estimate of the variance.

There are two commonly used derived measures based on the variance: the standard deviation and the coefficient of variation. By definition, the standard deviation (S) is the square root of the variance as follows:

(9.11)
$$S = \sqrt{\frac{1}{n-1}\left[\sum_{i=1}^{n} x_i^2 - \frac{1}{n}\left(\sum_{i=1}^{n} x_i\right)^2\right]}$$

It has the same units as both the underlying variable and the central tendency measures. Therefore, it is a useful descriptor of the dispersion or spread of a sample of data.

The coefficient of variation (cov or δ) is a normalized quantity based on the standard deviation and the mean. Because the standard deviation has the same units as the mean, the cov is dimensionless. The cov is defined as

(9.12)
$$\text{cov} = \frac{S}{\bar{X}}$$

It is also used as an expression of the standard deviation in the form of a percent of the average value. For example, consider $\bar{X}$ and S to be 50 and 20, respectively; therefore, $\text{cov}(X) = 0.4$ or 40%. In this case, the standard deviation is 40% of the average value.

Example 9.16: Dispersion Measures of Concrete Strength

A sample of five tests was taken to determine the compression strength (in ksi) of concrete. Test results are 2.5, 3.5, 2.2, 3.2, and 2.9 ksi. Compute the variance, standard deviation, and coefficient of variation of concrete strength. The mean value of concrete strength $\bar{X}$ is given by

(9.13)
$$\bar{X} = \frac{2.5 + 3.5 + 2.2 + 3.2 + 2.9}{5} = 2.86 \text{ ksi}$$

The variance of concrete strength is computed using Equation 9.10 as follows:

$$S^2 = \frac{2.5^2 + 3.5^2 + 2.2^2 + 3.2^2 + 2.9^2 - \dfrac{(2.5 + 3.5 + 2.2 + 3.2 + 2.9)^2}{5}}{5-1} = 0.273 \text{ ksi}^2$$

(9.14)

Therefore, the standard deviation is given by

$$(9.15) \qquad S = \sqrt{0.273} = 0.522 \text{ ksi}$$

The coefficient of variation can be computed as follows:

$$(9.16) \qquad \delta \text{ or } \text{cov}(X) = \frac{S}{\overline{X}} = \frac{0.522}{2.86} = 0.183$$

The relatively large coefficient of variation, 18.3%, suggests that the average value is not reliable and that additional measurements might be needed. If, with additional measurements, the coefficient of variation remains large, relatively large factors of safety should be used for projects that use the concrete.

9.5.3 PERCENTILES

A p percentile value (x_p) for a parameter or variable based on a sample is the value of the parameter such that p% of the data is less or equal to x_p. On the basis of this definition, the median value is considered to be the 50th percentile value. It is common in engineering to have interest in the 10th, 25th, 50th, 75th, or 90th percentile values for a variable.

Example 9.17: Operation of a Marine Vessel

The annual number of operation hours of a high-speed Coast Guard patrol boat was discussed in Example 9.9. The annual number of hours was recorded for 10 years. An analytical model is used to fit the data; this is shown in Figure 9.9 as a solid line representing an assumed frequency model for the population. The model is also shown in Figure 9.17. Also shown in this figure is the 25th percentile value, which is 702 hours. The 75th, 90th, and 10th percentile values are 1850 hours, 2244 hours, and 578 hours, respectively. The

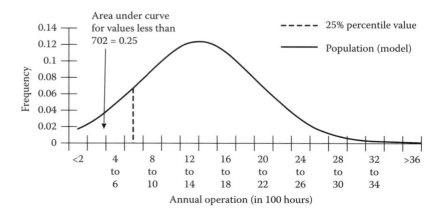

FIGURE 9.17 Percentile values for the operational profile.

calculation of the percentile values, in this example, requires the knowledge of the equation of the model and probabilistic analysis. The model used in Figure 9.17 is called the normal probability density function. The area under this model is a measure of likelihood or probability. The total area under the model is 1. Therefore, the 25th percentile value is the x value such that the area under the model up to this value is 0.25. This value is 702 hours, as shown in Figure 9.17.

9.5.4 BOX-AND-WHISKER PLOTS

A box-and-whisker plot, which is sometimes referred to as a boxplot, is a graphical method for showing the distribution of sampled data, including the central tendency (mean and median), dispersion, percentiles (10th, 25th, 75th, and 90th percentiles), and extremes (minimum and maximum). Additionally, it can be used to show the bias about the standard value and, if the figure includes multiple plots for comparison, the relative sample size.

To construct a box-and-whisker plot, the following characteristics of a data set need to be computed:

1. Mean and median of the sample
2. Minimum and maximum values in the sample
3. The 90th, 75th, 25th, and 10th percentile values

The plot consists of a box, the upper and lower boundaries of which define the 75th and 25th percentiles, and upper and lower whiskers, which extend from the ends of the box to the extremes, as shown in Figure 9.18. At the 90th and 10th percentiles, bars, which are one-half of the width of the box, are placed perpendicular to the whiskers. The mean and median are indicated by solid and dashed lines that are the full width of the box, respectively.

Figure 9.18 shows the box-and-whisker plot for the maximum daily ozone concentration. The mean and median values are 59 and 52 ppb (parts per billion), respectively. The 10, 25, 75, and 90th percentile points are 24, 36, 79, and 97 ppb, respectively.

If a figure includes more than one box-and-whisker plot and the samples from which each plot is derived are of different sizes, then the width of the box can be used to indicate the sample size, with the width of the box increasing as the sample sizes increase. Figure 9.19 shows box-and-whisker plots of samples of a toxic chemical analyzed at four different laboratories. Since the samples were of different concentrations, the distributions are presented as the differences between the true concentration and the concentration reported by the laboratory (i.e., error). The difference between the line for zero error and the solid line for the mean in each plot represents the corresponding lab's bias. Lab 1 tends to overpredict by almost 10 ppb. While lab 2 shows a slight positive bias, it shows a skewed distribution of results, with many values underestimating the true value. Laboratories 1, 2, 3, and 4 processed the same number of samples each. Therefore, the widths of the boxes are the same, reflecting the

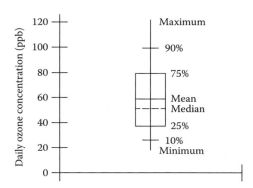

FIGURE 9.18   Box-and-whisker plot for maximum daily ozone concentration (ppb).

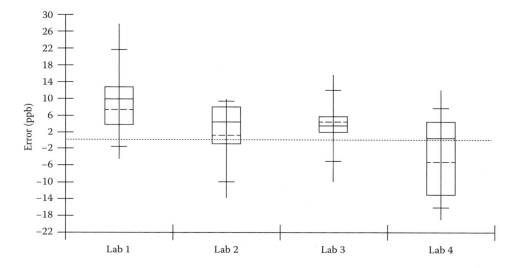

FIGURE 9.19   Display of multiple box-and-whisker plots.

equal sample sizes. However, it should be noted that, in general, greater accuracy of data descriptors can be expected for the descriptors based on larger sample sizes.

9.6 APPLICATIONS

9.6.1 TWO RANDOM SAMPLES

The two random samples shown in Table 9.2 are used to illustrate the meaning of dispersion in data. Both samples have the same mean value of 10. The mean value for sample 1 can be computed as

$$\text{mean of sample } 1 = \frac{15(9) + 15(11)}{30} = 10$$

TABLE 9.2 Two Random Samples

	Number of Occurrences	
Sampled Value	Sample 1	Sample 2
5	0	3
6	0	2
7	0	4
8	0	0
9	15	3
10	0	4
11	15	2
12	0	6
13	0	2
14	0	2
15	0	2

Similarly, the mean for sample 2 is

$$\text{mean of sample } 2 = \frac{3(5) + 2(6) + 4(7) + \ldots + 2(13) + 2(14) + 2(15)}{30} = 10$$

Although both samples have the same mean and the same sample size, they are different in their levels of scatter or dispersion. Figure 9.20 shows the histograms for the two samples, which show clearly the different levels of scatter. The variances, standard deviations, or coefficients of variation can be computed and used to measure the dispersion. The variance for sample 1 can be computed as

$$\text{variance of sample } 1 = \frac{15(9-10)^2 + 15(11-10)^2}{30-1} = 1.034$$

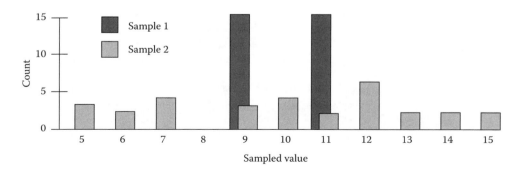

FIGURE 9.20 Histograms for two samples.

Similarly, the variance for sample 2 is

variance of sample 2

$$= \frac{3(5-10)^2 + 2(6-10)^2 + 4(7-10)^2 + \ldots + 2(13-10)^2 + 2(14-10)^2 + 2(15-10)^2}{30-1} = 9.379$$

Therefore, the variance of sample 2 is about nine times the variance of sample 1. The standard deviations for samples 1 and 2 are 1.017 and 3.063, respectively, and the coefficients of variations are 0.10 and 0.31, respectively. The larger scatter in sample 2 could reflect natural variation in the value of the variable or a difficulty in making precise measurements. In some cases, variation is not desirable in a sample of data if it reflects a lack

TABLE 9.3   Stage and Discharge for the Little Patuxent River, Guilford, Maryland

Year	Stage (m)	Discharge (cms)	Year	Stage (m)	Discharge (cms)
1933	3.81	119.2	1962	2.78	36.2
1934	2.86	41.9	1963	2.25	23.2
1935	2.35	25.9	1964	2.7	32.8
1936	2.74	37.4	1965	2.35	24.8
1937	3.14	56.6	1966	2.72	34
1938	3.08	51.5	1967	2.71	33.4
1939	2.41	27.4	1968	2.86	39.9
1940	3.51	77.6	1969	2.02	19.4
1941	1.87	15.9	1970	2.02	19.4
1942	3.25	61.7	1971	3.49	86.9
1943	2.79	39.6	1972	5.6	351.1
1944	3.19	58.3	1973	3.07	53.2
1945	3.72	107.9	1974	2.65	31.4
1946	2.52	30.6	1975	4.08	152
1947	2.44	23.1	1976	3.08	54.1
1948	3.02	43.9	1977	2.14	21.3
1949	2.22	19.9	1978	3.63	103.9
1950	1.95	15.9	1979	3.91	132.5
1951	3.19	55.8	1980	2.65	31.4
1952	4.04	150.1	1981	2.5	28
1953	3.11	56.6	1982	2.18	22
1954	1.91	18.3	1983	3.33	78.7
1955	3.69	107.3	1984	2.65	42.2
1956	2.55	28.6	1985	2.6	40.5
1957	1.9	17.3	1986	1.69	19.3
1958	2.75	34.8	1987	2.42	34.8
1959	2.14	21.4	1988	2.42	34.8
1960	2.54	28.3	1989	4.01	143.5
1961	2.49	27.4			

of precision. In other cases, sample variation is necessary to ensure that the sample is representative of the full range of the underlying population.

9.6.2 STAGE AND DISCHARGE OF A RIVER

The stage of a river is defined as its flow depth, and the discharge is defined as the volume of flow. For the Little Patuxent River near Guilford, Maryland, the stage in meters (m) and the discharge in cubic meters per second (cms) were obtained as shown in Table 9.3. The table shows the stage and discharge for the years 1933 to 1989. Descriptive statistics of stage and discharge are shown in Table 9.4. The frequency histogram for the stage is shown in Figure 9.21. The average, which is indicated in Table 9.4, is 2.84 m. The median stage is 2.71 m. The standard deviation and coefficient of variation are 0.71 and 0.25 m, respectively. The maximum and minimum values (extreme values) are 5.60 and 1.69 m, respectively.

The frequency histogram for the discharge is shown in Figure 9.22. The average, which is indicated in Table 9.4, is 54.82 cms. The median discharge is 34.8 cms. The standard deviation and coefficient of variation are 53.78 and 0.98 cms, respectively. The maximum and minimum values (extreme values) are 351.10 and 15.90 cms, respectively.

By comparing the two histograms in Figures 9.21 and 9.22, it can be observed that the relative dispersion in discharge is larger than that in stage. The coefficient of variation of discharge is 0.98, which is four times larger than the coefficient of variation of stage (0.25). Also, it can be observed that the frequency histogram for stage is almost bell shaped, whereas the histogram for discharge is highly skewed and appears as an exponential decay function. The closeness of the average stage (2.84 m) to the median stage (2.71 m) is an indication of a symmetric histogram. On the other hand, the average discharge (54.82 cms) is considerably different from the median discharge (34.8 cms), indicating a lack of symmetry in the discharge measurements.

TABLE 9.4   Descriptive Statistics of Stage and Discharge for the Little Patuxent River, Guilford, Maryland

Parameter	Stage (m)	Discharge (cms)
Average	2.84	54.82
Median	2.71	34.8
Mode	2.65	34.8
Standard deviation	0.71	53.78
Sample variance	0.505	2893
Coefficient of variation	0.25	0.98
Range	3.91	335.2
Minimum	1.69	15.9
Maximum	5.6	351.1
Count	57	57

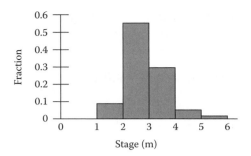

FIGURE 9.21   Frequency histogram of the stage of a river.

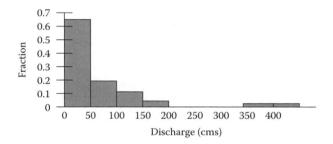

FIGURE 9.22   Frequency histogram of the discharge of a river.

PROBLEMS

9.1   For each of the measurement scales, identify five variables that are measured with the scale.

9.2   Using age as a variable of interest, identify a function that would be measured on each of the scales: ordinal, interval, and ratio.

9.3   Using the copper content of steel as a variable of interest, identify one function that would be measured for each of the four measurement scales.

9.4   Using energy as a variable of interest, identify a specific variable that would be appropriate for each of the four scales.

9.5   Using the game of football as a variable of interest, identify a specific variable that would be appropriate to represent each of the four scales.

9.6   For the eight methods of graphical analysis given in Section 9.3, develop a classification system for distinguishing among them. The classification system should center on basic, yet important, characteristics.

9.7   Using the data from Problem 9.21 and the following population totals, construct an area chart that shows the total and the breakdown of the total population in rural, suburban, and central city.

Year	1900	1910	1920	1930	1940	1950	1960	1970
Total ($\times 10^6$)	76	92	106	123	132	151	179	203

9.8 The following data show the mean monthly days of various depths of rainfall. Provide an area chart that illustrates the data.

Month	J	F	M	A	M	J	J	A	S	O	N	D	Total
Rain $P \geq 0.5$ in.	2	0	3	1	1	0	3	3	2	2	2	1	20
$0.25 < P < 0.5$	2	3	3	3	2	4	4	6	5	4	3	1	40
$0.01 \leq P \leq 0.25$	1	2	1	2	2	3	3	4	3	3	1	2	27
$P = 0$ (dry day)	26	23	24	24	26	23	21	18	20	22	24	27	278

9.9 Based on 1 year of data, the following data show the mean number of weekly visits to the emergency room of a hospital in the semirural community. Provide an area chart that illustrates the data.

Day	Su	M	Tu	W	Th	F	Sa
Broken bone	2.2	0	1.4	0.8	1.6	3.5	4.7
Heart concern	0.8	1.3	0.6	0.5	1.1	1.8	2.2
Alcohol/drug abuse	3.1	1.6	0.4	0.3	0.7	1.4	2.3
Other	11.6	5.3	2.4	1.9	2.3	8.6	15.4

9.10 The following data are the solid waste (millions of tons) produced annually in the United States:

Municipal trash and garbage	150
Industrial	350
Mining	1700
Agriculture	2300

Construct a pie chart to present the data. Discuss the merits of presenting the data in a pie chart versus the tabular summary given.

9.11 Using the percentages of the total visits to the emergency room given by the data in Problem 9.9, construct a pie chart, Also display the data in tabular form and discuss the benefits of both displays.

9.12 The following data give the age distribution of US citizens as a function of age group. Select a method for graphing the results. Interpret the results. Then combine the values into three meaningful groups and provide a graphical analysis of the distribution. How do the two analyses differ in their information content and emphasis?

Age group	<5	5–9	10–14	15–19	20–24	25–29	30–34
Percentage	7.2	7.4	8.1	9.3	9.4	8.6	7.8

Age group	35–39	40–44	45–49	50–54	55–59	60–64	≥65
Percentage	6.2	5.2	4.9	5.2	5.1	4.5	11.3

9.13   Researchers at an agricultural college wanted to study the perseverance of acid rain in soils. They irrigated plots with water at four levels of acidity (pH of 3.0, 3.5, 4.0, and 4.5) and measured the pH of the soil at a depth of 25 cm over a period of a week, with the following results:

pH of Irrigated Water				
Day	**3.0**	**3.5**	**4.0**	**4.5**
1	6.2	6.4	6.7	6.8
3	5.4	5.7	6.2	6.5
5	4.7	5.2	6.0	6.2
7	5.1	5.5	6.3	6.4

a.   Present the data using four bar charts, one for each day.
b.   Present the data using a combination chart.

9.14   A local highway department compiled the following percentages from accident records according to traffic control method (flashing red light, two-way stop sign, or four-way stop sign) and accident severity (loss of life, major damage, minor damage):

Traffic Control Method	**Loss of Life**	**Major Damage**	**Minor Damage**	**Total**
Flashing red light	23	41	36	100
Two-way stop sign	18	39	43	100
Four-way stop sign	12	21	67	100

a.   Present the data using three pie charts, one for each of the traffic control methods.
b.   Present all the data in a bar chart in a way that emphasizes differences between accident severity.
c.   Present all the data in a bar chart in a way that emphasizes differences between the traffic control methods.
d.   Present the data as a column chart.
e.   Discuss the advantages and disadvantages of each of the preceding graphical methods with respect to these data.

9.15   Compare the graphical analyses of Figures 9.3, 9.5a and b and identify the circumstances under which each would be the most appropriate for making decisions.

9.16   The following data provide the estimated remaining strippable resources and reserves of bituminous coal in the United States (billions of short tons). Create a bar chart to display the data.

Alaska	0.9
Rocky Mountains and Northern Great Plains	1.1
Interior and Gulf provinces	32
Eastern province	27

9.17 The following data include the glacial ice coverage G (sq) of the world separated by land area. Create a bar chart to display the data:

Land Area	$G\,(\times 10^3)$
Continental Europe	4
Continental Asia	43
Continental North America	31
Continental South America	10
South Polar Region	5000
North Polar Region	700
Other	<1

9.18 The following data are the volumes V (mi³), water surface area A (mi²), and shoreline length L (mi) for the five Great Lakes:

Lake	V	A	L
Superior	2900	32,000	2700
Michigan	1200	22,000	1600
Huron	850	23,000	3800
Erie	120	9900	900
Ontario	400	7300	700

a. Construct three pie charts, one for each of the variables (V, A, L).
b. Construct a column chart that includes a column for each of the three variables. In each column, show the percentage for each of the five lakes.

9.19 The following data are the sources of sediment discharged into surface water in a region, recorded as percentages:

Cropland	38
Stream bank	25
Pasture	14
Forest	6
Urban	3
Mining	1
Other	13
Total	100%

a. Construct a pie chart to display the data.
b. Display the data using a column chart.
c. Use a bar chart to display the data.
d. Briefly comment on the best graphical display of the data.

9.20   Create column charts to present the following estimates of US production of bituminous and lignite coal from surface and underground mines:

Year	Surface Mines	Underground Mines
1940	43	418
1950	123	393
1960	131	285
1970	264	339

Use one column chart to emphasize the temporal variation of total production. Use a second column chart to emphasize the proportion produced from surface mines. Use a third column chart to emphasize the total production from surface versus underground mines.

9.21   The following percentages indicate the change in rural, suburban, and central city US populations from 1900 to 1970. Present the data graphically to emphasize the decline in the proportion of the population living in rural areas. Present the data graphically to illustrate any association between the change in central city population and the increase in the proportion living in suburban areas.

Year	Rural	Suburban	Central City
1900	58	16	26
1910	55	17	28
1920	51	18	31
1930	46	20	34
1940	46	22	32
1950	45	24	31
1960	40	30	30
1970	34	38	28

9.22   Plot the data shown in the following table using a column chart:

| Superstructure Type | Number of Constructed Bridges by Year | | |
	1989	1990	1991
Steel	5	10	12
Concrete	10	6	7
Prestressed concrete	4	6	5
Total	19	22	24

9.23   Using the bridge data in Problem 9.22, show the number of bridges as a function of year and by bridge type using a line chart, a pie chart, and three-dimensional surface chart.

9.24   The following concrete strength data (in ksi) were collected using an ultrasonic nondestructive testing method at different locations of an existing structure: 3.5, 3.2, 3.1, 3.5, 3.6, 3.2, 3.4, 2.9, 4.1, 2.6, 3.3, 3.5, 3.9, 3.8, 3.7, 3.4, 3.6, 3.5, 3.5, 3.7, 3.6, 3.8, 3.2, 3.4, 4.2, 3.6, 3.1, 2.9, 2.5, 3.5, 3.4, 3.2, 3.7, 3.8, 3.4, 3.6, 3.5, 3.2, 3.6, and 3.8. Plot a histogram and a frequency diagram for concrete strength.

9.25   The following data are the annual maximum flow rates (ft^3/sec) in the Floyd River at James, Iowa (1935–1973): 1460, 4050, 3570, 2060, 1300, 1390, 1720, 6280, 1360, 7440, 5320, 1400, 3240, 2710, 4520, 4840, 8320, 13900, 71500, 6250, 2260, 318, 1330, 970, 1920, 15100, 2870, 20600, 3810, 726, 7500, 7170, 2000, 829, 17300, 4740, 13400, 2940, 5660.

   a.   Construct a histogram of the data. Use Equation. 9.5 to determine the number of cells in the histogram.

   b.   Compute the base 10 logarithm of each flow and construct a histogram of the data. Use the same number of cells.

   c.   Briefly comment on the two histograms.

9.26   For the data of Problem 9.24, determine the central tendency measures—that is, the average value, median, and mode.

9.27   For the data of Problem 9.25 determine the three measures of the central tendency. Compare the resulting values.

9.28   For the data of Problem 9.25, determine the three measures of dispersion. Compare the resulting values. Then compute the three measures using the logarithms of the data. Contract them with the measures of the raw data.

9.29   For the data of Problem 9.24, determine the dispersion measures, that is, the variance, standard deviation and coefficient of variation.

9.30   Create a box-and-whisker plot of the data in Problem 9.32. Also create a frequency histogram of the data. Discuss and compare the information content of the two graphical analyses.

9.31   Using the data of Problem 9.25, develop a box-and-whisker plot. Also create a frequency histogram and a bar graph of the data. Briefly compare the three results.

9.32   The following data are the maximum daily ozone concentrations for the months indicated. Graph the data in a way that will emphasize the monthly variation in the concentration. Also graph the data in a way that emphasizes the annual variation. Does the effect of monthly variation make it more difficult to assess the importance of the annual variation? Explain.

Year	Feb	Apr	June	Aug	Oct	Dec
1980	61	72	77	83	64	55
1981	63	71	78	87	66	58
1982	64	72	78	86	66	59
1983	68	72	80	89	72	63
1984	74	73	85	94	68	66
1985	73	74	86	92	70	64
1986	67	73	85	90	67	64
1987	66	72	83	86	63	59
1988	62	68	81	87	65	57
1989	59	66	82	86	62	51
1990	56	65	80	86	60	50

Curve Fitting and Regression Analysis

10

10.1 INTRODUCTION

In many engineering problems, variation in the value of a variable is associated with variation in one or more additional variables. By establishing the relationship between two or more variables, it may be possible to reduce the uncertainty in an estimate of the variable of interest. Thus, before making a decision based on the analysis of observations on a random variable alone, it is wise to consider the possibility of systematic associations between the variable of interest and other variables that have a causal relationship with it. These associations among variables can be understood using correlation and regression analyses.

To summarize these concepts, the variation in the value of a variable may be either random or systematic in its relationship with another variable. Random variation represents uncertainty. If the variation is systematically associated with one or more other variables, the uncertainty in estimating the value of a variable can be reduced by identifying the underlying relationship. Correlation and regression analyses are important methods to achieve these objectives. They should be preceded by a graphical analysis to determine (1) if the relationship is linear or nonlinear, (2) if the relationship is direct or indirect, and (3) if any extreme events might control the relationship.

10.2 CORRELATION ANALYSIS

Correlation analysis provides a means of drawing inferences about the strength of the relationship between two or more variables. That is, it is a measure of the degree to which the values of these variables vary in a systematic manner. Thus it provides a quantitative index of the degree to which one or more variables can be used to predict the values of another variable.

It is just as important to understand what correlation analysis cannot do as it is to know what it can do. Correlation analysis does not provide an equation for predicting the value of a variable. Also, correlation analysis does not indicate whether a relationship is causal; that is, it is necessary for the investigator to determine whether a cause-and-effect relationship exists between the variables. Correlation analysis only indicates whether the degree of common variation is significant.

Correlation analyses are most often performed after an equation relating one variable to another has been developed. This happens because the correlation coefficient, which is a quantitative index of the degree of common linear variation, is used as a measure of the goodness of fit between the prediction equation and the data sample used to derive the prediction equation. But correlation analyses can be very useful in model formulation, and every model development should include a correlation analysis.

10.2.1 GRAPHICAL ANALYSIS

The first step in examining the relationship between variables is to perform a graphical analysis. Visual inspection of the data can identify the following information:

1. The degree of common variation, which is an indication of the degree to which the two variables are related
2. The range and distribution of the sample data points
3. The presence of extreme events
4. The form of the relationship between the two variables
5. The type of relationship

Each of these factors is of importance in the analysis of sample data and in decision making.

When variables show a high degree of association, we assume that a causal relationship exists. If a physical reason to suspect, a causal relationship can be proposed, then the association demonstrated by the sample data provides empirical support for the assumed relationship. Common variation implies that, when the value of one of the variables is changed, the value of the other variable will change in a systematic manner. For example, an increase in the value of one variable occurs when the value of the other variable increases. If the change in the one variable is highly predictable from a given change in the other variable, then a high degree of common variation exists. Figure 10.1 shows graphs of different samples of data for two variables having different degrees of common variation. The common variation is measured in the figure using the correlation coefficient (R). In Figure 10.1a and e the degree of common variation is very high; thus the variables are said to be correlated. In Figure 10.1c there is no correlation between the two variables because, as the value of X is increased, it is not certain whether Y will increase or decrease. In Figure 10.1b and d, the degree of correlation is moderate; in Figure 10.1b, evidently Y will increase as X is increased, but the exact change in Y for a change in X is difficult to estimate from just looking at the graph. A more quantitative description of the concept of common variation is needed.

It is important to use a graphical analysis to identify the range and distribution of the sample data points so that the stability of the relationship can be assessed and so that we can assess the ability of the data sample to represent the distribution of the population. If the range of the data is limited, a computed relationship may not be stable; that is, it may not apply to the distribution of the population. The data set of Figure 10.2 shows a

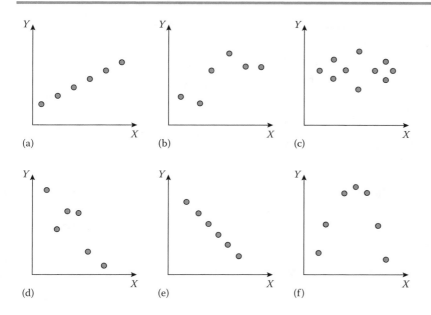

FIGURE 10.1   Different degrees of correlation between variables (X and Y):
(a) $R = 1.0$, (b) $R = 0.5$, (c) $R = 0.0$, (d) $R = -0.5$, (e) $R = -1.0$, and (f) $R = 0.3$.

case where the range of the sample is much smaller than the expected
range of the population. If an attempt is made to use the sample to project
the relationship between the two random variables, a small change in the
slope of the relationship causes a large change in the predicted estimate of
Y for values of X at the extremes of the range of the population. A graph of
two variables might alert the investigator to a sample in which the range
of the sample data may cause stability problems in a derived relationship
between two random variables, especially when the relationship is pro-
jected beyond the range of the sample data.

It is important to identify extreme events in a sample of data for several reasons. First, extreme events can dominate a computed relationship
between two variables. For example, in Figure 10.3a, the extreme point
suggests a high correlation between X and Y; in this case, the cluster of

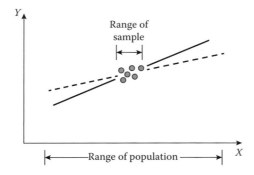

FIGURE 10.2   Instability in the relationship between random variables.

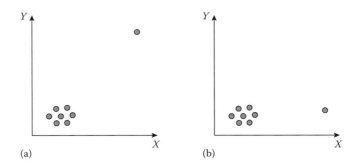

FIGURE 10.3   Effect of an extreme event in a data sample on the correlation: (a) high correlation and (b) low correlation.

points acts like a single observation. In Figure 10.3b, the extreme point causes a poor correlation between the two random variables; since the mean of the cluster of points is nearly the same as the value of Y of the extreme point, the data of Figure 10.3b suggest that a change in X is not associated with a change in Y. A correlation coefficient is more sensitive to an extreme point when the sample size is small. An extreme event may be due to (1) errors in recording or plotting the data or (2) a legitimate observation in the tail of the distribution. Therefore, an extreme event must be identified and the reason for the event determined. Otherwise, it is not possible to properly interpret the results of the correlation analysis.

Relationships can be linear or nonlinear. Since the statistical methods to be used for the two forms of a relationship differ, it is important to identify the form. Additionally, the most frequently used correlation coefficient depends on a linear relationship existing between the two random variables; thus low correlation may result for a nonlinear relationship even when a strong relationship is obvious. For example, the bivariate relationship of Figure 10.1f suggests a very predictable trend in the relationship between Y and X; however, the correlation is poor and is certainly not as good as that in Figure 10.1a, even though the two plots suggest equal levels of predictability.

Graphs relating pairs of variables can be used to identify the type of the relationship. Linear trends can be either direct or indirect, with an indirect relationship indicating a decrease in Y as X increases. This information is useful for checking the rationality of the relationship, especially when dealing with data sets that include more than two variables. A variable that is not dominant in the physical relationship may demonstrate a physically irrational relationship with another variable because of the values of the other variable affecting the physical relationship.

10.2.2 BIVARIATE CORRELATION

Correlation is the degree of association between the elements of two samples of data—that is, between observations on two variables. Correlation coefficients provide a quantitative index of the degree of linear association.

Examples of variables that are assumed to have a causal relationship and significant correlation are (1) the cost of living and wages and (2) the volumes of rainfall and flood runoff. However, examples that do not have a cause-and-effect relationship but may have significant correlation are (1) the crime rate and the sale of chewing gum over the past two decades and (2) annual population growth rates in nineteenth century France and annual cancer death rates in the United States in the twentieth century.

Many indexes of correlation exist. The method that is used most frequently is the *Pearson product-moment* correlation coefficient. Nonparametric correlation indexes include the contingency coefficient, the Spearman rank correlation coefficient, and the Kendall rank correlation coefficient. Only the Pearson correlation coefficient is considered in this chapter.

10.2.3 SEPARATION OF VARIATION

A set of observations on a variable Y has a certain amount of variation, which may be characterized by the variance of the sample of size n. The variance equals the sum of squares of the deviations of the observations from the mean of the observations divided by the degrees of freedom $(n - 1)$. Ignoring the degrees of freedom, the variation in the numerator can be separated into two parts: (1) variation associated with a second variable X and (2) variation not associated with X. That is, the total variation (TV), which equals the sum of the squares of the sample data points about the mean of the data points, is separated into the variation that is explained by variation in the second variable (EV) and the variation that is not explained—that is, the unexplained variation, UV. Thus TV can be expressed as

(10.1)
$$TV = EV + UV$$

Using the general form of the variation of a random variable, each of the three terms in this equation can be represented by a sum of squares as follows:

(10.2)
$$\sum_{i=1}^{n}(Y_i - \bar{Y})^2 = \sum_{i=1}^{n}(\hat{Y}_i - \bar{Y})^2 + \sum_{i=1}^{n}(\hat{Y}_i - Y_i)^2$$

where Y = an observation on the random variable, $\hat{Y}$ = the value of Y estimated from the best linear relationship with the second variable X, and $\bar{Y}$ = the mean of the observations on Y. These variation components are shown in Figure 10.4; the dashed lines illustrate the deviations of Equation 10.2. Figure 10.4a shows the variation of the observations about the mean value; this represents the variation that exists in the sample and that may potentially be explained by the variable X; it reflects the left side of Equation 10.2. The variation in Figure 10.4b represents the variation of the line about the mean and thus corresponds to the first term of the

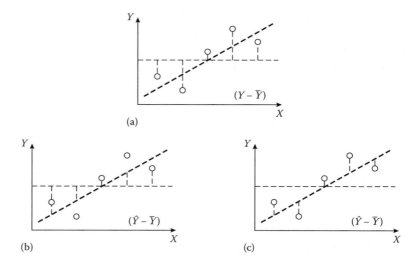

FIGURE 10.4   Separation of variation: (a) total variation, (b) explained variation, and (c) unexplained variation.

right side of Equation 10.2. If all the sample points fall on the line, the explained variation will equal the total variation. The variation in Figure 10.4c represents the variation of the points about the line; therefore, it corresponds to the second term on the right side of Equation 10.2. This is the variation that is not explained by the relationship between Y and X. If this unexplained variation equals the total variation, the line does not explain any of the variation in the sample, and the relationship between X and Y is not significant.

10.2.4 CORRELATION: FRACTION OF EXPLAINED VARIATION

The separation of variation concept is useful for quantifying the correlation coefficient by Pearson. Specifically, dividing both sides of Equation 10.1 by the total variation TV gives

(10.3)
$$1 = \frac{EV}{TV} + \frac{UV}{TV}$$

The ratio EV/TV represents the fraction of the total variation that is explained by the linear relationship between Y and X; this is called the *coefficient of determination* and is given by

(10.4)
$$R^2 = \frac{EV}{TV} = \frac{\sum_{i=1}^{n}(\hat{Y}_i - \bar{Y})^2}{\sum_{i=1}^{n}(Y_i - \bar{Y})^2}$$

The square root of the ratio is the correlation coefficient, R. If the explained variation equals the total variation, the correlation coefficient will equal 1. If the relationship between X and Y is inverse, as in Figure 10.1e, and the explained variation equals the total variation in magnitude, R will equal −1. These represent the extremes, but both values indicate a perfect association, with the sign only indicating the direction of the relationship. If the explained variation equals zero, R equals zero. Thus a correlation coefficient of zero, which is sometimes called the null correlation, indicates the lack of a linear association between the two variables X and Y.

10.2.5 COMPUTATIONAL FORM FOR CORRELATION COEFFICIENT

While Equation 10.4 provides the means to compute a value of the correlation coefficient, it can be shown that Equation 10.4 can be rearranged to the following form using a linear model for $\hat{Y}$ that minimizes the unexplained variation (UV):

$$(10.5) \quad R = \frac{\sum_{i=i}^{n} X_i Y_i - \frac{1}{n}\left(\sum_{i=1}^{n} X_i\right)\left(\sum_{i=1}^{n} Y_i\right)}{\sqrt{\sum_{i=1}^{n} X_i^2 - \frac{1}{n}\left(\sum_{i=1}^{n} X_i\right)^2}\sqrt{\sum_{i=1}^{n} Y_i^2 - \frac{1}{n}\left(\sum_{i=1}^{n} Y_i\right)^2}}$$

This equation is used most often because it does not require prior computation of the means, and thus the computational algorithm is easily programmed for a computer. The linear model $\hat{Y}$ that minimizes UV is based on the principle of least squares, as discussed in Section 10.4.

Example 10.1: Computational Illustration for Correlation Coefficient

A sample of five observations, which is given as the values of X and Y in Table 10.1, is used to illustrate the computational methods

TABLE 10.1   Calculation of the Correlation Coefficient

X	Y	X − X̄	Y − Ȳ	(X − X̄)(Y − Ȳ)	Ŷ	Ŷ − Ȳ	(Y − Ŷ)²	(Y − Ȳ)²	XY	X²	Y²
1	2	−2	−1	2	1.2	−1.8	3.24	1	2	1	4
2	1	−1	−2	2	2.1	−0.9	0.81	4	2	4	1
3	3	0	0	0	3.0	0.0	0.00	0	9	9	9
4	4	1	1	1	3.9	0.9	0.81	1	16	16	16
5	5	2	2	4	4.8	1.8	3.24	4	25	25	25
15	15			9			8.10	10	54	55	55

outlined herein. The last row of Table 10.1 contains column summations. Using Equation 10.4, the correlation is

(10.6)
$$R = \sqrt{\frac{8.10}{10}} = 0.9$$

The predicted values, $\hat{Y}$, were obtained by fitting a regression equation to the data points; the regression method is discussed in the remaining sections of this chapter. Therefore, Equation 10.5 yields

(10.7)
$$R = \frac{54 - \dfrac{15(15)}{5}}{\sqrt{55 - \dfrac{15(15)}{5}}\sqrt{55 - \dfrac{15(15)}{5}}} = 0.9$$

Thus 81% (based on $0.9^2 = 0.81$) of the total variation in Y is explained by variation in X.

10.3 INTRODUCTION TO REGRESSION

In many engineering projects, estimates of variables are required. For example, in the design of irrigation projects, it is necessary to provide estimates of evaporation. If we are fortunate to have a past record of daily evaporation rates from a pan, the mean of these observations multiplied by some proportionality constant (a pan coefficient) may be our best estimate. However, the error in such an estimate can be significant. Because the standard deviation of the observations is a measure of the variation in the past, it is our best estimate of the future variation and thus the error in a single measurement. The error in a mean value is also a function of the standard deviation.

If a variable can be related to other variables, it may be possible to reduce the error in a future estimate. For example, evaporation is a function of the temperature and humidity of the air mass that is over the water body. Thus, if measurements of the air temperature and the relative humidity are also available, a relationship, or model, can be developed. The relationship may provide a more accurate estimate of evaporation for the conditions that may exist in the future.

The process of deriving a relationship between a random variable and measured values of other variables is called *optimization*, or *model and calibration*. It makes use of the fundamental concepts of calculus and numerical analysis. The objective of optimization is to find the values of a vector of unknowns that provide the minimum or maximum value of some function. Before examining regression analysis, it is necessary to discuss the fundamentals of statistical optimization.

10.3.1 ELEMENTS OF STATISTICAL OPTIMIZATION

In statistical optimization, the function to be optimized is called the *objective function,* which is an explicit mathematical function that describes what is considered the optimal solution. There is a mathematical model that relates a random variable, called the *criterion* or *dependent variable,* to a vector of unknowns, called regression coefficients, and a vector of *predictor variables*, which are sometimes called *independent variables.* The predictor variables are usually variables that have a causal relationship with the criterion variable. For example, if we are interested in using measurements of air temperature and humidity to predict evaporation rates, the evaporation would be the criterion variable and the temperature and relative humidity would be the predictor variables. While the objective function of statistical optimization corresponds to the explicit function, the vector of unknowns in statistical optimization corresponds to the unknowns in the explicit function. The unknowns are the values that are necessary to transform values of the predictor variable(s) into a predicted value of the criterion variable. It is important to note that the objective function and the mathematical model are two separate explicit functions. The third element of statistical optimization is a data set. The *data set* consists of measured values of the criterion variable and the predictor variable(s). In summary, statistical optimization requires (1) an objective function, which defines what is meant by best fit; (2) a mathematical model, which is an explicit function relating a criterion variable to vectors of unknowns and predictor variable(s); and (3) a matrix of measured data.

As an example, we may attempt to relate evaporation (E), the criterion variable, to temperature (T), which is the predictor variable, using the equation

(10.8)
$$\hat{E} = b_0 + b_1 T$$

in which b_0 and b_1 = the unknown coefficients, and $\hat{E}$ = the predicted value of E. To evaluate the unknowns b_0 and b_1, a set of simultaneous measurements on E and T would be made. For example, if we were interested in daily evaporation rates, we may measure both the total evaporation for each day in a year and the corresponding mean daily temperature; this would give us 365 observations from which we could estimate the two regression coefficients. An objective function would have to be selected to evaluate the unknowns—for example, regression that minimizes the sum of the squares of the differences between the predicted and measured values of the criterion variable. Other criterion functions, however, may be used.

10.3.2 ZERO-INTERCEPT MODEL

Assume that the criterion variable Y is related to a predictor variable X using the linear model

(10.9)
$$\hat{Y} = bX$$

in which b = the unknown and $\hat{Y}$ = the predicted value of Y. The objective function, F, is defined as minimizing the sum of the squares of the differences (e_i) between the predicted values ($\hat{Y}_i$) and the measured values (Y_i) of Y as follows:

$$(10.10) \qquad F = \min \sum_{i=1}^{n} e_i^2 = \min \sum_{i=1}^{n} (\hat{Y}_i - Y_i)^2$$

where n = the sample size.

To derive a value for b, differentiate $\sum_{i=1}^{n} e_i^2$ with respect to the unknown, set the derivative equal to zero, and solve for the unknown. Substituting Equation 10.9 into Equation 10.10 yields

$$(10.11) \qquad F = \min \sum_{i=1}^{n} (\hat{Y}_i - Y_i)^2 = \min \sum_{i=1}^{n} (bX_i - Y_i)^2$$

The derivative of $\sum_{i=1}^{n} e_i^2$ with respect to the unknown b is

$$(10.12) \qquad \frac{\partial \sum_{i=1}^{n} e_i^2}{\partial b} = 2 \sum_{i=1}^{n} (bX_i - Y_i)(X_i)$$

Setting Equation 10.12 to 0 and solving for b yields

$$(10.13) \qquad b = \frac{\sum_{i=1}^{n} X_i Y_i}{\sum_{i=1}^{n} X_i^2}$$

The value of b can be computed using the two summations, $\sum_{i=1}^{n} X_i Y_i$ and $\sum_{i=1}^{n} X_i^2$.

Example 10.2: Fitting a Zero-Intercept Model

The following table gives the data that can be used for fitting the value of b:

i	X_i	Y_i
1	3	2
2	5	3
3	6	4

The following table shows the necessary computations:

i	X_i	X_i^2	Y_i	X_iY_i
1	3	9	2	6
2	5	25	3	15
3	6	36	4	24
Column summation		70		45

Therefore, the value of b equals 45/70 or 0.64286. The following equation could then be used to derive a future estimate of Y for a future value of X:

$$(10.14) \qquad \hat{Y} = 0.64286X$$

It is important to note that Equation 10.13 is valid only for the linear model of Equation 10.9 that does not include an intercept; Equation 10.13 would not be valid for Equation 10.8, since Equation 10.8 includes an intercept coefficient (b_0).

10.3.3 REGRESSION DEFINITIONS

The objective of regression is to evaluate the coefficients of an equation relating the criterion variable to one or more other variables, which are called the *predictor variables*. The predictor variables are variables in which their variation is believed to cause or agree with variation in the criterion variable. A predictor variable is often called an *independent variable*; this is a misnomer in that independent variables are usually neither independent of the criterion variable nor independent of other predictor variables.

The most frequently used linear model relates a criterion variable Y to a single predictor variable X by the equation

$$(10.15) \qquad \hat{Y} = b_0 + b_1X$$

in which b_0 = the intercept coefficient and b_1 = the slope coefficient; b_0 and b_1 are called *regression coefficients* because they are obtained from a regression analysis. Because Equation 10.15 involves two variables, Y and X, it is sometimes referred to as the *bivariate model*. The intercept coefficient represents the value of Y when X equals zero. The slope coefficient represents the rate of change in Y with respect to change in X. Whereas b_0 has the same dimensions as Y, the dimensions of b_1 equal the ratio of the dimensions of Y to X.

The linear multivariate model relates a criterion variable to two or more predictor variables:

$$(10.16) \qquad \hat{Y} = b_0 + b_1X_1 + b_2X_2 + \ldots + b_pX_p$$

in which p = the number of predictor variables, X_i = the ith predictor variable, b_i = the ith slope coefficient, and b_0 is the intercept coefficient, where i = 1, 2, ..., p. The coefficients b_i are often called *partial regression coefficients* and have dimensions equal to the ratio of the dimensions of Y to X_i. Equation 10.15 is a special case of Equation 10.16.

10.4 PRINCIPLE OF LEAST SQUARES

10.4.1 DEFINITIONS

The values of the slope and intercept coefficients of Equations 10.15 and 10.16 can be computed using the principle of least squares, which is a process of obtaining *best* estimates of the coefficients; it is referred to as a regression method. Regression is the tendency for the expected value of one of two jointly correlated random variables to approach more closely the mean value of its set than any other. The principle of least squares is used to regress Y on either X or the X_i values of Equations 10.15 and 10.16, respectively. To express the principle of least squares, it is important to define the error, e, or residual, as the difference between the predicted and measured values of the criterion variable:

$$(10.17) \qquad e_i = \hat{Y}_i - Y_i$$

in which $\hat{Y}_i$ = the ith predicted value of the criterion variable, Y_i = the ith measured value of Y, and e_i = the ith error. It is important to note that the error is defined as the measured value of Y subtracted from the predicted value. Some computer programs use the measured value minus the predicted value. This definition is avoided because it indicates that a positive residual implies underprediction. With Equation 10.17 a positive residual indicates overprediction, while a negative residual indicates underprediction. The objective function for the principle of least squares is to minimize the sum of the squares of the errors:

$$(10.18) \qquad F = \min \sum_{i=1}^{n} (\hat{Y}_i - Y_i)^2$$

in which n = the number of observations on the criterion variable—that is, the sample size.

10.4.2 SOLUTION PROCEDURE

After inserting the model of $\hat{Y}$, the objective function of Equation 10.18 can be minimized by taking the derivatives with respect to each unknown, setting the derivatives equal to zero, and then solving for the unknowns. The solution requires the model for predicting Y_i to be substituted into the objective function. It is important to note that the derivatives are taken with respect to the unknowns b_i and not the predictor variables X_i.

To illustrate the solution procedure, the model of Equation 10.15 is substituted into the objective function of Equation 10.18, which yields

$$(10.19) \qquad F = \min \sum_{i=1}^{n} (b_0 + b_1 X_i - Y_i)^2$$

The derivatives of the sum of squares of the errors with respect to the unknowns b_0 and b_1 are, respectively,

$$(10.20) \qquad \frac{\partial \sum_{i=1}^{n} (\hat{y}_i - y_i)^2}{\partial b_0} = 2 \sum_{i=1}^{n} (b_0 + b_1 X_i - Y_i) = 0$$

$$(10.21) \qquad \frac{\partial \sum_{i=1}^{n} (\hat{y}_i - y_i)^2}{\partial b_1} = 2 \sum_{i=1}^{n} (b_0 + b_1 X_i - Y_i) X_i = 0$$

Dividing each equation by 2, separating the terms in the summations, and rearranging yield the set of normal equations:

$$(10.22) \qquad n b_0 + b_1 \sum_{i=1}^{n} X_i = \sum_{i=1}^{n} Y_i$$

$$(10.23) \qquad b_0 \sum_{i=1}^{n} X_i + b_1 \sum_{i=1}^{n} X_i^2 = \sum_{i=1}^{n} X_i Y_i$$

All the summations in Equations 10.22 and 10.23 are calculated over all values of the sample. Sometimes the index values of the summation are omitted when they refer to summation over all elements in a sample of size n. Also, the subscripts for X and Y can be omitted but are inferred. The two unknowns, b_0 and b_1, can be evaluated by solving the two simultaneous equations as follows:

$$(10.24) \qquad b_1 = \frac{\sum_{i=1}^{n} X_i Y_i - \frac{1}{n} \sum_{i=1}^{n} X_i \sum_{i=1}^{n} Y_i}{\sum_{i=1}^{n} X_i^2 - \frac{1}{n} \left(\sum_{i=1}^{n} X_i \right)^2}$$

$$(10.25) \qquad b_0 = \overline{Y} - b_1 \overline{X} = \frac{\sum_{i=1}^{n} Y_i}{n} - \frac{b_1 \sum_{i=1}^{n} X_i}{n}$$

TABLE 10.2   Least-Squares Computations

i	X_i	Y_i	X_i^2	Y_i^2	X_iY_i	e_i	e_i^2
1	12	4	144	16	48	−0.2	0.04
2	11	3	121	9	33	−0.1	0.01
3	10	2	100	4	20	0.0	0.00
4	9	0	81	0	0	1.1	1.21
5	8	1	64	1	8	−0.8	0.64
Column summation	50	10	510	30	109	0.0	1.90

Example 10.3: Least-Squares Analysis

The data of Table 10.2 can be used to illustrate the solution of the least-squares principle for a data set of five observations on two variables. Substituting the sums of Table 10.2 into the normal equations of Equations 10.22 and 10.23 gives

(10.26) $$5b_0 + 50b_1 = 10$$

(10.27) $$50b_0 + 510b_1 = 109$$

Solving Equations 10.26 and 10.27 for the unknowns (b_0 and b_1) yields the model

(10.28) $$\hat{Y} = -7 + 0.9X$$

The errors, which are shown in Table 10.2, can be computed by substituting the measured values of X into Equation 10.28. In solving a linear model using the least-squares principle, the sum of the errors always equals zero, which is shown in Table 10.2. The sum of the squares of the errors equals 1.9, which is a minimum.

Note that if a line is passed through the first three observations of Table 10.2, the error for each of these three observations is zero. Such a model would be

(10.29) $$\hat{Y} = -8 + 1.0X$$

While three of the residuals are zero, the sum of the squares of the errors for Equation 10.29 equals 2.0, which is greater than the corresponding sum for Equation 10.28. Thus the least-squares solution always provides the straight line with the smallest sum of squares of the errors.

10.5 RELIABILITY OF THE REGRESSION EQUATION

Having evaluated the coefficients of the regression equation, it is of interest to evaluate the reliability of the regression equation. The following selected criteria can be assessed in evaluating the model: (1) the correlation coefficient, (2) the standard error of estimate, and (3) the rationality of the coefficients and the relative importance of the predictor variables, both of which can be assessed using the standardized partial regression coefficients.

10.5.1 CORRELATION COEFFICIENT

As suggested in Section 10.2, the correlation coefficient (R) is an index of the degree of linear association between two random variables. The magnitude of R indicates whether the regression provides accurate predictions of the criterion variable. Thus R is often computed before the regression analysis is performed in order to determine whether it is worth the effort to perform the regression. However, R is always computed after the regression analysis because it is an index of the goodness of fit. The correlation coefficient measures the degree to which the measured and predicted values agree and is used as a measure of the accuracy of future predictions. It must be recognized that if the measured data are not representative of the population—that is, data that will be observed in the future—the correlation coefficient cannot be indicative of the accuracy of future predictions.

The square of the correlation coefficient (R^2) equals the percentage of the variance in the criterion variable that is explained by the predictor variable. Because of this physical interpretation, R^2 is a meaningful indicator of the accuracy of predictions.

10.5.2 STANDARD ERROR OF ESTIMATE

In the absence of additional information, the mean is the best estimate of the criterion variable; the standard deviation S_y of Y is an indication of the accuracy of prediction. If Y is related to one or more predictor variables, the error of prediction is reduced from S_y to the standard error of estimate, S_e. Mathematically, the standard error of estimate equals the standard deviation of the errors, has the same units as Y, and is given by

$$(10.30) \qquad S_e = \sqrt{\frac{1}{\nu} \sum_{i=1}^{n} (\hat{Y}_i - Y_i)^2}$$

in which ν = the degrees of freedom, which equals the sample size minus the number of unknowns estimated by the regression procedure. For the bivariate model of Equation 10.15, $p = 1$ and $\nu = n - 2$. For the general linear model with an intercept, Equation 10.16, there are ($p + 1$) unknowns; thus $\nu = n - p - 1$. In terms of the separation of variation concept discussed previously, the standard error of estimate equals the square root of the ratio of the unexplained variation to the degrees of freedom. It is important to note that S_e is based on ($n - p - 1$) degrees of freedom, while the error S_Y is based on $(n - 1)$ degrees of freedom. Thus in some cases S_e may be greater than S_Y. To assess the reliability of the regression equation, S_e should be compared with the bounds of zero and S_Y. If S_e is near S_Y, the regression has not been successful. If S_e is much smaller than S_Y and is near zero, the regression analysis has improved the reliability of prediction.

The standard error of estimate is sometimes computed using the following relationship:

$$(10.31) \qquad\qquad S_e = S_Y \sqrt{1 - R^2}$$

Equation 10.31 must be considered as only an approximation to Equation 10.30 because R is based on n degrees of freedom and S_Y is based on $(n - 1)$ degrees of freedom.

Using the separation-of-variation concept, the exact relationship between S_e, S_Y, and R can be computed as follows:

$$(10.32) \qquad\qquad \text{TV} = \text{EV} + \text{UV}$$

The total variation is related to the variance of Y by

$$(10.33) \qquad\qquad S_y^2 = \frac{\text{TV}}{n - 1}$$

The square of the correlation coefficient is the ratio of the explained variation (EV) to the total variation (TV):

$$(10.34) \qquad\qquad R^2 = \frac{\text{EV}}{\text{TV}}$$

The standard error of estimate is related to the unexplained variation (UV) by

$$(10.35) \qquad\qquad S_e^2 = \frac{\text{UV}}{n - p - 1}$$

Equation 10.33 can be solved for TV, which can then be substituted into both Equations 10.32 and 10.34. Equation 10.34 can then be solved for EV, which can be substituted into Equation 10.32. Equation 10.35 can be solved for UV, which is also substituted into Equation 10.32. Solving for S_e^2 from Equation 10.32 yields

$$(10.36) \qquad\qquad S_e^2 = \frac{n - 1}{n - p - 1} S_y^2 (1 - R^2)$$

Thus Equation 10.36 is a more exact relationship than Equation 10.31. However, for large sample sizes the difference between the two estimates is small.

Although S_e may actually be greater than S_Y, in general, S_e is within the range from zero to S_Y. When a regression equation fits the data points exactly, S_e equals zero; this corresponds to a correlation coefficient of one. When the correlation coefficient equals zero, S_e equals S_Y. As was indicated previously, S_e may actually exceed S_Y when the degrees of freedom have a significant effect. The standard error of estimate is often preferred to the correlation coefficient because S_e has the same units as the criterion variable and its magnitude is a physical indicator of the error; the correlation coefficient is only a standardized index and does not properly account for degrees of freedom lost because of the regression coefficients.

10.5.3 STANDARDIZED PARTIAL REGRESSION COEFFICIENTS

Because the partial regression coefficient, b_1 of Equation 10.15, is dependent on the units of both Y and X, it is often difficult to use its magnitude to measure the rationality of the model. The partial regression coefficient can be standardized by

(10.37)
$$t = \frac{b_1 S_X}{S_Y}$$

in which t is called the standardized partial regression coefficient, and S_X and S_Y are the standard deviations of the predictor and criterion variables, respectively. Because the units of b_1 are equal to the ratio of the units of Y to units of X, t is dimensionless. For the bivariate regression model, t equals the correlation coefficient of the regression of the standardized value of Y on the standardized value of X. This suggests that t is a measure of the relative importance of the corresponding predictor variable. The value of t has the same sign as b_1. Therefore, if the sign of t is irrational, we can conclude that the model is not rational. Although it is more difficult to assess the rationality of the magnitude of a regression coefficient than it is to assess the rationality of the sign, the magnitude of the standardized partial regression coefficient can be used to assess the rationality of the model. For rational models, t must vary in the range $-1 \le t \le 1$. Because t equals R for the bivariate model of Equation 10.15, values of t outside this range should not occur; this is not true for multivariate models, such as Equation 10.16.

Example 10.4: Errors in Regression Models

The data of Table 10.2 were previously used to compute the standard error of estimate and the correlation coefficient; they can also be used to illustrate the computation of the standardized partial regression coefficient, which is computed using Equation 10.37 as

$$(10.38) \qquad t = 0.9 \sqrt{\dfrac{510 - \dfrac{(50)^2}{5}}{30 - \dfrac{(10)^2}{5}}} = 0.9$$

Using the equations given previously for computing the correlation coefficient, it is easy to show that t equals R.

10.5.4 ASSUMPTIONS UNDERLYING THE REGRESSION MODEL

The principle of least squares assumes that the errors—that is, the differences between the predicted and measured values of the criterion variable—(1) are independent of each other, (2) have zero mean, (3) have a constant variance across all values of the predictor variables, and (4) are normally distributed. If any of these assumptions is violated, we must assume that the model structure is not correct. Violations of these assumptions are easily identified using statistical analyses of the residuals.

When the sum of the residuals does not equal zero, it reflects a bias in the model. The regression approach, when applied analytically, requires the sum of the residuals for a linear model to equal 0. For a nonlinear model, the sum of the errors may not equal 0, which suggests bias. When an inadequate number of significant digits is computed for the regression coefficients, the sum of the residuals may not equal 0, even for a linear model. However, a model may be biased even when the sum of the residuals equals zero. For example, Figure 10.5 shows an X–Y plot in which the trend of the data is noticeably nonlinear, with the linear regression line also shown. For cases like Figure 10.5, the incorrect model structure yields a model with local biases. If the errors e_i are computed and plotted versus the corresponding values of the predictor variable, a noticeable trend appears (Figure 10.6). The errors are positive for both low and high values of X, while the errors show a negative bias of prediction for intermediate values of X. While the sum of the residuals equals 0, the trends

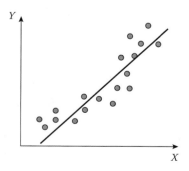

FIGURE 10.5 Biased linear regression model.

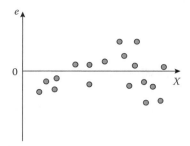

FIGURE 10.6   Residual plot for a biased linear regression model.

in the residuals suggest a biased model that should be replaced by a model that has a different structure. Figure 10.7 shows data with a trend in the error variance; specifically, as X increases, the error variance increases.

The residuals (i.e., $\hat{Y}_i - Y_i$) are an important criterion in assessing the validity of a bivariate regression equation. Plots of the residuals may indicate an incorrect model structure, such as a nonlinear relationship between the criterion variable and one of the predictor variables. Additionally, the residuals may suggest a need to consider another predictor variable that was not included in the model, requiring the use of multiple regression (see Section 10.8). A plot of the residuals should be made to assess the assumptions of zero mean, constant error variation, independence, and normality. A plot of the residuals versus the predicted value of the criterion variable (i.e., e_i versus $\hat{Y}_i$) may identify either a violation of the assumption of constant variance or a lack of independence of the observations.

The fourth assumption deals with the independence of the observations. This is not usually a problem except when time or space is the predictor variable. In such cases, the measurement of Y for one time period may not be completely independent of the measurement for an adjacent time period. For example, if X is the distance measured from the top of a slope and Y is the soil moisture measured at the distance X, then the residuals may not be independent, which would violate one of the basic

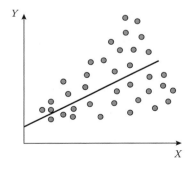

FIGURE 10.7   Model with a nonconstant error variance.

assumptions. In such a case, the real degrees of freedom may be much less than the actual sample size.

10.6 CORRELATION VERSUS REGRESSION

Before discussing other topics in regression, it is important to emphasize the differences between correlation and regression. Regression is a means of calibrating the unknown coefficients of a prediction equation, whereas correlation provides a measure of goodness of fit. Thus regression is a method for model calibration, while correlation would have usefulness in model formulation and model assessment. When using regression, it is necessary to specify which variable is the criterion and which is the predictor; when using correlation, it is not necessary to make such a distinction. The distinction is necessary with regression because a regression equation is not transformable, unless the correlation coefficient equals 1.0. That is, if Y is the criterion variable when the equation is calibrated, the equation cannot be rearranged algebraically to get an equation for predicting X. Specifically, Equation 10.16 can be rearranged to predict X as

$$(10.39) \qquad Y = b_0 + b_1 X$$

$$(10.40) \qquad Y - b_0 = b_1 X$$

Therefore,

$$(10.41) \qquad X = \frac{Y - b_0}{b_1}$$

or

$$(10.42) \qquad X = -\frac{b_0}{b_1} + \frac{1}{b_1} Y$$

$$(10.43) \qquad X = a_0 + a_1 Y$$

If X is regressed on Y, the resulting coefficients will not be the same as the coefficients obtained by regressing Y on X (Equation 10.39) and then setting $a_0 = -b_0/b_1$ and $a_1 = 1/b_1$. The correlation for the regression of Y on X is the same as the correlation for the regression of X on Y.

Computationally, the correlation coefficient is a function of the explained and total variation; the slope coefficient of a regression equation is related to the correlation coefficient by

$$(10.44) \qquad b_1 = \frac{RS_Y}{S_X}$$

in which S_X and S_Y are the standard deviations of X and Y, respectively.

After a regression equation is calibrated, it is very important to examine the rationality of the regression coefficients; specifically, is the predicted value of Y rational for all reasonable values of X? Since the slope coefficient of the regression equation represents the rate of change of Y with respect to X, is the effect of X on Y rational? A regression equation that is not considered rational should be used with caution, if at all. In addition to checking for rationality, the goodness-of-fit statistics (R and S_e) should be computed. If the expected accuracy is not acceptable, we may elect to collect more data or to develop a model that uses other predictor variables.

10.7 APPLICATIONS OF BIVARIATE REGRESSION ANALYSIS

10.7.1 ESTIMATING TRIP RATE

A traffic planner needs estimates of trip rates for residential areas. The planner conducts studies in 10 residential areas of different densities of development, with the following measured rates:

X	3	4	8	8	13	15	18	22	24	27
Y	4.5	3.3	3.5	2.3	3.8	2.6	2.7	1.6	1.9	1.7

where X = residential density (households per acre) and Y = trip rate (daily trips per household). The regression of trip rate on density yields the following linear equation:

$$(10.45) \qquad Y = 4.100 - 0.09226X$$

The negative slope coefficient indicates that the trip rate decreases by 0.092 daily trips per household for each unit increase in the number of households per acre. The standard error of estimate is 0.603, which is an improvement over the standard deviation of 0.965 ($S_e/S_y = 0.625$). The correlation coefficient is -0.808, which yields a normalized explained variance (R^2) of 0.653. Figure 10.8 shows the measured data and the regression line; the moderate variation of the measured points about the line confirms the moderate correlation between residential density and trip rate.

10.7.2 BREAKWATER COST

A construction engineering firm has recently had a number of contracts for the construction of breakwaters in coastal environments. To reduce the effort in developing cost estimates for future clients, the company compiled a data record of 14 recent breakwater construction projects. The data consist of the length of the breakwater and the cost for rock,

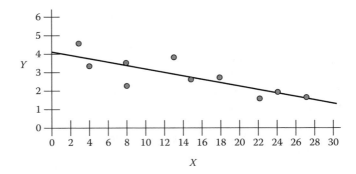

FIGURE 10.8   Bivariate regression of trip rate (Y, daily trips per household) on residential density (X, households per acre).

rock spurs, groins, marsh plants, and beach fill (see Table 10.3). The cost ($Y \times 10^3$) was regressed on the length of the breakwater (X):

(10.46) $\hat{Y} = 253.4 + 0.08295X$

The standard error of estimate is 149.1×10^3, which is slightly greater than the standard deviation of the measured costs ($S_y = \$145.5 \times 10^3$); that is, $S_e/S_y = 1.024$. Since S_e is greater than S_y, it would be more accurate to use the mean cost ($\$335.7 \times 10^3$) than the value estimated with the regression equation. If we use Equation 10.5 to compute the correlation coefficient, we get a value of 0.176; however, Equation 10.36 shows that we have essentially no reliability in the estimated values. Thus, for practical purposes, we have a correlation of zero. The adjusted correlation

TABLE 10.3   Bivariate Regression Analysis of Breakwater Cost

X	Y	$\hat{Y}$	e
450	230	291	61
625	490	305	−185
670	150	309	159
730	540	314	−226
810	280	321	41
880	120	326	206
1020	380	338	−42
1050	170	340	170
1100	530	345	−185
1175	290	351	61
1230	460	355	−105
1300	230	361	131
1350	470	365	−105
1510	360	379	19

computed with Equation 10.36 is a more realistic measure of reliability than is the value of R computed with the unadjusted, but more commonly used, equation (Equation 10.5), because Equation 10.36 accounts for differences in the degrees of freedom.

Why does the computed equation give such poor results? Certainly, breakwater length is a determinant of cost. However, it appears that other factors are more important and necessary to get accurate estimates of the cost.

10.7.3 STRESS–STRAIN ANALYSIS

Table 10.4 includes nine measurements of the observed stress and the resulting axial strain for an undisturbed sample of a Boston blue clay. A linear equation was computed by regressing the strain (Y) on the stress (X):

$$\hat{Y} = -8.29 + 0.6257X \tag{10.47}$$

The equation has a standard error of estimate of 1.04%, which is much smaller than the standard deviation of the measured strains; that is, $S_e/S_y = 0.35$. The correlation coefficient of 0.945 also suggests that the linear equation provides a reasonable fit to the measured data. An R^2 of 0.89 indicates that almost 90% of the total variation is explained by the linear regression; the standard error ratio of 0.35 suggests that the fit is not as good as suggested by the R^2. The predicted strains and residuals are given in Table 10.4 and shown in Figure 10.9. While most of the errors are small, there are local biases in the residuals, with negative residuals for small and large stresses. While the mean bias ($\bar{e}$) is zero, the local biases indicate that a nonlinear structure may be more appropriate. Another problem with the linear model is the negative estimates of strain that occur for stresses of less than 13.244 psi. The negative strain is physically not meaningful, so the model is not rational for small strains.

TABLE 10.4 Bivariate Regression Analysis of Stress–Strain Data

X	Y	$\hat{Y}$	e
10.8	0.1	−1.5	−1.6
14.1	0.4	0.5	0.1
16.3	1.1	1.9	0.8
17.5	1.4	2.7	1.3
18.6	2.7	3.4	0.7
19.8	3.6	4.1	0.5
22.1	5.7	5.5	−0.2
23.8	6.8	6.6	−0.2
24.3	8.3	6.9	−1.4

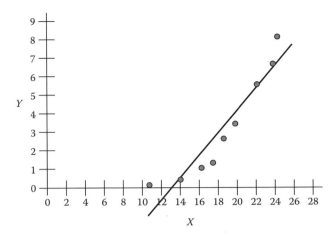

FIGURE 10.9   Axial strain (Y) of a Boston blue clay due to stress (X).

10.7.4 PROJECT COST VERSUS TIME

For large projects that require years to complete, the construction engineer needs to schedule project finances just as it is necessary to schedule construction activities. While costs vary with time, a construction engineer believes that a linear model can be used to represent the relationship between the value of work completed (Y) and the time since the start of the project (X). Since no work would be completed at the start of the project, a zero-intercept model would seem appropriate. The engineer compiled semiannual cost for a recent 6-year project; the data are given in Table 10.5. The linear, zero-intercept, bivariate model was fit to the data as follows:

$$(10.48) \qquad \hat{Y} = 0.50421X$$

where X = time in months from start of project, and $\hat{Y}$ = predicted value of work in $\$ \times 10^6$. The model is shown in Figure 10.10 and the residuals are given in Table 10.5. The model underpredicts for the middle part of the period, while it overpredicts for the early and latter parts of the time period. The standard error of estimate is $\$7.73 \times 10^6$, which gives $S_e/S_y =$ 0.757; this indicates that the model does not provide highly accurate estimates of Y. Additionally, the sum of the errors is not zero, so the model is biased.

Whereas a linear model with an intercept will provide an unbiased model, a zero-intercept model can be biased, as it is for this example with a mean error of $-\$0.52 \times 10^6$. This means that, on the average, Equation 10.48 tends to underpredict by $520,000. A correlation coefficient should not be computed for a biased model. The correlation coefficient assumes a linear model with an intercept, so it would be incorrect to use either Equation 10.5 or 10.36 to compute a correlation coefficient for Equation 10.48.

Because of the poor accuracy of the model of Equation 10.48, a linear model with an intercept was fit to the data of Table 10.5:

$$(10.49) \qquad Y = 2.36 + 0.4570X$$

TABLE 10.5   **Bivariate Regression Analyses of Project Cost (Y) versus Time (X)**

X	Y	Equation 10.48		Equation 10.49	
		$\hat{Y}$	e	$\hat{Y}$	e
6	1.5	3.03	1.53	5.10	3.60
12	4.3	6.05	1.75	7.84	3.54
18	11.2	9.08	−2.12	10.59	−0.61
24	15.5	12.10	−3.40	13.33	−2.17
30	19.1	15.13	−3.97	16.07	−3.03
36	21.8	18.15	−3.65	18.81	−2.99
42	24.0	21.18	−2.82	21.55	−2.45
48	26.2	24.20	−2.00	24.30	−1.90
54	26.9	27.23	0.33	27.04	0.14
60	27.8	30.25	2.45	29.78	1.98
66	30.1	33.28	3.18	32.52	2.42
72	33.8	36.30	2.50	35.26	1.46

Note: X = time (months) from start of project, Y = value of work completed ($\$ \times 10^6$), $\hat{Y}$ = predicted value, and e = error in predicted value = $\hat{Y} - Y$.

This model is also shown in Figure 10.10 and the residuals are given in Table 10.5. Equation 10.49 is unbiased and has a standard error of $\$2651 \times 10^6$, which gives $S_e/S_y = 0.26$. The model explains (R^2) 94% of the variation in Y and has a correlation coefficient of 0.969. In spite of the reasonably accurate estimates suggested by the goodness-of-fit statistics, the model still has local biases in the residuals (see Table 10.5), with under-predictions for the range $18 \leq X \leq 48$ months. However, even with the local biases, the model of Equation 10.49 may be sufficiently accurate for

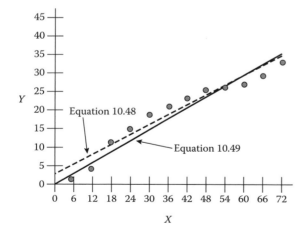

FIGURE 10.10   Regression of project cost (Y, 10^6) on time from start of project (X, months) for zero-intercept (Equation 10.48) and linear (Equation 10.49) models.

making project planning estimates. The nonzero intercept of Equation 10.48 suggests a cost of 2.36×10^6 at the time the project is initiated. While a value of $0 is expected, the nonzero intercept may suggest that the model may not be realistic for the initial part of a project.

In comparing the two models, the standard error ratio would suggest that the zero-intercept model is considerably less accurate than the nonzero-intercept model. However, in comparing the residuals of the two models, there appears to be very little difference. The averages of the absolute values of the residuals are 2.5 and 2.2 for Equations 10.48 and 10.49, respectively. The nonzero-intercept model is slightly better, but greater accuracy may be achieved with another model structure.

10.7.5 EFFECT OF EXTREME EVENT

Small samples very often contain an extreme event, which can greatly influence the resulting equation. Table 10.6 contains eight measurements of the concentration of ^{241}Am in soil samples and the distance of the soil sample from the nuclear facility. The eight measurements were used to regress the ^{241}Am concentration (Y) on the distance (X):

$$(10.50) \qquad \hat{Y} = 0.07366 - 0.0000123X$$

The negative slope coefficient indicates that the concentration decreases with distance. The standard error of estimate is 0.00694, which gives $S_e/S_y = 0.764$. This suggests poor prediction ability. The correlation coefficient of -0.706 means that 50% of the total variation in Y is explained by the regression equation. The predicted values of Y and the errors for Equation 10.50 are shown in Table 10.6. Except for the value for a distance of 640, the predicted values range from 0.046 to 0.050 picocuries per gram (pCi/g); however, the measured values for these distances ranged from

TABLE 10.6   Effect of Extreme Event on Regression

		Equation 10.50		Equation 10.51	
X	**Y**	**$\hat{Y}$**	**e**	**$\hat{Y}$**	**e**
640	0.063	0.066	0.003	–	–
1930	0.061	0.050	−0.011	0.058	−0.003
1980	0.057	0.049	−0.008	0.055	−0.002
2010	0.049	0.049	0.000	0.053	0.004
2080	0.043	0.048	0.005	0.049	0.006
2120	0.048	0.048	0.000	0.046	−0.002
2200	0.039	0.047	0.008	0.041	0.002
2250	0.042	0.046	0.004	0.038	−0.004

Note: X = distance from nuclear power facility, Y = concentration of ^{241}Am in soil samples, $\hat{Y}$ = predicted value of Y, and e = residual = $\hat{Y} - Y$.

0.039 to 0.061 pCi/g. These predicted values are not much different from the mean of 0.050 pCi/g. The batch of seven points acts as a single point.

A second model was developed using seven of the eight values. By eliminating the extreme event, the following model was fitted:

(10.51) $$\hat{Y} = 0.1755 - 0.0000611X$$

This model has a standard error of 0.00412 pCi/g ($S_e/S_y = 0.51$). The correlation coefficient in this case is −0.885, which means that 78.3% of the total variation in the seven measurements was explained by the regression equation of Equation 10.51. The residuals for Equation 10.51 are small, and the predicted values range from 0.038 to 0.058 pCi/g, which is much greater than the range for the model of Equation 10.50. But for a distance of 640, which is the measured value not included in fitting Equation 10.51, the value predicted with Equation 10.51 is 0.136 pCi/g; this is considerably higher than the measured value of 0.063 pCi/g.

In summary, in spite of the goodness-of-fit statistics for both models, neither model is an accurate representation of the system. This is a result of both the small sample size ($n = 8$) and the distribution of the data. It is undesirable to have a cluster of points and then one extreme event. Such a distribution of values makes it difficult to attain a model that is representative over a wide range of values of the predictor variable.

10.7.6 VARIABLE TRANSFORMATION

Studies have shown that fuel consumption (Y, gallons per mile) is related to the vehicle velocity (X, miles per hour) by

(10.52) $$\hat{Y} = b_0 + \frac{b_1}{X}$$

By creating a new variable, $W = 1/X$, a relationship between Y and X can be fitted. The data of Table 10.7 can be used to illustrate the fitting of a model that requires a transformation of the predictor variable.

A linear regression analysis of Y on W produces the following equation:

(10.53) $$\hat{Y} = 0.03939 + \frac{0.9527}{X}$$

The standard error of estimate for Equation 10.53 is 0.0141 gal/mi, which is 17.5% of the standard deviation of Y (i.e., $S_e/S_y = 0.175$). Accurate estimates are also suggested by the correlation coefficient of 0.986 ($R_2 = 0.972$). The residuals (see Table 10.7) are small and a trend is not apparent.

In this case, only the predictor variable was transformed, which does not create problems. Problems can arise when it is necessary to transform the criterion variable in order to put the relationship into a linear form.

TABLE 10.7   Bivariate Regression of Fuel Consumption (Y) on Speed (X)

X	W	Y	$\hat{Y}$	e
3.4	0.294	0.32	0.319	−0.001
8.1	0.123	0.14	0.157	0.017
16.4	0.0610	0.13	0.0975	−0.0325
23.7	0.0422	0.094	0.0796	−0.0144
31.3	0.0319	0.062	0.0698	0.0078
38.8	0.0258	0.066	0.0640	−0.0020
42.3	0.0236	0.051	0.0619	0.0109
49.0	0.0204	0.056	0.0588	0.0028
54.6	0.0183	0.055	0.0568	0.0018
58.2	0.0172	0.049	0.0558	0.0068
66.1	0.0151	0.051	0.0538	0.0028

Note: X = fuel consumption (gallons per mile), $W = 1/X$, and Y = mean speed (miles per hour).

10.8 MULTIPLE REGRESSION ANALYSIS

What options are available to improve the accuracy of predictions if the accuracy from a bivariate regression is still not sufficient for the design problem? Multivariate systems are a possible solution to this problem. Where one predictor variable may not be adequate, several predictor variables may provide sufficient prediction accuracy. Thus one reason for using multivariate models is to reduce the standard error of estimate.

Multivariate data analyses are also of value when a goal of the analysis is to examine the relative importance of the predictor variables. If a predictor variable is not included in the model, a regression analysis cannot determine the sensitivity of the criterion variable to variation in that predictor variable. Also, if a predictor variable that has not been included in the model is correlated with another predictor variable that is included in the model, the regression coefficient for the former variable may reflect the effects of both the former and latter predictor variables. Thus predictor variables that have only slight correlation with the criterion variable are often included in the model so that a complete analysis may be made.

How does a multivariate analysis compare with a bivariate analysis? Actually, the two are very similar. First, the same least-squares objective function is used to calibrate the regression coefficients. Second, bivariate correlations are still computed. Third, the data should be properly screened using graphical analyses prior to selecting a model structure. The major difference between multivariate and bivariate analyses is the necessity to account for the interdependence (i.e., correlation) of the predictor variables.

10.8.1 CORRELATION MATRIX

The first step in a multivariate analysis is to perform a graphical analysis. This includes developing plots of the criterion variable versus each of the predictor variables in the data set. Also, plots of each pair of predictor variables should be made; such plots will help to identify the characteristics of the interdependence between predictor variables. It will become evident that the interdependence is extremely important in the analysis of multivariate systems. All the plots should be examined for the characteristics identified previously for bivariate analyses—for example, extreme events, nonlinearity, and random scatter.

After a graphical analysis, the bivariate correlation coefficients should be computed for each pair of variables; this includes the correlation between the criterion variable and each predictor variable, as well as the correlation between each pair of predictor variables. The correlations are best presented in matrix form. The correlation matrix is a means of presenting in an organized manner the correlations between pairs of variables in a data set; it appears as

(10.54)

$$
\begin{array}{c|cccccc}
 & X_1 & X_2 & X_3 & \cdots & X_p & Y \\
\hline
X_1 & 1.0 & r_{12} & r_{13} & \cdots & r_{1p} & r_{1Y} \\
X_2 & & 1.0 & r_{23} & \cdots & r_{2p} & r_{2Y} \\
X_3 & & & 1.0 & \cdots & r_{3p} & r_{3Y} \\
\vdots & & & & \cdots & \vdots & \vdots \\
X_p & & & & & 1.0 & r_{pY} \\
Y & & & & & & 1.0
\end{array}
$$

Note that the correlation matrix, which includes p predictor variables $(X_i, i = 1, 2, ..., p)$ and the criterion variable (Y), is shown in upper-triangular form because the matrix is symmetric (i.e., $r_{ij} = r_{ji}$); also, the elements on the principal diagonal equal 1.0 because the correlation between a variable with itself is unity. The matrix is $(p + 1) \times (p + 1)$.

Example 10.5: Sediment Yield Data

The sediment database includes four predictors and a criterion variable, the sediment yield. The ultimate objective would be to develop a model for estimating the sediment yield from small watersheds. The predictor variables are (1) the precipitation/temperature ratio (X_1), which reflects the vegetation potential of the area; (2) the average watershed slope (X_2), which reflects the erosion potential due to the momentum of surface runoff; (3) the soil particle size index (X_3), which reflects the coarse-particle composition of the soil; and (4) the soil aggregation index (X_4), which reflects the dispersion characteristics of the small particles. Since vegetation retards

---

TABLE 10.8   Correlation Matrix of the Sediment Database

	X_1	X_2	X_3	X_4	Y
X_1: Precipitation/temperature ratio	1.000	0.340	−0.167	−0.445	−0.297
X_2: Average slope		1.000	−0.051	−0.185	0.443
X_3: Soil particle size			1.000	0.069	−0.253
X_4: Soil aggregation index				1.000	0.570
Y: Sediment yield					1.000

---

erosion, we would expect r_{1y} to be negative. Similarly, coarse particles are more difficult to transport, so r_{3Y} should also be negative. As the slope increases, the water will move with a greater velocity and more soil can be transported, so r_{2Y} should be positive. Erosion should increase as the dispersion characteristics of small particles increase, so r_{4y} should be positive.

The correlation matrix for 37 measurements of these five variables is given in Table 10.8. In general, the intercorrelations are low, with the largest in absolute value being for r_{14} (−0.445); this suggests that more than one predictor variable will probably be used to obtain accurate estimates of the criterion variable. Many of the intercorrelations are very low, less than 0.2; this is usually a desirable characteristic. The predictor–criterion correlations are low to moderate; the largest is 0.570. Using the square of the predictor–criterion correlations, the fraction of variance explained ranges from 0.062 to 0.325. These low values would also suggest that more than one predictor variable will be necessary to accurately estimate the sediment yield.

Example 10.6: Evaporation Data

The evaporation data set includes four predictor variables and a criterion variable, which is the pan evaporation from a small pond in southeastern Georgia. The predictor variables are (1) the mean daily temperature in degrees Fahrenheit (°F), X_1; (2) the wind speed in miles per day, X_2; (3) the radiation in equivalent inches, X_3; and (4) the vapor pressure deficit, X_4. The wind speed attempts to define the rate at which the drier air is moved over the pond to replace the moister air. The temperature and the radiation measure the potential energy of the water molecules and their potential for escaping from the water surface. The vapor pressure deficit represents the potential for the overlying air mass to accept water molecules.

The correlation matrix, which was based on 354 daily measurements, is shown in Table 10.9. The correlation matrix provides a sharp contrast to that from the sediment data because the matrix of Table 10.9 is characterized by high intercorrelation and moderate predictor–criterion correlations. The high intercorrelations suggest that only one or two predictor variables will be necessary to estimate evaporation rates. The moderate predictor–criterion

TABLE 10.9   Correlation Matrix for the Evaporation Database

	X_1	X_2	X_3	X_4	Y
X_1: Temperature (°F)	1.000	−0.219	0.578	0.821	0.581
X_2: Wind speed (miles/per day)		1.000	−0.261	−0.304	−0.140
X_3: Radiation (equivalent inches)			1.000	0.754	0.578
X_4: Vapor pressure deficit				1.000	0.635
Y: Pan evaporation (inches)					1.000

correlations suggest that the potential accuracy of predictions is only moderate. The fraction of the variance explained by X_4 is 0.403. Because most of the intercorrelations are high, it is unlikely that the variation explained by the four predictor variables will be much greater than the 40% explained by just one of the predictor variables.

10.8.2 CALIBRATION OF THE MULTIPLE LINEAR MODEL

The three components of regression analysis are the model, the objective function, and the data set. The data set consists of a set of n observations on p predictor variables and one criterion variable, where n should be, if possible, at least four times greater than p. The data set can be viewed as a matrix having dimensions of n by $(p + 1)$. The principle of least squares is used as the objective function. The model, in raw score form, is

$$(10.55) \qquad \hat{Y} = b_0 + b_1 X_1 + b_2 X_2 + \ldots + b_p X_p$$

in which X_j ($j = 1, 2, \ldots, p$) are the predictor variables, b_j ($j = 1, 2, \ldots, p$) are the partial regression coefficients, b_0 is the intercept coefficient, and Y is the criterion variable. Using the least-squares principle and the model of Equation 10.55, the objective function becomes

$$(10.56) \qquad F = \min \sum_{i=1}^{n} e_i^2 = \min \sum_{i=1}^{n} \left(b_0 + \sum_{j=1}^{p} b_j X_{ij} - Y_i \right)^2$$

in which F is the value of the objective function. It should be noted that the predictor variables include two subscripts, with i indicating the observation and j the specific predictor variable.

The method of solution is to take the $(p + 1)$ derivatives of the objective function, Equation 10.56, with respect to the unknowns, b_j ($j = 0, 1, \ldots, p$), setting the derivatives equal to zero and solving for the unknowns. A set of $(p + 1)$ normal equations is an intermediate result of this process.

As an example, consider the case where $p = 2$; thus Equation 10.55 reduces to

(10.57) $$\hat{Y} = b_0 + b_1 X_1 + b_2 X_2$$

Also, the objective function, Equation 10.56, is given by

(10.58) $$F = \min \sum_{i=1}^{n} e_1^2 = \min \sum_{i=1}^{n} (b_0 + b_1 X_{i1} + b_2 X_{i2} - Y_i)^2$$

The resulting derivatives are

(10.59) $$\frac{\partial \sum_{i=1}^{n} e_i^2}{\partial b_0} = 2 \sum_{i=1}^{n} (b_0 + b_1 X_{i1} + b_2 X_{i2} - Y_i)(1) = 0$$

(10.60) $$\frac{\partial \sum_{i=1}^{n} e_i^2}{\partial b_1} = 2 \sum_{i=1}^{n} (b_0 + b_1 X_{i1} + b_2 X_{i2} - Y_i)(X_{i1}) = 0$$

(10.61) $$\frac{\partial \sum_{i=1}^{n} e_i^2}{\partial b_2} = 2 \sum_{i=1}^{n} (b_0 + b_1 X_{i1} + b_2 X_{i2} - Y_i)(X_{i2}) = 0$$

Rearranging Equations 10.59 through 10.61 yields a set of normal equations:

(10.62) $$n b_0 + b_1 \sum_{i=1}^{n} X_{i1} + b_2 \sum_{i=1}^{n} X_{i2} = \sum_{i=1}^{n} Y_i$$

(10.63) $$b_0 \sum_{i=1}^{n} X_{i1} + b_1 \sum_{i=1}^{n} X_{i1}^2 + b_2 \sum_{i=1}^{n} X_{i1} X_{i2} = \sum_{i=1}^{n} X_{i1} Y_1$$

(10.64) $$b_0 \sum_{i=1}^{n} X_{i2} + b_1 \sum_{i=1}^{n} X_{i1} X_{i2} + b_2 \sum_{i=1}^{n} X_{i2}^2 = \sum_{i=1}^{n} X_{i2} Y_1$$

The solution of the three simultaneous equations yields values of b_0, b_1, and b_2.

10.8.3 STANDARDIZED MODEL

When the approach of the previous section is used and the means and standard deviations of the predictor and criterion variables are significantly different, round-off error, which results from the inability to maintain a sufficient number of significant digits in the computations, may cause the partial regression coefficient to be erroneous. Thus most multiple regression analyses are made using a standardized model:

$$(10.65) \qquad Z_y = t_1 Z_1 + t_2 Z_2 + \ldots + t_p Z_p$$

in which t_j (j = 1, 2, ..., p) are called standardized partial regression coefficients, and Z_y and the Z_j (j = 1, 2, ..., p) are the criterion variable and the predictor variables, respectively, expressed in standardized form; specifically, for i = 1, 2, ..., n, they are computed by

$$(10.66) \qquad Z_Y = \frac{Y_i - \bar{Y}}{S_Y}$$

and

$$(10.67) \qquad Z_j = \frac{X_{ij} - \bar{X}_j}{S_j}$$

in which S_y is the standard deviation of the criterion variable and S_j (j = 1, 2, ..., p) are the standard deviations of the predictor variables. It can be shown that the standardized partial regression coefficients (i.e., the t_j's) and the partial regression coefficients (i.e., the b_j's) are related by

$$(10.68) \qquad b_j = \frac{t_j S_Y}{S_j}$$

The intercept coefficient can be computed by

$$(10.69) \qquad b_0 = \bar{Y} - \sum_{j=1}^{p} b_j \bar{X}_j$$

Thus the raw score model of Equation 10.55 can be computed directly from the standardized model, Equation 10.65, and Equations 10.68 and 10.69.

Although the differences between regression and correlation have been emphasized, the correlation matrix can be used in solving for the standardized partial regression coefficients. The solution is represented by

$$(10.70) \qquad \mathbf{R}_{11}\, \mathbf{t} = \mathbf{R}_{12}$$

in which $\mathbf{R}_{11}$ is the $p \times p$ matrix of intercorrelations between the predictor variables, $\mathbf{t}$ is a $p \times 1$ vector of standardized partial regression coefficients, and $\mathbf{R}_{12}$ is the $p \times 1$ vector of predictor–criterion correlation coefficients. Since $\mathbf{R}_{11}$ and $\mathbf{R}_{12}$ are known while $\mathbf{t}$ is unknown, it is necessary to solve the matrix equation, Equation 10.70, for the $\mathbf{t}$ vector. This involves premultiplying both sides of Equation 10.70 by $\mathbf{R}_{11}^{-1}$ (i.e., the inverse of $\mathbf{R}_{11}$) and using the matrix identities, $\mathbf{R}_{11}^{-1} \cdot \mathbf{R}_{11} = \mathbf{I}$ and $\mathbf{I} \cdot \mathbf{t} = \mathbf{t}$, where $\mathbf{I}$ is the unit matrix, as follows:

$$(10.71) \qquad \mathbf{R}_{11}^{-1} \cdot \mathbf{R}_{11} \cdot \mathbf{t} = \mathbf{R}_{11}^{-1} \cdot \mathbf{R}_{12}$$

$$(10.72) \qquad \mathbf{I} \cdot \mathbf{t} = \mathbf{R}_{11}^{-1} \cdot \mathbf{R}_{12}$$

$$(10.73) \qquad \mathbf{t} = \mathbf{R}_{11}^{-1} \cdot \mathbf{R}_{12}$$

It is important to recognize from Equation 10.73 that the elements of the $\mathbf{t}$ vector are a function of both the intercorrelations and the predictor–criterion correlation coefficients. If $\mathbf{R}_{11} = \mathbf{I}$, then $\mathbf{R}_{11}^{-1} = \mathbf{I}$ and $\mathbf{t} = \mathbf{R}_{21}$; this suggests that since the t_j values serve as weights on the standardized predictor variables, the predictor–criterion correlations also reflect the importance, or "weight," that should be given to a predictor variable. However, when $\mathbf{R}_{11}$ is very different from $\mathbf{I}$ (i.e., when the intercorrelations are significantly different from zero), the t_j values will provide considerably different estimates of the importance of the predictors than would be indicated by the elements of $\mathbf{R}_{12}$.

10.8.4 INTERCORRELATION

It is evident from Equation 10.73 that intercorrelation can have a significant effect on the t_j values and thus on the b_j values. In fact, if the intercorrelations are significant, the t_j values can be irrational. It should be evident that irrational regression coefficients can lead to irrational predictions. Thus it is important to assess the rationality of the coefficients.

The irrationality results from the difficulty in taking the inverse of the $\mathbf{R}_{11}$ matrix that is necessary for Equation 10.73; this corresponds to the round-off-error problem associated with the solution of the normal equations of Equations 10.62 through 10.64. A matrix in which the inverse cannot be evaluated is called a singular matrix. A near-singular matrix is one in which one or more pairs of the standardized normal equations are nearly identical (i.e., linearly dependent).

The determinant of a square matrix, such as a correlation matrix, can be used as an indication of the degree of intercorrelation. The determinant is a unique scalar value that characterizes the intercorrelation of the $\mathbf{R}_{11}$ matrix; that is, the determinant is a good single-valued representation of the degree of linear association between the normal equations.

Consider the following four matrices:

(10.74)
$$A_1 = \begin{bmatrix} 1 & 0 & 0 \\ 0 & 1 & 0 \\ 0 & 0 & 1 \end{bmatrix}$$

(10.75)
$$A_2 = \begin{bmatrix} 1.0 & 0.5 & 0.5 \\ 0.5 & 1.0 & 0.5 \\ 0.5 & 0.5 & 1.0 \end{bmatrix}$$

(10.76)
$$A_3 = \begin{bmatrix} 1.0 & 0.9 & 0.9 \\ 0.9 & 1.0 & 0.9 \\ 0.9 & 0.9 & 1.0 \end{bmatrix}$$

(10.77)
$$A_4 = \begin{bmatrix} 1.0 & 0.9 & 0.0 \\ 0.9 & 1.0 & 0.0 \\ 0.0 & 0.0 & 1.0 \end{bmatrix}$$

The determinants of these matrices are

(10.78)
$$|A_1| = 1$$

(10.79)
$$|A_2| = 0.5$$

(10.80)
$$|A_3| = 0.028$$

(10.81)
$$|A_4| = 0.19$$

These values indicate that (1) the determinant of a correlation matrix will lie between 0.0 and 1.0; (2) if the intercorrelations are zero, the determinant of a correlation matrix equals 1.0; (3) as the intercorrelations become more significant, the determinant approaches zero; and (4) when any two rows of a correlation matrix are nearly identical, the determinant approaches zero.

10.8.5 CRITERIA FOR EVALUATING A MULTIPLE REGRESSION MODEL

After a multiple regression model has been calibrated, we may ask how well the linear model represents the observed data. The following criteria should be used in answering this question: (1) the rationality of the coefficients; (2) the coefficient of multiple determination (R^2); (3) the standard error of estimate, which is usually compared with S_Y; (4) the relative importance of the predictor variables; and (5) the characteristics of the residuals. The relative importance of each of these five criteria may vary with the problem, as well as with the analyst.

The coefficient of multiple determination equals the fraction of the variation in the criterion variable that is explained by the regression equation. Thus it is the square of the correlation coefficient. Mathematically, it is defined as

(10.82)
$$R^2 = \frac{\sum_{i=1}^{n} (\hat{Y}_i - \bar{Y})^2}{\sum_{i=1}^{n} (Y_i - \bar{Y})^2}$$

It may be computed by

(10.83)
$$R^2 = \sum_{j=1}^{p} t_j r_{jY}$$

in which r_{jY} is the predictor–criterion correlation for predictor j. The value of R^2 is always in the range from 0 to 1.0, with a value of 0 indicating that Y is not related to any of the predictor variables. The coefficient of multiple determination must be at least as large as the square of the largest predictor–criterion correlation; if the intercorrelations are high, there will be little difference between the two, which would indicate that including more than one predictor variable in the regression equation does little to improve the accuracy of predictions. If the intercorrelations are low, including more than one predictor variable will probably provide greater accuracy than a bivariate model.

The standard error of estimate (S_e) is defined as the square root of the sum of the squares of the errors divided by the degrees of freedom (v):

(10.84)
$$S_e = \left[\frac{1}{v} \sum_{i=1}^{n} (\hat{Y}_i - Y_i)^2 \right]^{0.5}$$

Previously, the degrees of freedom used to compute S_e for a bivariate model were ($n - 2$); however, in general, the degrees of freedom are defined as

(10.85)
$$v = n - q$$

in which q is the number of unknowns. For the case where the equation includes p partial regression coefficients and one intercept coefficient, $q = p + 1$ and $v = n - p - 1$. The S_e has the same units as the criterion variable and should be compared with S_Y in assessing the accuracy of prediction.

If we use Equation 10.68 to compute t_i for the bivariate case (i.e., one predictor variable), it will be evident that t, the standardized partial regression coefficient, equals the correlation coefficient. Even in the case where $p > 1$, t_i is still a measure of the common variation. However, Equation 10.73 indicates that the t_i values are a function of both the intercorrelation and the predictor–criterion correlations. When the intercorrelations are significant (i.e., $\mathbf{R}_{11}$ of Equation 10.73 is significantly different from $\mathbf{I}$, the identity matrix), the t_i values may be irrational; thus they would not be valid measures of the relative importance of the corresponding predictor variable. Since the t value corresponds to a correlation coefficient in the bivariate case, it is not unreasonable to provide a similar interpretation for a multiple regression; thus t_i values should be considered irrational in magnitude if they are greater than 1.0 in absolute value. Even values of t near 1.0 may indicate irrationality. If a t value is irrational, the corresponding b value must be considered irrational, since b is a scaled value of t.

10.8.6 ANALYSIS OF RESIDUALS

The residuals (i.e., $\hat{Y}_i - Y_i$) were shown to be an important criterion in assessing the validity of a bivariate regression equation; the same concepts apply in the analysis of the residuals for a multiple regression equation. Plots of the residuals may indicate an incorrect model structure, such as a nonlinear relationship between the criterion variable and one of the predictor variables. Additionally, the residuals may suggest a need to consider another predictor variable that was not included in the model. Plots of the residuals should be made to assess the assumptions of zero mean, constant error variance, independence, and normality. A plot of the residuals versus the predicted value of the criterion variable (i.e., e_i versus $\hat{Y}_i$) may identify either a violation of the assumption of constant variance or a lack of independence of the observations.

10.8.7 COMPUTATIONAL EXAMPLE

Although we rarely perform a multiple regression without the aid of a computer, it is instructive to see the manual computations at least one time. Table 10.10 provides values for the criterion variable (Y) and two predictor variables (X_1 and X_2); the sample consists of six observations

TABLE 10.10   Database for Example of a Two-Predictor Model

Y	X_1	X_2	X_1^2	X_1X_2	X_1Y	X_2^2	X_2Y
2	1	2	1	2	2	4	4
2	2	3	4	6	4	9	6
3	2	1	4	2	6	1	3
3	5	5	25	25	15	25	15
5	4	6	16	24	20	36	30
6	5	4	25	20	30	16	24
Column summation 21	19	21	75	79	77	91	82

(i.e., $n = 6$). Figure 10.11 shows the graphical relationship between each pair of variables; in general, linear associations are evident, although the scatter about the apparent linear relationships is significant. The correlation between the two predictor variables appears to be high relative to the degree of linearity between the criterion variable and the predictor variables.

After examining the plots, the correlation matrix ($\mathbf{R}$) should be computed:

(10.86)
$$
\mathbf{R} = \begin{array}{c|ccc}
 & X_1 & X_2 & Y \\
\hline
X_1 & 1.000 & 0.776 & 0.742 \\
X_2 & & 1.000 & 0.553 \\
Y & & & 1.000
\end{array}
$$

The correlation coefficient of 0.776 between X_1 and X_2 indicates that the intercorrelation is high. The determinant of the intercorrelation matrix

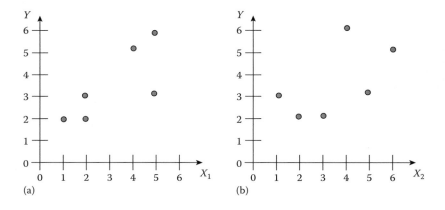

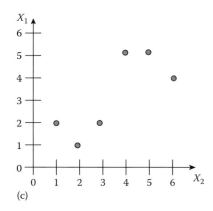

FIGURE 10.11 Graphs of variables for computational example: (a) Y versus X_1, (b) Y versus X_2, and (c) X_1 versus X_2.

equals 0.398, which should be considered as moderate; that is, it suggests that we should carefully examine the resulting regression equation for rationality.

The summations shown in Table 10.10 can be used to develop the normal equations using Equations 10.62 through 10.64, as follows:

(10.87) $$19b_0 + 75b_1 + 79b_2 = 77$$

(10.88) $$21b_0 + 79b_1 + 91b_2 = 82$$

(10.89) $$6b_0 + 19b_1 + 21b_2 = 21$$

The normal equations can be solved to obtain values for the coefficients of the following equation:

(10.90) $$\hat{Y} = 1.30 + 0.75X_1 - 0.05X_2$$

Similarly, the correlation matrix could be used to develop the normal equations for the standardized model:

(10.91) $$\begin{bmatrix} 1.000 & 0.776 \\ 0.776 & 1.000 \end{bmatrix} \begin{bmatrix} t_1 \\ t_2 \end{bmatrix} = \begin{bmatrix} 0.742 \\ 0.553 \end{bmatrix}$$

which yields

(10.92) $$t_1 + 0.776t_2 = 0.742$$

(10.93) $$0.776t_1 + t_2 = 0.553$$

Solving for t_1 and t_2 yields the following standardized equation:

(10.94) $$Z_y = 0.786Z_1 - 0.057Z_2$$

Both the raw score equation (Equation 10.90) and the standardized equation (Equation 10.94) indicate that as X_2 increases Y decreases; this is in conflict with both the correlation coefficient and the graphical relationship shown in Figure 10.11. This irrationality is the result of the high intercorrelation between the predictor variables.

The correlation coefficient can be computed from either Equation 10.82 or 10.83. Using Equation 10.83 yields

(10.95) $$R^2 = t_1 r_{1Y} + t_2 r_{2Y}$$

(10.96) $$= 0.786(0.742) + (-0.057)(0.553) = 0.552$$

Therefore, R equals 0.743. Predictor X_1 explained 55.1% (i.e., 0.742^2) of the variation in Y by itself. Thus the use of X_2 is not justified because it does not make a significant contribution to the explained variation and, even more important, it results in an irrational model. The standard error of estimate is

$$(10.97) \qquad S_e = S_Y \sqrt{\frac{n-1}{n-p-1}(1-R)^2} = 1.643 \sqrt{\frac{5}{3}(1-0.552)^2}$$

$$(10.98) \qquad = 1.643(0.864) = 1.420$$

Thus the two-predictor model reduced the error variance by 25.3%; the standard error of estimate is about 86% of the standard deviation of the criterion variable. For the one-predictor model with X_1 as the predictor variable, the standard error of estimate would be 1.231, which is less than the standard error of estimate for the two-predictor model; this occurs because the ratio $(n-1)/(n-p-1)$ is much less for the one-predictor model. This example illustrates the importance of computing the standard error of estimate when the sample size is small. The standard error provides a more realistic measure of the goodness of fit than the correlation coefficient does.

10.9 REGRESSION ANALYSIS OF NONLINEAR MODELS

In most empirical analyses, linear models are attempted first because of the relative simplicity of linear analysis. Also, linear models are easily applied, and the statistical reliability is easily assessed.

Linear models may be rejected because of either theoretical considerations or empirical evidence. Specifically, theory may suggest a nonlinear relationship between a criterion variable and one or more predictor variables; for example, biological growth curves used by environmental engineers are characterized by nonlinear forms. When a model structure cannot be identified by theoretical considerations, empirical evidence can be used in model formulation and may suggest a nonlinear form. For example, the hydrologic relationship between peak discharge and the drainage area of a watershed has been found to be best represented by a log–log equation, which is frequently referred to as a power model. A nonlinear form may also be suggested by the residuals that result from a linear analysis; that is, if a linear model produces nonrandomly distributed residuals (i.e., a bias is apparent), the underlying assumptions are violated. A nonlinear functional form may produce residuals that satisfy the assumptions that underlie the principle of least squares.

One very common use of nonlinear models is the fitting of existing design curves that have significant nonlinearity. The equations are then used as part of computer software that performs a design procedure that, in the past, was solved manually.

10.9.1 COMMON NONLINEAR ALTERNATIVES

Linear models were separated into bivariate and multivariate; the same separation is applicable to nonlinear models. It is also necessary to separate nonlinear models on the basis of the functional form. Although polynomial and power models are the most frequently used nonlinear forms, it is important to recognize that other model structures are available and may actually provide the correct structure. In addition to the power and polynomial structures, forms such as square root, exponential, and logarithmic may provide the best fit to a set of data. Since the polynomial and power forms are so widely used, it important to identify their structure:

1. Bivariate
   a. Polynomial

(10.99) $$\hat{Y} = b_0 + b_1 X + b_2 X^2 + \ldots + b_p X^p$$

   b. Power

(10.100) $$\hat{Y} = b_0 X^{b_1}$$

2. Multivariate
   a. Polynomial

(10.101) $$\hat{Y} = b_0 + b_1 X_1 + b_2 X_2 + b_3 X_1^2 + b_4 X_2^2 + b_5 X_1 X_2$$

   b. Power

(10.102) $$\hat{Y} = b_0 X_1^{b_1} X_2^{b_2} \ldots X_p^{b_p}$$

In these relationships, Y is the criterion variable, X is the predictor variable in the bivariate case, X_i is the ith predictor variable in the multivariate case, b_j ($j = 0, 1, \ldots, p$) is the jth regression coefficient, and p is either the number of predictor variables or the order of the polynomial. The bivariate forms are just special cases of the multivariate forms.

10.9.2 CALIBRATION OF POLYNOMIAL MODELS

The power and polynomial models are widely used nonlinear forms because they can be transformed in a way that makes it possible to use the principle of least squares. Although the transformation to a linear structure is desirable from the standpoint of calibration, it has important consequences in terms of assessing the goodness-of-fit statistics.

The bivariate polynomial of Equation 10.99 can be calibrated by forming a new set of predictor variables:

(10.103) $$W_i = X^i \quad \text{for} \quad i = 1, 2, \ldots, p$$

This results in the model

(10.104) $$\hat{Y} = b_0 + b_1 W_1 + b_2 W_2 + \ldots + b_p W_p$$

The coefficients b_j can be estimated using a standard linear multiple regression analysis.

The multivariate polynomial of Equation 10.101 can also be solved using a multiple regression analysis for a set of transformed variables. For the model given by Equation 10.101, the predictor variables are transformed as follows:

$$(10.105) \qquad W_1 = X_1$$

$$(10.106) \qquad W_2 = X_2$$

$$(10.107) \qquad W_3 = X_1^2$$

$$(10.108) \qquad W_4 = X_2^2$$

$$(10.109) \qquad W_5 = X_1 X_2$$

The revised model has the form

$$(10.110) \qquad \hat{Y} = b_0 + \sum_{i=1}^{5} b_i W_i$$

It is important to note that the polynomial models do not require a transformation of the criterion variable.

The model given by Equation 10.101 has only two predictor variables (i.e., X_1 and X_2) and is a second-order equation. In practice, a model may have more predictor variables and may be of higher order. In some cases the interaction terms (i.e., $X_i X_j$) may be omitted to decrease the number of coefficients that must be calibrated. However, if the interaction terms are omitted when they are actually significant, the goodness-of-fit statistics may suffer, unless the variation is explained by the other terms in the model; when this occurs, the regression coefficients may lack physical significance.

10.9.3 FITTING A POWER MODEL

The bivariate power model of Equation 10.100 can be calibrated by forming the following set of variables:

$$(10.111) \qquad \hat{Z} = \ln \hat{Y}$$

$$(10.112) \qquad c = \ln b_0$$

$$(10.113) \qquad W = \ln X$$

These transformed variables form the following linear equation, which can be calibrated using bivariate linear regression analysis:

$$\hat{Z} = c + b_1 W \tag{10.114}$$

After values of c and b_1 are obtained, the coefficient b_0 can be determined using

$$b_0 = e^c \tag{10.115}$$

Base 10 logarithms can also be used; in such a case, Equations 10.111 and 10.113 will use the base 10 logarithm and Equation 10.115 will use a base 10 rather than e.

The multivariate power model of Equation 10.102 can be evaluated by making a logarithmic transformation of both the criterion and the predictor variables:

$$\hat{Z} = \ln \hat{Y} \tag{10.116}$$

$$c = \ln b_0 \tag{10.117}$$

$$W_i = \ln X_i, \quad \text{for} \quad i = 1, 2, ..., p \tag{10.118}$$

The resulting model has the form

$$\hat{Z} = c + \sum_{i=1}^{p} b_i W_i \tag{10.119}$$

The coefficients of Equation 10.119 can be evaluated using a multiple regression analysis. The value of b_0 can be determined by making the transformation of Equation 10.115. Again, a base 10 transformation can be used rather than a natural log transformation.

10.9.4 GOODNESS OF FIT

Since most computer programs that perform multiple regression analyses include goodness-of-fit statistics as part of the output, it is important to recognize the meaning of these statistics. For nonlinear models in which the criterion variable $\hat{Y}$ is not transformed, the goodness-of-fit statistics are valid indicators of the reliability of the model; however, when the criterion variable is transformed, such as is necessary for the power model form, the principle of least squares is applied in the log–log space. As a result, the residuals that are used to compute the standard error of estimate—and therefore the correlation coefficient—are measured in the domain of the logarithm of $\hat{Y}$ and not the $\hat{Y}$ domain. Therefore, the goodness-of-fit statistics are not necessarily a reliable indicator of model reliability, especially since engineering decisions are made to predict Y and not $\log(Y)$. Therefore, when a model requires a transformation of the criterion variable $\hat{Y}$ in order to calibrate the coefficients, the goodness-of-fit statistics that are included with the multiple regression output should not be used as measures of reliability. Instead, values for the goodness-of-fit statistics should be recomputed using values of $\hat{Y}$ rather than $\log(\hat{Y})$. The correlation coefficient and the standard error of estimate should be computed using Equations 10.82 and 10.84, respectively. In summary, when an equation is calibrated with the criterion variable transformed, such as the log–log space, the least-squares concepts apply only to the transformed space, not to the measurement nontransformed space. The sum of the residuals based on the logarithms of Y will equal zero; however, the sum of the residuals of the transformed values of Y may not equal zero, and the sum of the squares of the errors may not be a minimum in the Y domain even though it is in the $\log(Y)$ domain. Furthermore, the assumption of a constant variance may also not be valid. Because these basic assumptions of regression are not valid in the X–Y space, many practitioners object to data transformations. However, in spite of these theoretical considerations, the transformations may provide reasonable estimates in the X–Y space.

10.9.5 ADDITIONAL MODEL FORMS

In addition to the polynomial and power models, other forms can provide good approximations to the underlying population relationship between two or more variables. An exponential model has the form

$$(10.120) \qquad \hat{Y} = b_0 e^{b_1 X}$$

in which b_0 and b_1 are coefficients requiring values by fitting to data. Values for b_0 and b_1 can be obtained by linear bivariate regression after taking the logarithms of both sides of Equation 10.120 as follows:

$$(10.121) \qquad \ln \hat{Y} = \ln b_0 + b_1 X$$

In this case, $\hat{Y}$ is transformed; thus the correlation coefficient and standard error of estimate apply to the $\ln \hat{Y}$ and not to $\hat{Y}$. Also, the intercept coefficient obtained from the regression analysis for Equation 10.121 must be transformed for use with Equation 10.120.

A logarithmic curve can also be used to fit values of Y and X:

$$(10.122) \qquad \hat{Y} = b_0 + b_1 \ln X$$

In this case, $\hat{Y}$ is not transformed, so the values of the correlation coefficient and the standard error of estimate are valid indicators of the goodness of fit in the $\hat{Y}$ space.

10.10 APPLICATIONS

10.10.1 ONE-PREDICTOR POLYNOMIAL OF SEDIMENT YIELD VERSUS SLOPE

First-, second-, and third-order polynomials that regress sediment yield on watershed slope were fitted. The correlation matrix is given in Table 10.11. It is evident that the intercorrelation coefficients are very significant and should lead to irrational regression coefficients; furthermore, because of the high intercorrelation it is unlikely that the explained variance will improve significantly as the order of the equation increases. Table 10.12 provides a summary of the correlation coefficients, the standardized partial regression coefficients, and the partial regression coefficients. The second- and third-order models do not provide a significant increase in the explained variation when compared with the linear model.

10.10.2 ONE-PREDICTOR POLYNOMIAL OF EVAPORATION VERSUS TEMPERATURE

The evaporation data were used to derive polynomials for the regression of evaporation on temperature. The correlation matrix, which is shown in Table 10.13, shows very high intercorrelation. Table 10.14 provides a summary of the regression analyses. Whereas the linear model explains approximately 34% of the variation in evaporation, the higher order terms provide only a marginal increase in the explained variation. The quadratic model increases the explained variance by about 8%. The marginal improvement is also evident from the small decrease in the standard error of estimate. In judging the rationality of a model, it is of value to examine

TABLE 10.11   Correlation Matrix: Sediment Yield (Y) versus Slope (S)

	S	S^2	S^3	Y
S	1.000	0.908	0.791	0.443
S^2		1.000	0.972	0.443
S^3			1.000	0.428
Y				1.000

TABLE 10.12   Summary of Polynomial Analyses for Sediment Yield versus Slope Models

Model	R	Intercept b_0	Partial Regression Coefficients b_1	b_2	b_3	Standardized Partial Regression Coefficients t_1	t_2	t_3
First order	0.443	0.220	0.0263	–	–	0.443	–	–
Second order	0.454	0.323	0.0136	2.31×10^{-4}	–	0.229	0.235	–
Third order	0.499	0.042	0.0865	–0.0031	3.43×10^{-5}	1.456	–3.164	2.35

TABLE 10.13   Correlation Matrix: Evaporation (E) versus Temperature (T)

	T	T^2	T^3	E
T	1.000	0.995	0.981	0.581
T^2		1.000	0.996	0.602
T^3			1.000	0.627
E				1.000

TABLE 10.14   Summary of Regression Analyses

Model	R	S_e^a	t_1	t_2	b_3	b_0	b_1	b_2	b_3
First order	0.581	0.0763	0.581	–	–	−0.114	0.00383	–	–
Second order	0.648	0.0715	2.16	2.76	–	0.425	−0.0142	0.000143	–
Third order	0.654	0.0711	3.11	−8.44	5.98	−0.239	0.0205	−0.000439	3.14×10^{-6}

Note: $S_y = 0.0936$ in./day.

the predicted values of evaporation for values of the predictor variable that are likely to be observed. The three models were used to estimate the daily evaporation depth for selected temperatures (Table 10.15). Because water freezes at 32°F, evaporation rates near zero are expected; only the linear model gives a rational estimate at 32°F. Estimates made using the second-order model decrease with increases in temperature for temperatures up to about 50°F. Both the second- and third-order models provide especially high rates at 100°F. In summary, the high intercorrelations lead to irrational estimates when the higher order forms are used.

10.10.3 SINGLE-PREDICTOR POWER MODEL

A data set that consists of five observations will be used to illustrate the fitting of a power model that includes a single predictor variable. The data set is given in Table 10.16. The values were converted to natural logarithms and the coefficients of Equation 10.114 were calibrated using least squares:

$$(10.123) \qquad \ln \hat{Y} = -0.1 + 1.9 \ln X$$

TABLE 10.15   Predicted Evaporation Rates (in./day)

Model	Temperature (°F)			
	32	50	75	100
First order	0.009	0.078	0.173	0.269
Second order	0.117	0.073	0.164	0.435
Third order	0.070	0.081	0.154	0.561

TABLE 10.16   Database for Single-Predictor Power Model

X	Y	ln X	ln Y	Using Equation 10.125		Using Equation 10.124	
				$\hat{Y}$	e	ln $\hat{Y}$	e_e
2.718	7.389	1	2	−214.54	−221.929	6.048	−1.341
7.389	20.086	2	3	53.18	33.095	40.445	20.359
20.086	403.429	3	6	780.93	377.499	270.427	−133.002
54.598	2980.96	4	8	2759.03	−221.927	1807.96	−1173.00
148.413	8103.08	5	9	8136.00	32.923	12087.4	3984.31
					$\Sigma e = 0$		$\Sigma e_e \neq 0$

A correlation coefficient of 0.9851 was computed for Equation 10.123. Equation 10.123 can be retransformed to the X–Y coordinate space:

$$(10.124) \qquad \hat{Y} = 0.9048X^{1.9}$$

A linear analysis of the values of X and Y provides a correlation coefficient of 0.9975 and the following regression equation:

$$(10.125) \qquad \hat{Y} = -370.3301 + 57.3164X$$

The residuals for Equations 10.124 and 10.125 are given in Table 10.16. Whereas the residuals for Equation 10.125 have a sum of zero, the sum of the residuals for Equation 10.124 is not equal to zero; however, the sum of the residuals for Equation 10.123 will equal zero in the natural-log space.

The correlation coefficient and standard error of estimate that are computed for the model in the natural-log space are not valid indicators of the accuracy of the estimates in the domain of the nontransformed criterion variable. For example, the correlation coefficient of 0.9851 computed with Equation 10.123 is not a measure of the accuracy of Equation 10.124. Values of Y that were estimated with Equation 10.124 are given in Table 10.16, and the resulting residuals are denoted as e_e. If the separation of variation concept is applied to Equation 10.124, the components are

$$(10.126) \qquad \sum_{i=1}^{n}(Y_i - \bar{Y})^2 = \sum_{i=1}^{n}(Y_i - \hat{Y}_i)^2 + \sum_{i=1}^{n}(\hat{Y}_i - \bar{Y})^2$$

$$(10.127) \qquad 48{,}190{,}499.4 \neq 17{,}276{,}427.6 + 100{,}525{,}199.1$$

The separation of variation is not valid for Equation 10.124 because the model was calibrated by taking logarithms. The separation of variation does, however, apply to the logarithms. It is evident that, if the correlation coefficient were computed as the ratio of the explained variation to the total variation (Equation 10.82), then from the values of Equation 10.127 the correlation coefficient would be greater than 1.0, which is not rational.

Thus the standard error of estimate, rather than the correlation coefficient, should be used as a measure of accuracy:

(10.128)
$$S_e = \left(\frac{17,276,427.6}{5-2} \right)^{0.5} = 2399.75$$

The standard deviation of the criterion variable is 3470.97; therefore, the power model is not highly accurate, and the computed correlation for the log model (i.e., 0.9851) certainly does not reflect the accuracy of the model. The small sample and large variation in both the X and Y contribute to the problem.

10.10.4 MULTIVARIATE POWER MODEL

Sediment yield data will be used to illustrate a power model that includes more than one predictor variable. The small particle size index cannot be included as a predictor variable because some of the observations equal zero; thus the model regresses sediment yield (Y) on three predictor variables: the precipitation/temperature ratio (P), the slope (S), and the soil index (I). The data are given in Table 10.17. The resulting regression equation in logarithm form is

(10.129) $\ln \hat{Y} = -4.563 - 1.123 \ln P + 0.8638 \ln S - 0.1089 \ln I$

Transforming Equation 10.129 to the X–Y space yields

(10.130) $\hat{Y} = 0.01043 P^{-1.123} S^{0.8638} I^{-0.1089}$

A computer program provided a correlation coefficient of 0.725 and a standard error estimate of 1.019; however, these are for Equation 10.129 and may not be accurate indications of the accuracy of Equation 10.130. The standard error of estimate is expressed in logarithmic units. The standard error of estimate as measured in the X–Y space was computed as 0.595, which can be compared with a standard deviation of 0.817 for the 37 observations of sediment yield.

For a comparison, the following linear model was calibrated using the same three predictor variables:

(10.131) $\hat{Y}_{\text{linear}} = 0.8920 - 1.428P + 0.03664S - 0.01308I$

This resulted in a multiple correlation coefficient of 0.720 and a standard error of estimate of 0.592. Thus, for this case, there is no significant difference in the expected accuracy of the linear and power models.

TABLE 10.17   Sediment Yield Data

P	S	I	Y
0.135	4.000	40.000	0.140
0.135	1.600	40.000	0.120
0.101	1.900	22.000	0.210
0.353	17.600	2.000	0.150
0.353	20.600	1.000	0.076
0.492	31.000	0.000	0.660
0.466	70.000	23.000	2.670
0.466	32.200	57.000	0.610
0.833	41.400	4.000	0.690
0.085	14.900	22.000	2.310
0.085	17.200	2.000	2.650
0.085	30.500	15.000	2.370
0.193	6.300	1.000	0.510
0.167	14.000	3.000	1.420
0.235	2.400	1.000	1.650
0.448	19.700	3.000	0.140
0.329	5.600	27.000	0.070
0.428	4.600	12.000	0.220
0.133	9.100	58.000	0.200
0.149	1.600	28.000	0.180
0.266	1.200	15.000	0.020
0.324	4.000	48.000	0.040
0.133	19.100	44.000	0.990
0.133	9.800	32.000	0.250
0.133	18.300	19.000	2.200
0.356	13.500	58.000	0.020
0.536	24.700	62.000	0.030
0.155	2.200	27.000	0.036
0.168	3.500	24.000	0.160
0.673	27.900	22.000	0.370
0.725	31.200	10.000	0.210
0.275	21.000	64.000	0.350
0.150	14.700	37.000	0.640
0.150	24.400	29.000	1.550
1.140	15.200	11.000	0.090
1.428	14.300	1.000	0.170
1.126	24.700	44.000	0.170

Source: E. M. Flakman, "Predicting Sediment Yield in Western United States," *Journal of the Hydraulic Division*, ASCE, 98(HY12), pp. 2073–2085, December 1972.

Note: Y = sediment yield, X_1 = climate variable, X_2 = watershed slope variable, and X_3 = soil aggregation variable.

10.10.5 ESTIMATING BREAKWATER COSTS

In Section 10.7.2, a bivariate model between the construction cost of a breakwater and length of the breakwater was developed. It was found to be inaccurate, with the suggestion given that other variables associated with cost were probably better determinants of the cost.

Table 10.18 includes the original measurements of length and cost along with a second predictor variable, the average depth of the breakwater. The correlation matrix is

	X_1	X_2	Y
X_1: Length	1.000	−0.229	0.176
X_2: Depth		1.000	0.905
Y: Cost			1.000

While length and depth are poorly correlated, depth and cost have a high correlation, with R^2 equal to 0.819.

A regression analysis yields the following equation:

$$(10.132) \qquad \hat{Y} = -199.8 + 0.1904X_1 + 58.72X_2$$

The multiple correlation coefficient equals 0.987; thus $R^2 = 0.974$, which indicates that the two-predictor model provides accurate estimates of the cost. The standard error of estimate is 25.3×10^3, with $S_e/S_y = 0.174$.

The standardized partial regression coefficients are $t_1 = 0.405$ and $t_2 = 0.998$. While t_2 is below 1, it is still very large.

The intercept is also very large in magnitude. While none of the predicted costs is negative, it is still possible because the intercept is negative.

TABLE 10.18   Data for Breakwater Cost Estimation

Length, X_1 (ft)	Depth, X_2 (ft)	Cost, Y ($\$ \times 10^3$)
450	6.3	230
625	9.6	490
670	3.1	150
730	10.1	540
810	5.4	280
880	3.5	120
1020	6.5	380
1050	2.9	170
1100	8.8	530
1175	3.9	290
1230	7.4	460
1300	3.3	230
1350	7.2	470
1510	4.6	360

For example, at a length of 450 ft and depth of 2 ft, the estimated cost would be negative. Equation 10.132 should be used cautiously beyond the range of the data: $450 < X_1 < 1510$ ft and $2.9 < X_2 < 10.1$ feet.

It is interesting to compare the two-predictor multiple regression model with both one-predictor models. The length versus cost model was discussed in Section 10.7.2. It provided very poor accuracy, $R^2 = 0.031$. The depth versus cost model would explain 81.9% of the variation in cost (0.905^2). Since the multiple regression model explains 97.4%, adding the length to the model increases the explained variance by 15.5% (97.4 − 81.9); this is a significant increase, so the two-predictor model is preferable to either of the one-predictor models. Also, when the depth is added to the equation, the importance of the length increases. The t for the bivariate model is 0.176, and $t_1 = 0.405$ for the multiple regression model. This is a positive effect of the intercorrelation between the predictor variables.

10.10.6 TRIP GENERATION MODEL

For a regional-planning study, a transportation engineer collects data on traffic. The data of Table 10.19 give measurements of nonwork, home-based-trip-production rates for suburban, medium-density households (Y) as a function of the number of persons per household (X_1) and the number of vehicles per household (X_2). In this case, the predictor variables are integers.

The correlation matrix for the sample of 15 is given by

	X_1	X_2	Y
X_1	1.000	0.000	0.931
X_2		1.000	0.264
Y			1.000

TABLE 10.19   Regression Analysis

X_1	X_2	Y	$\hat{Y}$	e
1	0	0.9	0.42	−0.48
2	0	1.8	1.87	0.07
3	0	2.4	3.32	0.92
4	0	4.3	4.76	0.46
5+	0	6.6	6.21	−0.39
1	1	1.9	1.13	−0.77
2	1	2.7	2.58	−0.12
3	1	3.6	4.03	0.43
4	1	5.4	5.47	0.07
5+	1	7.7	6.92	−0.78
1	2+	2.1	1.84	−0.26
2	2+	3.1	3.29	0.19
3	2+	3.7	4.74	1.04
4	2+	5.9	6.18	0.28
5+	2+	8.3	7.63	−0.67

The intercorrelation between the two predictor variables is zero because the data were collected on a grid. One advantage of having no intercorrelation is that the potential for irrational regression coefficients is minimized. The predictor–criterion correlation coefficient for the number of persons per household is large ($R^2 = 0.868$), while that for the number of vehicles per household is small ($R^2 = 0.070$). This may suggest that the bivariate equation $\hat{Y} = f(X_1)$ may be as accurate as the multiple regression model $\hat{Y} = f(X_1, X_2)$.

A multiple regression analysis produced the following equation:

$$(10.133) \qquad \hat{Y} = -1.023 + 1.4467X_1 + 0.7100X_2$$

The correlation coefficient is 0.968, which gives $R^2 = 0.937$. Since the intercorrelation between X_1 and X_2 is zero, the sum of the squares of the two predictor–criterion correlation coefficients equals the coefficient of multiple determination:

$$(10.134) \qquad R^2 = R_{1Y}^2 + R_{2Y}^2$$

$$(10.135) \qquad 0.937 = (0.931)^2 + (0.264)^2$$

This equality holds only when there is no intercorrelation. It also shows that the multiple regression model explains 7% more variation than the bivariate model based solely on X_1.

The standard error of estimate is 0.6153, which gives $S_e/S_y = 0.271$. This value supports the conclusion suggested by the correlation coefficient that the model is highly accurate.

The standardized partial regression coefficients are $t_1 = 0.931$ and $t_2 = 0.264$. Since the intercorrelation coefficient is 0, the t values equal the predictor–criterion correlation coefficients. Again, the number of persons per household is shown to be the more important predictor.

The intercept of Equation 10.133 is -1.023, which is irrational, since a value of 0 would be expected. If both X_1 and X_2 are 0, then Y should be 0. Furthermore, the magnitude of the intercept is large—that is, approximately 25% of the mean of Y and larger than the value of Y for $X_1 = 1$ and $X_2 = 0$. This suggests that another model structure may be needed to realistically represent the data.

The residuals are given in Table 10.19, and they also suggest that the linear model of Equation 10.133 is an incorrect structure. There are very noticeable local biases. For each value of X_2 the residuals for $X_1 = 1$ and $X_1 = 5$ are negative; additionally, for each value of X_2 the residuals for $X_1 = 3$ and $X_1 = 4$ are positive. The trends for each value of X_2 suggest that a nonlinear structure would be more appropriate.

10.10.7 ESTIMATION OF THE REAERATION COEFFICIENT

To estimate the biochemical oxygen demand of a stream, the environmental engineer needs an estimate of the reaeration coefficient. The reaeration coefficient varies with the stream velocity (X_1, in meters per second), the water depth (X_2, in meters), and the water temperature (X_3, in °C). Table 10.20 gives a set of 21 measurements of the three predictor variables and estimated reaeration coefficients (Y). The correlation matrix is

	X_1	X_2	X_3	Y
X_1	1.000	0.839	0.139	0.436
X_2		1.000	0.240	0.456
X_3			1.000	0.621
Y				1.000

The intercorrelation is high ($|\mathbf{R}_{11}| = 0.276$) because of the high correlation between X_1 and X_2. The water temperature is not highly correlated with either velocity or depth. The regression analysis of the 21 sample values yields

TABLE 10.20   Multiple Regression Analysis of the Stream Reaeration Coefficient (Y) on Stream Velocity,[a] Water Depth,[b] and Water Temperature[c]

X_1	X_2	X_3	Y	$\hat{Y}$	e
1.4	1.24	10	2.16	2.69	0.80
0.6	0.87	9	2.50	2.57	0.07
2.2	1.87	13	2.54	3.55	1.01
1.5	0.99	11	2.60	3.06	0.46
0.5	0.85	14	2.71	2.97	0.26
1.6	0.97	8	3.00	2.83	−0.17
1.3	1.12	19	3.02	3.70	0.68
0.8	1.21	11	3.03	2.85	−0.18
1.6	1.64	14	3.27	3.41	0.14
2.3	1.91	6	3.42	2.98	−0.44
1.0	0.70	15	3.42	3.21	−0.21
1.9	1.62	12	3.45	3.34	−0.11
1.3	1.01	17	3.49	3.51	0.02
1.8	1.43	15	3.58	3.54	−0.04
1.7	1.05	15	3.61	3.47	−0.14
0.7	0.62	18	3.69	3.36	−0.33
2.2	1.58	12	3.85	3.43	−0.42
1.4	1.21	15	3.88	3.39	−0.49
1.8	2.07	20	3.99	4.04	0.05
1.5	1.59	16	4.51	3.55	−0.96
2.8	2.38	24	4.74	4.75	0.01

[a] X_1, m/sec.
[b] X_2, m.
[c] X_3, °C.

$$(10.136) \qquad \hat{Y} = 1.508 + 0.3315X_1 + 0.09731X_2 + 0.08672X_3$$

All the regression coefficients are rational in sign. Equation 10.136 results in a correlation coefficient of 0.715, which means that the equation explains 51.2% of the total variation. The standard error of estimate is 0.497 ($S_e/S_y = 0.758$), which suggests that the equation provides estimates of Y that are not highly accurate.

Of the three predictor variables, the water temperature (X_3) would provide the highest R^2 for a bivariate model ($R^2 = 0.386$). The multiple regression model of Equation 10.136 explains 51.2%. Thus the addition of X_1 and X_2 increases the explained variance by 12.6%. This means that more than one predictor variable should be used, but it does not necessarily mean that all three predictor variables are needed. It may be worthwhile developing two-predictor models: $\hat{Y} = f(X_1, X_3)$ and $\hat{Y} = f(X_2, X_3)$. Because X_1 and X_2 are highly correlated, both are probably not needed, so a two-predictor model may provide estimates that are just as accurate as the three-predictor model.

The standardized partial regression coefficients are $t_1 = 0.299$, $t_2 = 0.070$, and $t_3 = 0.563$. These suggest that X_3 is most important, followed by X_1 and finally by X_2, which appears to be unimportant. This differs somewhat from the predictor–criterion correlation coefficients, which suggest that X_3 is most important, but that X_1 and X_2 are of equal importance. Since velocity is probably physically more important than depth, the t values probably give a more realistic assessment of the relative importance of the predictor variables than are given by the predictor–criterion correlation coefficients.

10.10.8 ESTIMATING SLOPE STABILITY

Slope stability is very important to engineers who design with unrestrained slopes. The stability number is an index used by geotechnical engineers to assess the safety of a slope. Among others, the slope stability is a function of the angle of friction of the soil (X_1, in degrees) and the slope angle (X_2, in degrees). Twelve laboratory measurements were made of the stability number for selected values of the angle of friction and the slope angle (see Table 10.21). The correlation matrix for the twelve tests is

	X_1	X_2	Y
X_1	1.000	0.520	0.415
X_2		1.000	−0.543
Y			1.000

The intercorrelation is moderate; the predictor–criterion correlation coefficients are not high, with the two values only representing 17.2% and 29.5% of the total variation in the stability number.

The least-squares regression equation is

$$(10.137) \qquad \hat{Y} = 0.04234 + 0.002842X_1 - 0.005093X_2$$

TABLE 10.21 Multiple Regression Analysis of Slope Stability Values

X_1	X_2	Y	$\hat{Y}$	e
10	5	0.053	0.045	−0.008
10	10	0.005	0.020	0.015
20	5	0.081	0.074	−0.007
20	12	0.036	0.038	0.002
20	18	0.012	0.008	−0.004
30	6	0.102	0.097	−0.005
30	13	0.058	0.061	0.003
30	21	0.021	0.021	0.000
40	9	0.105	0.110	0.005
40	14	0.079	0.085	0.006
40	21	0.052	0.049	−0.003
40	26	0.027	0.024	−0.003

Note: X_1 = friction angle (degrees), X_2 = slope angle (degrees), and Y = stability number.

The multiple correlation coefficient for Equation 10.137 is 0.981, which means that the equation explains 96.2% of the variation in the measured values of the stability number. The standard error of estimate is 0.0072, which gives S_e/S_y = 0.214. The goodness-of-fit statistics suggest that Equation 10.137 provides accurate estimates of the stability number.

Note that, in spite of the relatively low predictor–criterion correlation coefficients (i.e., 0.415 and −0.543), the total R^2 is very high. The intercorrelation actually enhances the reliability of the equation.

The goodness-of-fit statistics are revealing, but they do not give an entirely accurate assessment of the model. The standardized partial regression coefficients for X_1 and X_2 are t_1 = 0.96 and t_2 = −1.04. Because the latter is greater than 1, it is considered irrational and suggests that b_2 gives an unrealistic indication of the rate of change of Y with respect to X_2. The high accuracy indicated by the goodness-of-fit statistics suggests that the model is acceptable for estimation; however, the t_2 value suggests that the model may not be reliable upon extrapolation or for assessing model sensitivity.

The residuals (see Table 10.21) should be evaluated. Figure 10.12 shows a plot of the residuals as a function of the angle of friction (ordinate) and the slope angle (abscissa). While the sample size is probably too small to be very conclusive, there appears to be a region of positive residuals between friction angles of 9° and 15° and two regions of negative residuals, friction angles less than 6° and greater than 18°. This would suggest that the structure of the linear model may be incorrect.

The intercept also appears to be a problem. If both X_1 and X_2 are zero, then the stability number equals 0.042, which does not seem reasonable.

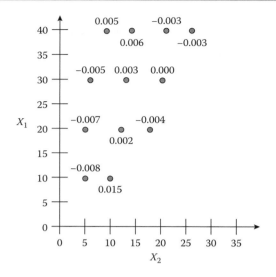

FIGURE 10.12   Residuals of a multiple regression model versus predictor variables X_1 and X_2.

PROBLEMS

10.1   Plot the following data, assess the degree of systematic variation, and assess the appropriateness of using a linear model:

X	0.7	1.3	2.8	5.1	7.3	8.9	11.5
Y	1.1	3.8	7.2	7.9	9.7	8.8	9.6

10.2   Plot the following data and assess the quality of the data for
prediction:

X	0.7	0.8	1.3	1.2	1.7	1.9	4.1
Y	1.0	1.8	1.2	2.1	1.0	2.8	5.7

10.3   Graph the values of Y versus X, sketch a straight line through
the mass of points, and qualitatively assess the degree of
fit. Using the graph, approximate the intercept and slope.
Discuss their meaning and rationality.

X	2.1	2.7	3.5	3.8	4.4	4.9
Y	9.6	18	19	27	37	29

10.4   Graph the data characterizing the growth of a microorganism (Y) over time (T, hours). Discuss the shape of the relations and implication of using a linear model to represent the data.

T	1.6	2.4	4.0	5.3	5.9	6.8	7.7	8.1	9.6	10.8	11.7	13.1
Y	0.03	0.08	0.08	0.13	0.21	0.28	0.32	0.43	0.45	0.43	0.47	0.46

10.5   Using the following values of X and Y, show that the total variation (TV) equals the sum of the explained variation (EV) and the unexplained variation (UV):

X	2	5	7	6	2
Y	1	3	5	7	9

Assume that the regression equation relating Y and X is given by

$$\hat{Y} = 4.5849 + 0.09434X$$

10.6   Using the following data, compute TV, EV, and UV. Use EV and TV to compute the correlation coefficient. The regression line is $Y = 0.1395 + 0.9419 * X$

X	1	2	1	2	6
Y	1	1	2	2	6

10.7   Using the following data sets, compute TV, EV, and UV. Compute R^2. Discuss the results.

a.

X	2.0	2.6	4.2	5.4	7.8	8.2
Y	2.2	1.1	3.6	1.6	4.4	3.2

b.

X	2.0	2.6	4.2	5.4	7.8	8.2
Y	4.4	2.2	7.2	3.2	8.8	6.4

10.8   Using the following values of X and Y, show that the TV equals the sum of the EV and the UV. Assume that the regression equation relating Y and X is given by

$$\hat{Y} = 1.2 + 0.657X$$

Compute the correlation coefficient.

X	1	2	3	4	5	6
Y	2	2	3	5	4	5

10.9 For the following observations on X and Y, compute the correlation coefficient. Discuss the potential accuracy of the estimated value of Y.

X	−3	−2	−1	0	1	2
Y	2	2	3	0	−2	−1

10.10 For the data of Problem 10.6, compute and assess the correlation coefficient using the standard normal transformation.

10.11 For the data given in Problem 10.5, compute and assess the correlation coefficient using the standard normal transformation definition of Equation 10.5.

10.12 Develop the relation between Y and X ($Y = a + bX$) using the standard normal transforms $Z_y = tZ_x$.

10.13 Compute the correlation coefficient for the data of Problem 10.2.

10.14 Derive Equation 10.5 from Equation 10.4.

10.15 Given the following observations on X and Y, (a) construct a graph of X versus Y, and (b) compute the Pearson correlation coefficient.

X	1	2	3	4	5
Y	1	1	2	4	4

10.16 Given the following paired observations on variables X and Y and the values of the standardized variables Z_x and Z_y, compute the Pearson correlation coefficient between each of the following pairs: (a) X and Y; (b) Z_x and Z_y; (c) Z_x and X; (d) Z_x and Y; (e) X and Y − 3; and (f) X and 2Y.

X	Y	Z_x	Z_y
3	4	−1.5	−1.5
5	8	−0.5	0.5
6	7	0.0	0.0
7	6	0.5	−0.5
9	10	1.5	1.5

10.17   Given the paired observations on variables X and Y and the corresponding Z_x and Z_y values, compute the correlation coefficient between each of the following pairs: (a) X and Y; (b) Z_x and Z_y; (c) Z_x and X; (d) Z_x and Y; (e) X and Y + 2.0; and (f) X and Z_y.

X	Y	Z_x	Z_y
2	2	−1.095	−1.201
4	6	−0.5477	−0.2402
8	8	0.5477	0.2402
10	12	1.095	1.201

10.18   Using the data of Problem 10.2, fit the zero-intercept model of Equation 10.9.

10.19   Using the following data, fit the zero-intercept model of Equation 10.9:

Test 1	98	94	93	90	87	85	84
Test 2	96	94	91	85	93	86	90

10.20   Using the data of Problem 10.4, fit the zero-intercept model of Equation 10.9.

10.21   Using the data of Problem 10.3, fit the zero-intercept model of Equation 10.9.

10.22   Given the following four pairs of observations

X	1	1	2	2	?
Y	1	2	1	2	?

compute the correlation coefficient and the regression coefficients if (a) the fifth pair is (Y = 8, X = 8) and (b) the fifth pair is (Y = 2, X = 8). (c) Plot both sets of points and draw the regression lines.

10.23   Use least squares regression analysis to fit the coefficients of Equation 10.15. Compute the correlation coefficient and the sum of squares of the errors. Then assume the model $Y = 2 + X$ and recompute the sums of squares of the errors and the correlation coefficient. Briefly discuss the results.

X	2	2	3	4	4
Y	4	5	5	5	6

10.24   Show that Equation 10.29 results in a larger sum of squares of the errors for the data of Table 10.2 than the value obtained with Equation 10.28.

10.25 Given the following paired observations on Y and X, (a) graph Y versus X; (b) calculate the correlation coefficient; (c) determine the slope and intercept coefficients of the linear regression line; (d) show the regression line on the graph of part (a); (e) compute the predicted value of Y for each observed value of X; and (f) calculate Σ_e and S_e. (g) What is the predicted value of Y for $X = 5$? $X = 10$?

X	3	5	6	7	9
Y	4	8	7	6	10

10.26 Given the following paired observations on Y and X, (a) graph Y versus X; (b) calculate the correlation coefficient; (c) determine the slope and intercept coefficients of the linear regression line; (d) show the regression line on the graph of part (a); (e) compute the predicted value of Y for each observed value of X; and (f) calculate Σe and S_e. (g) What is the predicted value of Y for $X = 3$? $X = 10$?

X	1	2	4	5	7
Y	5	6	4	2	3

10.27 Given the following paired observations on Y and X, (a) graph Y versus X; (b) calculate the correlation coefficient; (c) determine the slope and intercept coefficients of the linear regression line; (d) show the regression line on the graph of part (a); (e) compute the predicted value of Y for each observed value of X; and (f) calculate Σe and S_e. (g) What is the predicted value of Y for $X = 3.5$? $X = 10$?

X	1	2	3	4	5	6	8
Y	3	6	6	4	5	3	4

10.28 Compare correlation analysis and regression analysis as tools in decision making.

10.29 Using the following data set, regress (a) Y on X and (b) X on Y. (c) Compute the correlation coefficient for each regression. (d) Transform the regression of Y on X into an equation for computing X and compare how these coefficients of parts (b) and (d) differ.

X	2.0	2.6	4.0	5.4	7.8	8.4
Y	2.2	1.5	3.1	2.4	4.7	3.4

10.30 Using the following data set, regress (a) Y on X and (b) X on Y. (c) Compute the correlation coefficient for each regression. (d) Transform the regression of Y on X into an equation for computing X and compare these coefficients with the coefficients for the regression of X on Y. (e) Why do the coefficients of parts (b) and (d) differ?

X	1	3	4	6	8
Y	6	7	5	3	4

10.31   Using the following data set, regress (a) Y on X and (b) X on Y. (c) Compute the correlation coefficient for each regression. (d) Transform the regression of Y on X into an equation for computing X and compare these coefficients with the coefficients for the regression of X on Y. (e) Why do the coefficients of parts (b) and (d) differ?

X	2	1	3	2
Y	1	2	4	5

10.32   The data shown are the results of a tensile-test of a steel specimen, where Y is elongation in thousandths of an inch that resulted when the tensile force was X thousands of pounds. Fit the data with a linear model and make a complete assessment of the results.

X	0.8	1.6	3.1	4.4	6.3	7.9	9.2
Y	2.8	4.9	6.5	8.1	8.8	9.1	8.9

10.33   The stream reaeration coefficient (Y) is related to the water temperature (T) in °C. Fit the following with a linear model and make a complete assessment of the results:

T	Y
11	2.69
11	3.69
13	2.41
14	2.89
15	3.56
16	3.05
17	4.17
18	3.35
19	3.06
20	3.78
21	4.08
23	4.20
23	4.18
24	3.30
25	3.58

10.34 The peak discharge ratio (q, ft^3/sec) is related to the drainage area (A, acres). Fit the following data with a linear model and make a complete assessment of the results:

A	13	23	28	49	67	96
q	11	21	39	38	42	52

10.35 Corrosion penetration (Y) is related to the years of exposure (X). Make a complete analysis of the data using a linear model.

X	0.5	1.0	1.5	2	4	8	16
Y	26	41	51	58	62	66	71

10.36 The dissolved chloride concentration (C, mg/L) varies with the discharge (Q, ft^3/sec). Using a linear model, make a complete analysis for predicting C.

Q	265,000	163,000	76,700	60,200	32,700	22,200
C	6.4	6.1	9.2	27.0	17.0	18.0

10.37 Using the principle of least squares, derive the normal equations for the following model:

$$\hat{Y} = b_1 X_1 + b_2 X_2 + b_3 X_3$$

10.38 Using the principle of least squares, derive the normal equations for the following model:

$$y = b_1 x_1 + b_2 x_1 x_2 + b_3 x_2^2$$

10.39 Using the method of least squares, derive the normal equations for the model $\hat{Y} = b_1 X + b_2 X^2$. Evaluate the coefficients using the data of Problem 10.15.

10.40 Using the method of least squares, derive the normal equations for the model $\hat{Y} = b_0 + b_1 \sqrt{X}$. Evaluate the coefficients using the data of Problem 10.15.

10.41 Starting with the general multivariate model

$$y = b_0 + b_1 x_1 + b_2 x_2 + b_3 x_3$$

develop the relations between the partial regression coefficient and the standardized partial regression coefficients.

10.42   The cost of producing power (P) in mills per kilowatt-hour is a function of the load factor (L) in percent and the cost of coal (C) in cents per million Btu. Perform a regression analysis to develop a regression equation for predicting P using the following data:

L	C	P
85	15	4.1
80	17	4.5
70	27	5.6
74	23	5.1
67	20	5.0
87	29	5.2
78	25	5.3
73	14	4.3
72	26	5.8
69	29	5.7
82	24	4.9
89	23	4.8

10.43   Using the first six sets of data of Problem 10.42 fit the multiple regression coefficients for Equation 10.57. Then fit the second six data sets. For each case compute S_e, S_e/S_4, R, and R^2. Compare the regression coefficients of the two analyses and discuss possible causes of differences.

10.44   Visitation to lakes for recreation purposes varies directly with the population of the nearest city or town and inversely with the distance between the lake and the city. Develop a regression equation to predict visitation (V) in person-days per year, using as predictors the population (P) of the nearby town and the distance (D) in miles between the lake and the city. Use the following data:

D	P	V
10	10,500	22,000
22	27,500	98,000
31	18,000	24,000
18	9,000	33,000
28	31,000	41,000
12	34,000	140,000
21	22,500	78,000
13	19,000	110,000
33	12,000	13,000

10.45   Using the data of Problem 10.32, fit a linear model and a quadratic polynomial and make a complete assessment of each. Which model is better for prediction? Support your choice.

10.46   Using the data of Problem 10.33, fit a linear model and a quadratic polynomial and make a complete assessment of each. Which model is better for prediction? Support your choice.

10.47   Using the data of Problem 10.4, fit linear and quadratic models and assess the goodness of fit of both. Which model is better for prediction? Support your choice.

10.48   Using the data of Problem 10.35, fit linear and quadratic models and assess the goodness of fit of both. Which model is better for prediction? Support your choice.

10.49   Using the data of Problem 10.3, fit linear and quadratic models and assess the goodness of fit of both. Which model is better for prediction? Support your choice.

10.50   Using the method of least squares, derive the normal equations for the model $\hat{Y} = ax^b$. Evaluate the coefficients using the data of Problem 10.15. (*Hint*: Convert the equation to linear form by making a logarithmic transformation.)

10.51   Using the data of Problem 10.32, fit a linear model and a power model. Make a complete assessment of each. Which model is better for prediction? Support your choice.

10.52   Using the data of Problem 10.33, fit a linear model and a power model. Make a complete assessment of each. Which model is better for prediction? Support your choice.

10.53   Using the data of Problem 10.42, fit a multiple power model. Make a complete assessment of the model.

10.54   Using the data of Problem 10.4, fit a power model and assess its goodness of fit.

10.55   Using the data of Problem 10.34, fit a power model and assess its goodness of fit.

10.56   Using the data of Problem 10.35, fit a power model and assess its goodness of fit.

10.57   Using the data of Problem 10.3, fit a power model and assess its goodness of fit.

Numerical Optimization

11.1 INTRODUCTION

Numerical optimization has the same objective as analytical optimization: to find the values of coefficients of a function that yield best estimates of some variable. Several types of models or functions can be solved analytically, as discussed in Chapter 10. The linear model of Equation 10.15 is the most basic. Multiple regression models (see Equation 10.55) and polynomial functions (see Equation 10.99) are easily solved analytically. A power model (see Equation 10.100) can be solved analytically using the logarithms of the variables. Similarly, the coefficients of an exponential decay function can be fitted analytically using a logarithmic transformation. Analytical solutions are generally easy to obtain, which is a desirable characteristic, but they are only available for simple functional forms.

It is more difficult to find the optimum set of coefficients of complex functions or functions that are subject to constraints. For example, a sine curve has four characteristic variables (i.e., the mean, the amplitude, the phase angle, and the frequency); these cannot be fitted analytically. For such models, a numerical analysis is a viable alternative. Analytical and numerical solutions have many common characteristics. They most often use the same objective function (i.e., least squares). The analytical solution is based on the analytical derivatives of the objective function with respect to the unknown coefficients. For a numerical solution, the derivatives are computed numerically.

A principal difference between the analytical and numerical solutions is that iteration is required for the numerical solution. Additionally, an analytical solution yields an exact solution, while a numerical solution generally only approaches the true solution. Numerical optimization requires initial estimates of the coefficients being fitted. If the initial estimates are highly erroneous, the solution procedure may not converge to an optimum solution.

Problems can be formulated to maximize or to minimize a function. For example, where profit is the objective, the optimization will likely be one of maximization; that is, the objective is to find the values of the function that maximize the profit. Some problems seek to find a minimum. The least squares problems of Chapter 10 are examples of minimization—that is, finding the values of the unknowns that minimize the objective function. Most of the problems of this chapter will involve minimization, but all of the concepts apply equally to problems of maximization.

11.2 THE RESPONSE SURFACE ANALYSIS

The solution of a numerical optimization problem can be graphically portrayed by graphing the value of the objective function F versus each of the unknowns. Therefore, a model with two unknown coefficients can be viewed as a three-dimensional graph (i.e., F vs. b_1 and b_2). Such graphs are referred to as response surfaces. If the least squares F was computed for many combinations of b_1 and b_2 and the values of F graphed as a function of the b values, the resulting graph would appear as a bowl. A specific set of coefficients, b_1 and b_2, would represent the values that produced the minimum value of F, which would appear as the low point of the bowl. This value of F would be the point that provides the smallest root mean square error, RMSE.

This visual representation of the problem is useful to understanding the procedure for numerically optimizing a function. In numerical optimization, a set of values of the unknowns serves as the starting point; this combination of values is referred to as the initial base point (i.e., a point on the surface of the bowl). The objective is to move from this base point to the bottom of the bowl, where the derivatives $\Delta F/\Delta b_i$ will be zero.

The complexity of the shape of the bowl depends on the complexity of the function being fitted. A simple function, such as the linear model of Equation 10.15, will produce a simple response surface (i.e., one that may look like a kitchen mixing bowl). However, a complex function, especially one subject to constraints, will produce a more complex response surface. For such surfaces, movement over the surface requires a more complex strategy and a better understanding of the numerical procedure.

Example 11.1: Response Surface for a Maximization Problem

Consider the function

(11.1)
$$F = -0.4b^2 + 2b + 2$$

The objective is to find the value of b that maximizes the value of F. The relation between F and b is shown in Figure 11.1. An analytical solution is easily found by setting the derivative dF/db equal to zero and solving for the value of b, which is found to be 2.5. This gives maximum F of 4.5. But what is the option if the function is not easily differentiated to find an analytical solution? Then a value of b can be found by starting at some point on the response function of Figure 11.1 and systematically moving to the peak.

Example 11.2: A Least Squares Response Surface

Least squares is a problem of minimization. Rather than a closed form function such as that of Equation 11.1, the procedure is used

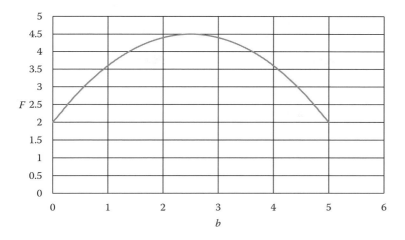

FIGURE 11.1   Response surface for a maximization problem.

to fit a model to a set of data. This was discussed in Chapter 10. For example, the zero-intercept model, which was illustrated in Section 10.7.4, relates a criterion variable Y and a predictor variable X:

(11.2) $$\hat{Y} = bX$$

The value of b is unique to the data set, such as the following data:

X	3	5	6
Y	2	3	4

The analytical least squares solution is

(11.3) $$\hat{Y} = 0.64286$$

The response surface for the model would appear as values of F versus b, not Y versus X. For the given data set, the response surface is shown in Figure 11.2, which appears as a two-dimensional bowl. From the graph it is evident that the optimum value of b lies between 0.6 and 0.7. The objective of a numerical analysis would be to start at some point (e.g., $b = 0.4$) and numerically move to the optimum.

11.3 NUMERICAL LEAST SQUARES

As with analytical optimization, the objective of the numerical procedure is to minimize the sum of the squares of the errors (F) relative to one or more unknown coefficients (b_i), where an error is the difference between a computed value of the criterion variable and the corresponding measured value. Numerical least squares require initial estimates of the unknown coefficients; these are inputs to the algorithm. The location of

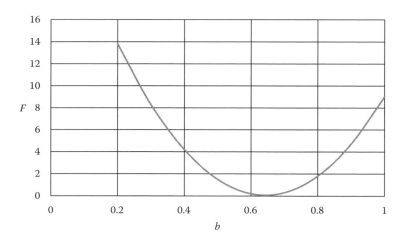

FIGURE 11.2   A least squares response surface.

the unknown coefficients is referred to as the base point, and the objective function value at the base point can be denoted by a subscript o (i.e., F_o). The values of the coefficients at the base point are denoted as b_{oi}, where i refers to the ith unknown coefficient.

To optimize the objective function, derivatives will be needed for each unknown. To compute the derivative of F for coefficient i, the base point value b_{oi} is incremented by an amount Δb_i, which is often referred to as the step size. The objective function F_{1i} at the incremented value of coefficient i is computed as a function of b_{1i}, where $b_{1i} = b_{oi} + \Delta b_i$. Thus the numerical estimate of the derivative d_i is

(11.4)
$$d_i = \frac{F_{1i} - F_o}{b_i}$$

When computing F_{1i} for coefficient i, all other coefficients are held at their base point values, b_{oi}.

Example 11.3: Computing Numerical Derivatives

If $b = 0.4$ is considered the base point for the function of Example 11.2, the derivative of F with respect to b can be estimated by computing the value of F at $b = 0.4$ and $b = 0.41$ and solving Equation 11.4. In this case $F_0 = 4.2$ and $F_1 = 3.867$. Therefore, the estimate of the derivative is

(11.5) $\Delta F/\Delta b = (F_1 - F_0)/(b_1 - b_0) = (3.867 - 4.2)/(0.41 - 0.40) = -33.3$

The minus sign indicates that the slope is negative at $b = 0.4$.

If $b = 0.9$ were selected as the base point, then the derivative would be

(11.6) $\Delta F/\Delta b = (F_1 - F_0)/(b_1 - b_0) = (5.067 - 4.7)/(0.91 - 0.90) = 36.7$

In this case, the derivative is positive, as is apparent from Figure 11.2.

11.4 STEEPEST DESCENT METHOD

The steepest descent method is a special case of the numerical least squares approach. The intent of this method is to find values for a set of coefficients that minimize the sum of the squares of the errors. The same procedure can be used to solve a maximization problem; this is called the steepest ascent algorithm. The inputs for the method include:

1. Initial estimates of the unknown coefficients
2. Step sizes Δc that are used to compute the derivatives for determining the incremental changes in the coefficients
3. The tolerance, T, used to assess convergence
4. The distance, D, for movement, which is also referred to as a step size
5. The sample size, n
6. The n values of the p predictor variable(s) and criterion variable, Y
7. The maximum number of iterations, i_{max}

The procedure is iterative in that it starts at a base point, which is defined by the initial estimates of the unknown coefficients, and progressively moves along the response surface toward the optimum value of F. The value of the objective function at the base point is denoted as F_o. Then the following iterative process is used to move from the base point across the response surface to a point near the optimum:

1. Compute the derivatives $d_j = \Delta F / \Delta b_j$ of the objective function F with respect to each of the coefficients; each of the d_j values represents the slope of the response surface in the direction of coefficient j.
2. Compute the weighted distance C, which is used to compute the adjustment to each coefficient for the iteration:

(11.7)
$$C = D \Big/ \left(\sum d_j^2 \right)^{0.5}$$

in which d_j is the derivative of the objective function with respect to coefficient b_j and is computed using Equation 11.4.
3. Adjust the most recent value of each coefficient using

(11.8)
$$b_{jt+1} = b_{jt} - C d_j$$

where the subscript j refers to the jth coefficient, and the subscript t refers to the tth iteration.
4. Compute the objective function F_{t+1} for iteration $t + 1$.
5. Check for convergence: objective function
   a. If $F_{t+1} < F_t$, then the new set of coefficients provide a better value of the objective function; proceed to step 7.

b.   If $F_{t+1} \geq F_t$, then the new set of coefficients did not improve the objective function; iteration can be discontinued and the previous set of coefficients are considered the optimum.

6. Check for convergence: tolerance

a.   If $|F_{t+1} - F_t| < T$, then assume convergence as the objective function did not change significantly; the most recent estimates of the coefficients are considered the optimum values.

b.   If $|F_{t+1} - F_t| > T$, then continue iterating; note that because of step 5a, the convergence cannot be assumed; return to step 1.

7. Check for discontinuance of iteration. If the number of iterations exceeds i_{max}, then assume the solution will not converge and that continuing the iteration is not likely to improve the objective function. If the number of iterations is less than i_{max}, return to step 1 using the most recent point as the new base point.

For a maximization problem, the negative sign in Equation 11.8 should be replaced with a plus sign. Note in Equation 11.8 that the change in a coefficient is directly proportional to the magnitude of the derivative. The steeper the slope of the response surface is, the more important the coefficient and any variable associated with that coefficient will be. Unimportant coefficients have shallow slopes, which results in slow movement across the response surface—that is, small changes in the term (Cd_j) for any one iteration. It can be expected that an unimportant coefficient may not approach its true value because the tolerance causes iteration to discontinue before the unimportant coefficient converges to its true value.

11.4.1 THE DISTANCE D

One input to the steepest descent algorithm is the distance to be moved for each iteration. The larger the value of D is, the faster the algorithm iterates to the solution; that is, a fewer number of iterations is required. However, the larger the value of D is, the sooner the tolerance may cause iteration to discontinue and thus the more distant the solution will be from the optimum. The implications of this are that (1) the selection of the value of D should be chosen in coordination with the value of the tolerance for convergence and (2) having reached an apparent optimum, the program should be tried again using a smaller tolerance or a smaller D to allow further movement toward the optimum solution. As additional runs are made, the step sizes Δb_j should be reduced to get more accurate estimates of the derivatives.

11.4.2 THE TOLERANCE FOR CONVERGENCE

As the set of coefficients approaches the point of zero gradient, the slope of the response surface decreases and movement causes very little change in the objective function F (i.e., $|F_{t+1} - F_t|$ decreases). Additional iterations do not result in a significantly lower sums of squares of the errors; that is, the optimum has been approached. To keep the number of iterations to a minimum, the algorithm for numerical least squares uses a tolerance, T, which can be used to discontinue the iterative process. Specifically, when the change in the objective function ΔF_j (where subscript j refers to the iteration number) from one iteration to the next ($\Delta F_j = |F_j - F_{j+1}|$) is less than the tolerance T, then iteration is discontinued and the values of the coefficients for iteration $j + 1$ are considered the optimum values.

The selection of the tolerance is an important decision. If the objective function is the sum of the squares of the errors, then the tolerance would have the same units. For example, if the criterion variable Y has units of force, then the least squares solution will have units of force squared, which will be the units of F. The sums of the squares will depend on the magnitude of the variables and the sample size. Therefore, the value of F will be quite variable, which makes it difficult to set a value of the tolerance. For example, if the mean value of Y is 10 and a set of b_i values produce predicted values $\hat{Y}$ of 15 on average, then the sums of squares for a sample of 10 would be 250, while a sample size of 20 would produce an F value of 500. This makes it difficult to estimate a good value of T, as it would be different for $F = 250$ versus $F = 500$.

One possible solution to this problem is to use the RMSE or the relative RMSE as the objective function. The relative RMSE is especially attractive as a decision criterion because it is dimensionless and somewhat insensitive to sample size. The RMSE is given by

$$(11.9) \qquad \mathrm{RMSE} = \left[\frac{1}{n-p} \sum e^2 \right]^{0.5} = \left[\frac{1}{n-p} \sum (\hat{Y} - Y)^2 \right]^{0.5}$$

where p is the number of fitted coefficients. The relative root mean square error is computed by

$$(11.10) \qquad S_e/S_Y = \mathrm{RMSE}/S_Y$$

where S_Y is the standard deviation of the criterion variable Y. Therefore, a tolerance of 1% or 0.1% could be applied to a wide variety of numerical solutions.

11.4.3 ADVANTAGES AND DISADVANTAGES

Numerical optimization has several advantages. First, complex functions can be calibrated. This includes both single-valued functions of complex form or multifunction (i.e., composite) forms. For example, a single-valued function such as a variable intercept power model with both ordinate and abscissa location parameters could not be fit analytically:

$$(11.11) \qquad \hat{Y} = b_1 + e^{b_2(X-X_o)}(X - X_o)^{b_3}$$

where b_1 is the intercept coefficient, X_o is the location parameter for the X axis, and b_2 and b_3 are empirical coefficients; four coefficients would need to be fitted (b_1, b_2, b_3, and X_o).

In many cases, data are characterized by different shapes for different regions of the variables. For example, one portion of the data may appear linear while for other values of X the variation may be nonlinear. For example, if the first part of the data appears as an increasing function with a decreasing slope and the second part of the data set—that is, the values are greater than X_C—then the following composite model would be appropriate:

$$
(11.12) \qquad \hat{Y} = \begin{cases} b_1 e^{b_2 X} X^{b_3} & \text{for} \quad X \leq X_C \\ b_4 + b_5 (X - X_C) & \text{for} \quad X > X_C \end{cases}
$$

(11.13)

Functions such as these are easily fitted with data using a numerical procedure.

Nonlinear models fitted using least squares can produce biased predictions (i.e., the sum of the errors is not equal to zero). An unbiased model is often preferred to a biased model. The bias should always be computed as it is a good measure of the systematic error (see Section 3.2 in Chapter 3). While methods of unbiasing analytical functions are available, they may produce unacceptable results. Unbiasing is easily handled using numerical optimization.

Equations 11.12 and 11.13 include a constraint on the range of X values for the two functions. Other types of constraints are sometimes desired. For example, a least-squares fitted model that produces positive predicted values may be needed, where unconstrained coefficients may lead to negative predicted values. The constraints needed to prevent negative predictions can be included in the numerical algorithm. For example, analytical least squares solution of a linear model, $\hat{Y} = b_0 + b_1 X$, often produces a negative intercept (b_0). This is easily handled in numerical optimization using a statement such as a command to set $b_0 = 0$ if the solution value of $\hat{Y}$ is negative; this, of course, will cause the slope coefficient (b_1) to differ from the unconstrained solution, but the objective of non-negative predictions would be met.

Numerical optimization has disadvantages. First, it requires more expertise than analytical optimization since it can provide an inexact solution that depends in part on the user's understanding of and experience with the convergence criterion. Convergence depends on both the error tolerance T and the step size D for movement. Specifically, it is the interdependence of these two factors that confounds the optimization. If the bottom of the response surface is relatively flat, then a small tolerance should be used; otherwise, iteration will stop prematurely. If the step size D is too large, then the optimum set of coefficients can be overshot (i.e., move from one side of the response surface, past the optimum, and to the other side). For important problems, it is necessary to try several different combinations of the step size D and the tolerance T to ensure that the optimum solution has been achieved.

A second disadvantage, but generally one of minimal significance, is that initial values of the unknowns need to be estimated. These can usually be approximated using mean values of the inputs. However, it is often advisable to make several trial runs using widely different, but feasible, estimates.

Example 11.4: Selection of T and D

Consider the response surface of Figure 11.2. If the base point of $b = 0.4$ were selected, with the optimum being 0.643, then it might seem reasonable to select a step size for movement of 0.02. If being within $|0.02|$ was the desired objective, then a tolerance of 0.01 may be needed. With a D of 0.02, it would take at least $(0.643 - 0.40)/0.02 = 12.15$ (or 13) iterations to converge. If at an iteration $b = 0.604$, then F_0 would equal 0.17712. Using a step size of Δb of 0.001, then F_1 would be 0.17175 and the derivative of Equation 11.4 would be

(11.14) $\Delta F/\Delta b = (0.17175 - 0.17712) / (0.605 - 0.604) = -5.37$

Therefore, the next estimate of C would be computed using Equation 11.7:

(11.15) $C = 0.02 / (-5.37) = -0.00372$

and the new estimate of b is obtained using Equation 11.8:

$b_1 = b_0 - Cd = 0.604 - (-0.00372)(5.37) = 0.6241$

which would give a value of 0.09606 for the objective function F. Thus, the objective function was reduced by 0.0811. Since this is greater than the tolerance of 0.01, iteration would continue.

11.5 ILLUSTRATING APPLICATIONS

This section includes numerous examples that illustrate the method of steepest descent. In general, the functions apply simple models, often those where an analytical solution is available. This enables comparisons to be made between analytical and numerical solutions. More detailed applications are given in Section 11.6.

Example 11.5: Minimize a Two-Variable Function and Show the Effect of the Step Size

This is a simple case of a second-order polynomial with two coefficients, b_1 and b_2:

(11.16) $F = 30b_1^2 + 2b_2^2 - b_1 + 2b_2 - 5$

TABLE 11.1   Example 11.5: A Two-Variable Function (True Solution $b_1 = 0.016667$, $b_2 = -0.5$, $F = -5.50833$)

	Initial			Final			
Run	b_1	b_2	D	b_1	b_2	F	n_i
1	1	1	0.100	0.01012	-0.4131	-5.4469	22
2			0.010	0.06265	-0.5007	-5.5075	224
3			0.001	0.011667	-0.5014	-5.5076	2238

where b_1 and b_2 need to be fitted. The analytical solution yields a minimum value when $b_1 = 1/60$ and $b_2 = -0.5$. The purpose of this example is solely to show the effect of the step size distance. The three runs started with the same initial estimates, $b_1 = b_2 = 1$, and incremental step sizes of 0.01 to compute the derivatives. Table 11.1 summarizes the results of the three analyses.

The results indicate that the value of D has a significant effect on the resulting solution. The number of iterations is an inverse function of the value of D. While b_2 for run 2 was closer to the true value of b_2 than with run 3, the value of b_1 for run 3 was much closer than for run 2. This is dependent on the initial parameters used.

Example 11.6: Minimization a Two-Variable Function and Show the Effect of the Step Size and Tolerance

This example is used to show the effect of the step size D and the tolerance T. The criterion is not least squares, but rather minimizing the function. Convergence is assumed when the difference in the values of the objective function from one iteration to the next becomes less than the input tolerance T. The function is

(11.17) $$F = 3b_1^2 + 2b_2^2 - b_1 + 2b_2 - 5$$

in which b_1 and b_2 are the unknown coefficients to be fitted. An analytical analysis yielded optimum values of $b_1 = 0.1667$ and $b_2 = -0.5$. A summary of the results is shown in Table 11.2. The first run uses the true values as inputs, with immediate convergence.

The second and third runs differ only in the tolerance, with values of 0.01 and 0.001, respectively. Both runs fail to converge. This indicates that, for the given value of D, the solution is insensitive to the tolerance; that is, convergence is not achieved because the step size D is too large.

Run 4 uses a smaller D (0.05) and a tolerance of 0.01. Convergence is completed after 35 iterations, with the values of b_1 and b_2 approaching the true values. The flatness of the response surface and the tolerance conspire to prevent movement closer to the optimum.

In run 5, the D was reduced from 0.05 (run 4) to 0.01, with the same tolerance (0.01). In this case, the final estimates were poorer even though the value of D was smaller. This shows that decreasing

TABLE 11.2   Example 11.6: Minimization of a Two-Coefficient Function

Run	Initial				Final			
	b_1	b_2	Tol	D	b_1	b_2	F	n_i
1	0.1667	−0.5	0.01	0.05	0.1667	−0.5000	−5.5833	1
2	1	1	0.01	0.20	0.3019	−0.5050	−5.5284	3333 max
3	1	1	0.001	0.20	0.3019	−0.5050	−5.5284	3333 max
4	1	1	0.01	0.05	0.1602	−0.5103	−5.5830	35
5	1	1	0.01	0.01	0.2134	−0.2679	−5.4690	150
6	1	1	0.001	0.05	0.2019	−0.5050	−5.5796	3333 max
7	2	2	0.01	0.05	0.1604	−0.4978	−5.5832	63

D may not always improve the fit. The decrease in D resulted in smaller step sizes and smaller changes in the objective function. Thus iteration was discontinued because the changes in the objective function were less than the tolerance. Therefore, the tolerance was too large for the given value of D.

Run 6 used the same D as run 4, but decreased the tolerance from 0.01 to 0.001. The solution for run 6 was not better than that of run 4; b_2 was closer to its optimum but b_1 was not as close to its true value. The value of F was slightly poorer (i.e., further from the true minimum).

In run 7, the initial parameter values were moved further from those of runs 2 to 6. The values of the tolerance and D were selected based on the best of runs 2 to 6, which was run 4. While more iterations were needed for run 6 than for run 4, the solution was actually improved, but very marginally. The solution process traversed the response surface differently, but suggested that the optimum could be achieved even if the initial coefficients are far from the optimum as long as the step size D and the tolerance T are appropriately selected.

Example 11.7: Zero-Intercept Model

This example illustrates the use of numerical analysis for a zero-intercept model. The analyses show the effects of the initial estimate, the step size, and the tolerance. The data of Table 11.3 suggest a linear trend that passes through the origin of the X–Y axes. An analytical analysis yielded the following model:

(11.18) $$\hat{Y} = 0.92647X$$

TABLE 11.3   Data for Zero-Intercept Model

X	1	2	3	4	5	6	7	8
Y	2	2	2	5	3	5	8	7

with a relative bias of −1.9%, a standard error ratio S_e/S_Y of 0.4746, and a correlation coefficient R of 0.8802. Overall, the analytical model provided for a good fit with the data.

The first run used the analytical solution as the initial estimate, which led to the same solution as found analytically. Convergence occurred on the first iteration.

Runs 2 to 6 used an initial estimate of 2 for the slope coefficient (see Table 11.4). The five analyses differed in the values of the step size D and the tolerance T. Run 2 used a large tolerance (0.001) and a large step size (0.05). It converged in 22 iterations to a coefficient that differed from the true value by only 2.5%. Reducing the tolerance from 0.001 to 0.0001 (run 2 to run 3) did not change the result. In this case, D was the controlling factor, as it was too large to allow movement.

In run 4, the step size was reduced from 0.05 (run 2) to 0.005. This increased the required number of iterations and yielded a slightly better fit. The slope coefficient was a 1.1% closer to the true value. In run 5, the step size D was decreased even more (from 0.005 of run 4 to 0.0005), but this resulted in a poorer fit even though the number of iterations increased significantly. The loss of accuracy was due to the movement being stopped because the small step size resulted in a step-to-step reduction in accuracy that was less than the tolerance. Thus the tolerance was the controlling factor. The standard error ratio became poorer (i.e., increased significantly, from 0.4756 in run 4 to 0.9985 in run 5).

In run 6, the tolerance T was reduced to 0.0001, with the step size D maintained at 0.0005 (same as run 5). Now the optimum coefficient was approached because the step size and tolerance were more compatible with each other. The number of iterations was large because both the tolerance and step size were small; thus movement was slow and differences in the objective function from iteration to iteration were allowed to be small.

For run 7, the initial parameter was much different from the true value and the initial values of runs 2 to 6. However, the solution did converge, with the coefficients within 1.4% of the true value. The model provided a high level of goodness of fit (i.e., low bias and good S_e/S_Y).

TABLE 11.4   Results for Zero-Intercept Model

| Run | Initial | | | Final | | | | |
	c_1	Tol.	D	c_1	$\bar{e}/\bar{Y}$	S_e/S_Y	R	n_i
1	0.9265	0.001	0.005	0.9265	−0.019	0.4746	0.8802	1
2	2.0	0.001	0.05	0.9500	0.006	0.4770	0.8786	22
3	2.0	0.0001	0.05	0.9500	0.006	0.4776	0.8786	22
4	2.0	0.001	0.005	0.9400	−0.005	0.4756	0.8797	212
5	2.0	0.001	0.0005	1.3131	0.390	0.9985	0.0543	1374
6	2.0	0.0001	0.0005	0.9446	−0.000	0.4764	0.8792	2111
7	5.0	0.001	0.005	0.9399	−0.005	0.4756	0.8797	812

Example 11.8: Fitting a Power Model

The next analysis fitted a simple power model. An analytical analysis based on a logarithmic transform of both Y and X was first fitted to the database of Table 11.5, which resulted in the following equation:

$$(11.19) \qquad \hat{Y} = 3.484 X^{0.3136}$$

The goodness-of-fit statistics were $S_e/S_Y = 0.0267$ and $R^2 = 0.9994$.

The objective of this analysis was to illustrate the effects of the step size, the tolerance for convergence, and the initial parameter estimates. Nine numerical analyses were made (see Table 11.6).

The first analysis showed that the solution converged immediately with the same goodness-of-fit statistics as the analytical solution. For runs 2 through 9, the initial estimates of the coefficients deviated from the true coefficients. The initial estimates of the two coefficients for runs 2 to 5 were not too far from the true values, but the step size and tolerance influenced the extent of convergence. Runs 2 and 5 used the same step size ($D = 0.001$) but different tolerances (0.0001 vs. 0.00005); the same solution resulted. In this case, the flatness of the response surface caused the variation of the tolerance to have no effect.

Runs 2 and 3 (Table 11.6) used the same initial parameter estimates and tolerance ($T = 0.0001$), but had a different step size D (0.001 vs. 0.01). In this case, the step size made a significant difference in the convergence. The smaller step size (run 2) allowed

---

TABLE 11.5   Data for the Power Model of Equation 11.19

X	1	2	3	4	5	6
Y	3.5	4.3	4.9	5.4	5.8	6.1

TABLE 11.6   Results for Example 11.8

Run	Initial Coefficients C_1	C_2	Inputs Tol.	D	Final Coefficients C_1	C_2	$\bar{e}/\bar{Y}$	Goodness of Fit S_e/S_Y	R	n_i
1	3.484	0.3136	0.0001	0.001	3.484	0.3136	0.000	0.02680	0.9994	1
2	4	0.5	0.0001	0.001	3.557	0.2998	0.004	0.05856	0.9973	670
3	4	0.5	0.0001	0.010	3.773	0.2548	0.009	0.20400	0.9667	72
4	4	0.5	0.0010	0.001	3.945	0.2322	0.028	0.3104	0.9229	274
5	4	0.5	0.00005	0.001	3.557	0.2998	0.004	0.5856	0.9973	670
6	10	0.5	0.0001	0.001	6.375	−0.1112	0.131	1.803	0.0000	4444
7	10	0.5	0.0001	0.010	3.775	0.2545	0.009	0.2049	0.9664	750
8	10	0.5	0.0001	0.050	4.583	0.0999	0.025	0.7546	0.5444	189
9	10	0.5	0.00005	0.010	3.775	0.2545	0.009	0.2049	0.9664	750

the procedure to get closer to the true solution. The standard error ratio improved from 0.204 (run 3) to 0.059 (run 2), which reflects a much better fit to the data. Values of both fitted coefficients were closer to the true values when the step size was smaller. In this case, the smaller tolerance of run 2 (vs. run 4) with the same step size D also allowed the true optimum to be approached more closely. The standard error ratio for run 4 was significantly poorer than for run 2 (i.e., a 430% increase).

Runs 5 to 9 used the same initial estimates of the coefficients, but the first coefficient was different from that used with runs 2 to 5 ($c_1 = 4$ vs. $c_1 = 10$). Run 6 used the same tolerance and step size as run 2, but the results were much poorer, as convergence was not achieved in run 6. The model showed a significant bias (13.1% over-prediction) and did not explain any variance ($R^2 = 0$). The iterations were stopped at the limit of 4444. The large difference in the initial estimate of c_1 prevented getting sufficiently close to the true value.

If runs 6 and 9, which have the same initial parameters, are compared, the analyses differ in the tolerances and step sizes. The larger step size of run 9 allows it to get closer to the true solution, but it is still not sufficiently close, which is reflected in the level of fit. The flatness of the response surface combined with the tolerance T caused the iterating to stop

Runs 7 and 8 differ only in the step size. With run 8, the larger step size (0.05 vs. 0.01) prevented the solution from converging. The larger step size of run 8 on the shallow response surface caused the derivative to be too small and the change in the objective function was less than the tolerance, so the iteration stopped. A large step size D is not always preferred.

Example 11.9: Fitting a Bivariate Power Model to Wide Data Range

While this example uses a simple power model, the data show considerable variation, which confounds the fitting. Example 11.8 also involved a power model, but the data covered a small range of X and Y values. In this example, the values of X vary from 7 to 8100 and the Y values from 2 to 148 (see Table 11.7). A logarithmic analysis was made, and the analytical solution produced the following biased model:

$$(11.20) \qquad \hat{Y} = 1.150 X^{0.5107}$$

This produced the following statistics: $\bar{e} | \bar{Y} = 7\%$, $S_e/S_Y = 0.357$, and $R^2 = 0.904$. The correlation coefficient in the log-space was 0.985, which is not reflective of the accuracy of predictions of Y, only log Y.

TABLE 11.7 Database for Example 11.9

X	7.389	20.087	403.429	2981	8103
Y	2.718	7.389	20.086	54.598	148.41

Because of the significant bias, the intercept coefficient of 1.150, which was based on the log–log fitting, was changed to 1.244, which resulted in the following unbiased (i.e., $\bar{e} = 0$) model:

(11.21) $Y = 1.244X^{0.5107}$

In addition to eliminating the bias, the S_e/S_Y improved to 0.310 and the R^2 increased to 0.928.

Run 1 (see Table 11.8) used the solution of Equation 11.20 as the initial conditions. Numerical optimization minimized the sum of the squares of the Y values, while the analytical solution of Equation 11.20 minimized the sum of squares of the logarithms of Y. Thus, the numerical solution is slightly different from the analytical solution. The overall goodness-of-fit statistics are better than those of the analytical solution because the analytical analysis optimizes in the logarithmic space, while the numerical solution is based on Y-space values.

Run 2 is a numerical analysis that uses the coefficients of Equation 11.21 as the initial estimates. Very little movement occurred because the response surface is relatively flat, so convergence was quickly reached (10 iterations).

In Run 3, the initial values of the coefficients were moved from the true values. Even after 99 iterations, the intercept coefficient c_1 had not moved toward the true value. This reflects the insensitivity of the model accuracy to c_1. The value of c_2 had moved close to its true value because the least squares criterion is more sensitive to the slope coefficient for this data set. Overall, the goodness-of-fit statistics are not too much different for this run from those of run 1 (i.e., a difference in R^2 of just 1.13%). This run shows that a numerical solution may converge but the solution may not be exactly at the optimum.

Run 4 used the same initial estimates for the two coefficients, but the step size was changed from 0.001 to 0.01. Convergence was achieved after only 11 iterations because the step size was not causing as much variation in the least squares sum as was allowed by the tolerance. The goodness-of-fit statistics for runs 3 and 4 were essentially identical.

TABLE 11.8   Results for Example 11.9

Run	Initial Coefficients C_1	Initial Coefficients C_2	Inputs Tol.	Inputs D	Final Coefficients C_1	Final Coefficients C_2	Goodness of Fit $\bar{e}/\bar{Y}$	Goodness of Fit S_e/S_Y	Goodness of Fit R^2	Goodness of Fit n_i
1	1.14994	0.5107	0.0001	0.001	1.1514	0.5286	0.070	0.2803	0.9411	19
2	1.24367	0.5107	0.0001	0.001	1.2442	0.5197	0.075	0.2974	0.9380	10
3	1.5	0.4	0.0001	0.001	1.507	0.4977	0.091	0.3059	0.9298	99
4	1.5	0.4	0.0001	0.010	1.507	0.4977	0.109	0.3072	0.9292	11
5	1.5	0.51	0.0001	0.010	1.499	0.5000	0.105	0.3063	0.9296	2
6	1.5	0.4	0.0001	0.010	1.5299	0.4989	0.118	0.3099	0.9280	41

In run 5, the initial estimate of c_2 was more deviant from the true value than the estimates for the other trials. The solution converged after 41 iterations. The final estimate of c_1 did not move closer to the true value as it has a very small derivative—that is, a shallow slope of the response surface. The increased value of c_1 caused the relative bias to increase slightly.

Example 11.10: Fitting a Variable-Intercept Power Model

This example illustrates the numerical optimization of a nonlinear equation, while showing the effects of the initial values of the coefficients, the step size, and the tolerance. A variable-intercept power model is fitted with values of a single predictor variable X:

$$ \hat{Y} = c_1 e^{c_2 X} X^{c_3} \tag{11.22} $$

The data set has values of Y that are skewed to the right (see Table 11.9). The analytical solution, which requires a logarithmic transformation of Y and X, yielded the following equation.

$$ \hat{Y} = 1.323 e^{0.2086 X} X^{0.3561} \tag{11.23} $$

Equation 11.23 was unbiased with a standard error ratio S_e/S_Y of 0.2228 and an R^2 of 0.9996.

In addition to a trial run using the coefficients of Equation 11.23 as initial estimates, six runs were made (see Table 11.10). Run 1 converged to the analytical solution as the true values were used as the initial estimates. Runs 2 and 3 differed only in the tolerance (0.005 vs. 0.0005). The change in tolerance did not make a significant difference. In both cases, the intercept did not converge, but the two exponents both converged toward their analytical values.

TABLE 11.9   Database for Example 11.10

X	1.1	1.6	2.2	2.9	3.7	4.1	4.8	5.8	7.3	9.2
Y	1.6	2.2	2.9	3.4	4.6	5.3	6.2	8.4	12.2	19.9

TABLE 11.10   Results for Example 11.10

Run	Initial Coefficients			Inputs		Final Coefficients			Goodness of Fit			
	c_1	c_2	c_3	Tol.	D	c_1	c_2	c_3	$\bar{e}/\bar{Y}$	S_e/S_Y	R^2	n_i
1	1.323	0.2806	0.3561	0.0050	0.001	1.323	0.2086	0.3561	0.000	0.2223	0.9996	1
2	1.5	0.150	0.3	0.0050	0.001	1.504	0.2032	0.3134	0.023	0.0532	0.9978	55
3	1.5	0.150	0.3	0.0005	0.001	1.503	0.2032	0.3140	0.024	0.0528	0.9978	56
4	1.5	0.150	0.3	0.0010	0.010	1.505	0.2080	0.3146	0.056	0.0862	0.9942	7
5	1.5	0.100	0.3	0.0050	0.001	1.508	0.1995	0.3253	0.024	0.0581	0.9974	103
6	1.5	0.010	0.3	0.0050	0.010	1.516	0.1934	0.3473	0.028	0.0663	0.9966	20
7	1.35	0.010	0.3	0.0050	0.010	1.368	0.2028	0.3500	−0.012	0.0528	0.9978	21

Run 4 differed from runs 2 and 3 in the tolerance and the step size. The step size D of run 4 was larger than for runs 2 and 3 (0.01 vs. 0.001), which resulted in faster convergence (7 iterations vs. 55 and 56). c_2 was closer to it true value than was c_3. Run 4 has a significant bias, 5.6% overprediction. While the R^2 is high, the S_e/S_Y is much larger than the value for run 1.

For run 5, the initial estimate of c_2 was moved further away from the true value. The step size D was relatively small and the tolerance T was relatively large. The smaller step size made more iterations necessary, but the solution was nearer to the true solution. Run 5 produced a moderate relative bias of 2.4%.

Run 6 differed from run 5 in the greater deviation of c_2 from the optimum and a larger step size. The iteration stopped (except for c_1) at only 20 iterations because of the larger step size. The goodness-of-fit statistics for runs 2 to 6 were comparable, but in all cases the value of the intercept coefficient did not approach the true value.

In run 7, the initial estimate of c_1 was moved closer to the true value. Relatively large values of D and the tolerance T were used. The solution converged quickly (21 iterations), with the value of c_3 being closer to the true value than any of the other runs. The value of c_1 actually moved further from the true value. The R^2 for runs 2 and 7 are the same even though the final estimates of c_1 are noticeably different. This indicates that the response surface for the intercept coefficient is flat, which means that the coefficient has little influence on the fitting accuracy.

Example 11.11: Fitting a Two-Predictor Model

This example uses a two-predictor, linear model (i.e., a typical multiple regression problem). The values of the two predictor variables are on a grid, which produces an intercorrelation of zero. The data of Table 11.11 were fitted analytically, with the following result:

$$(11.24) \qquad Y = 1.083 + 0.75X_1 + X_2$$

The unbiased model produced an S_e/S_Y of 0.1325 and $R^2 = 0.9868$. The numerical analyses are summarized in Table 11.12. Run 1 was used to verify the analytical model of Equation 11.24.

Runs 2 to 6 used the same initial parameter estimates but differed in the step sizes and tolerances. Run 2 used moderate values of D and T but did not converge to the optimum, with the final estimates of the intercept and slope coefficient for X_1 being far from their optimum values. Run 3 used a larger step size ($D = 0.01$) than

---

TABLE 11.11   Database for Example 11.11

X_1	1	1	1	3	3	3	5	5	5
X_2	1	3	5	1	3	5	1	3	5
Y	3	5	7	4	6	8	6	8	10

TABLE 11.12   Results for Example 11.4

Run	Initial Coefficients			Inputs		Final Coefficients			Goodness of Fit			
	c_1	c_2	c_3	Tol.	D	c_1	c_2	c_3	$\bar{e}/\bar{Y}$	S_e/S_Y	R^2	n_i
1	1.083	0.75	1.00	0.0010	0.001	1.083	0.743	0.993	0.000	0.1325	0.9869	1
2	2.0	1.00	1.50	0.0010	0.001	1.850	0.576	0.997	0.037	0.2459	0.9574	670
3	2.0	1.00	1.50	0.0010	0.010	1.667	0.662	0.912	0.009	0.1776	0.9763	94
4	2.0	1.00	1.50	0.0010	0.020	1.764	0.654	0.912	0.020	0.1923	0.9723	43
5	2.0	1.00	1.50	0.0001	0.010	1.372	0.704	0.954	0.002	0.1454	0.9841	156
6	2.0	1.00	1.50	0.0001	0.020	1.453	0.689	0.939	0.000	0.1545	0.9821	96
7	1.5	1.00	1.25	0.0001	0.010	1.362	0.706	0.956	0.002	0.1446	0.9843	54
8	1.5	1.00	1.25	0.0010	0.010	1.396	0.701	0.951	0.003	0.1473	0.9837	44
9	0.5	0.50	0.75	0.0001	0.010	0.805	0.787	1.037	−0.009	0.1414	0.9844	92

in run 2 ($D = 0.001$). Thus, run 3 converged faster than run 2 (94 vs. 670 iterations) and provided better goodness of fit (i.e., $\Delta R^2 = 0.0189$, less bias, and significantly smaller S_e/S_Y. However, the estimation of the intercept was still poor and was only in moderate agreement with the two slope coefficients. The smaller step size of run 2 limited the rate of movement, which caused smaller changes in the objective function, which allowed the tolerance to prematurely discontinue iteration.

Run 4 used a larger step size (i.e., 0.01 vs. 0.01 for run 3 to 0.02 for run 4). The larger step size required fewer iterations (43 vs. 94) but resulted in poorer accuracy; that is, the S_e/S_Y increased from 0.1776 to 0.1923 and the bias increased from 0.009 to 0.020. Even with the same tolerance, the larger step size would not allow the procedure to approach the optimum. The final estimates of c_3 were the same, but for run 4 both c_1 and c_2 were further from the optimum values than in run 3.

For run 5, when compared to run 3, the tolerance was lower, so more iterations were performed and the procedure was able to get closer to the true solution. The bias for run 5 (0.2%) was closer to zero than for run 3 and the S_e/S_Y and R^2 were significantly improved compared to those for run 3—that is, a decrease in the standard error ratio of more than 3%. This shows the importance of the tolerance; however, it does not necessarily imply that a smaller tolerance is better.

For run 6, the step size was increased from 0.01 (run 5) to 0.02. Fewer iterations were required (96 vs. 156) but the solution did not converge closer to the optimum. Both c_2 and c_3 for run 6 were further from the optimum than for run 5. Again, the larger step size D resulted in the procedure not being able to approach the optimum.

If the results for runs 3 to 6 are compared, the effects of tolerances and step sizes are evident. Run 5, with the smaller tolerance and step size, produced the most accurate results, but at the expense of the larger number of iterations. These results should not be viewed as implying that smaller tolerances and step sizes are best. It does, however, imply that various combinations of tolerance and step size should be tried.

For run 7, the initial values of c_1 and c_3 were moved closer to the true values, with the preferred values of the tolerance and step size suggested by the results of run 5. The goodness-of-fit statistics for run 7 were essentially identical to those of run 5, which indicates that, for this case, the initial estimates of the unknown coefficients were not as critical as the tolerance and step size.

Run 8 used a larger tolerance than run 7. The solution converged faster but provided marginally poorer goodness of fit.

Run 9 used initial estimates of the coefficients that were on the other side of the optimum compared to run 5. The accuracy of fit was essentially the same—that is, slightly less biased and little change in R^2 (only 0.03%).

Example 11.12: Fitting a Two-Predictor Linear Model and Examining Convergence

The objectives of this example are (1) to show the effect of the initial estimates of the coefficients and (2) to evaluate the influence of the tolerance on convergence. The model is a two-predictor linear equation with an intercept that can be solved analytically using least squares. The data set is shown in Table 11.13. In this case, the two predictor variables, X_1 and X_2, are uncorrelated because their values are on the nodes of a grid. The resulting analytical solution is

$$(11.25) \qquad \hat{Y} = 1.417 + 0.75X_1 + 3X_2$$

The model is unbiased, with an R^2 of 0.9978 and a standard error ratio of 0.0538.

A series of numerical analyses were made. The analytical derived coefficient was used for the first run (see Table 11.14). The solution converged on the first iteration because the initial estimates were the true values. Runs 2 through 5 used a different set of initial parameter estimates, with none of the solutions converging to the true solution. The two slope coefficients were reasonably close to the optimum values, but the intercept showed significant variation. Changing the tolerance had little effect on the fitted coefficients (see Table 11.14) or the goodness-of-fit statistics. For run 6, the initial estimate of the intercept coefficient was set closer to the true value as compared to runs 2 to 5, but the algorithm did not move closer to the true value because movement was dominated by the slope coefficients. The slope of the response surface for the intercept coefficient is very shallow, which makes it difficult to converge to the true value.

TABLE 11.13   Database for Example 11.12

X_1	1	1	1	3	3	3	5	5	5
X_2	1	3	5	1	3	5	1	3	5
Y	5	11	17	7	13	19	8	14	20

TABLE 11.14   Results for Example 11.12

Run	Initial Coefficients			Inputs		Final Coefficients			Goodness of Fit			
	C_1	C_2	C_3	Tol.	D	C_1	C_2	C_3	$\bar{e}/\bar{Y}$	S_e/S_Y	R^2	n_i
1	1.417	0.75	3.0	0.00100	0.01	1.417	0.750	3.000	0.000	0.0538	0.9978	1
2	1	0.50	2.5	0.00100	0.01	1.142	0.815	3.014	−0.003	0.0599	0.9973	68
3	1	0.50	2.5	0.00010	0.01	1.182	0.787	3.037	−0.001	0.0574	0.9975	74
4	1	0.50	2.5	0.00010	0.01	1.142	0.818	3.011	−0.003	0.0603	0.9973	673
5	1	0.50	2.5	0.00005	0.01	1.182	0.787	3.037	−0.001	0.0574	0.9975	75
6	1.5	0.50	2.5	0.00010	0.01	1.565	0.727	2.977	0.001	0.0553	0.9977	64

11.6 APPLICATIONS

The examples of Section 11.5 were used to introduce and illustrate some of the issues associated with numerical optimization. The strength of techniques like the basic steepest descent method is that it allows complex models to be fitted to data using the same least squares criterion used in analytical optimization. However, the complexity interjects some difficulties, most notably the need for good initial parameter estimates, the difficulty in selecting a step size D, and the problem created by the interaction between the step size D and the convergence tolerance T.

Two applications are provided in this section. First, a set of actual data is used to show that the introduction of a constraint can help overcome an irrational model. Additionally, this model uses a two-function form, with one part using a quadratic model and the second part using a linear model. The coefficients were fitted such that the two models have the same magnitude and the same slope at the intersection point.

The second application illustrates the use of numerical optimization in fitting a multicomponent algorithm that uses time series as inputs and outputs. The fitting of the model parameters must address many of the issues associated with the selection of the initial parameters, the step size, and the tolerance.

11.6.1 COMPOSITE REGRESSION OF SNOWMELT RUNOFF DATA

In some regions, snowmelt runoff is a major source of water for irrigation, power, and recreation. Estimates of runoff volumes that are expected to occur during the growing season are needed in the spring for agricultural planning. Forecast models use the runoff volume as the criterion variable. The measured snow water equivalent, which is the melted water content of the snowpack, is used as the predictor variable, with the measurements often made on the first day of a month such as March or April. The location of the station where snow water measurements are made can influence the shape of the relationship between runoff volumes (V) and the snow water equivalents (S). For example, in the case of a watershed that has a large elevation range, the relationship between V and S for snow water data from a station at a relatively low elevation may be characterized by a linear trend for large volumes of V and a parabolic trend that approaches some nonzero volumes as S approaches zero. That is, even in years when the measured snow water is near zero at the low elevation station, large volumes of snow may accumulate at the upper elevations, and this snow can produce reasonably large volumes of run-off. This situation would produce data for which a composite regressions model would be appropriate.

The data chosen for the composite regression analysis are from the Upper Sevier River above Hatch in south-central Utah. The Sevier River above Hatch has a drainage area of 340 square miles, with elevations ranging from about 6500 to 11,000 ft and about 70% of the land area being above 9000 ft. The natural stream flow regime consists of low flows from August through March and high snowmelt flows between April and August. The average yearly discharge is about 94,000 acre-feet with about 75% occurring between April and August. While the snow-melt runoff between April and August varies considerably from year to year, the low flow shows little variation from year to year. The mean monthly precipitation in April and May is small. Snow water equivalent measurements are available at the Duck Creek station, which is at an elevation of 8700 ft. The measurements made on April 1 of each year will be used as the predictor variable, while the April 1 to July 31 flow will be used as the criterion variable. A record of 25 years (1952–1976) is used. The mean and standard deviation of the flow for this period was 39.276×10^3 acre-feet and 25.931×10^3 acre-feet, respectively. The mean and standard deviation of the April 1 Duck Creek snow water measurements were 13.324 in. and 8.448 in., respectively. The data are given in Table 11.15. For years in which the April 1 snow water equivalent at Duck Creek is near zero, the runoff is about 20×10^3 acre-feet; specifically, the runoff values decrease linearly as the snow water measurements decrease until about 10 in. With one exception (1967) the flow is between 18 and 30 ($\times 10^3$ acre-feet) for snow water values less than 12 in. Thus the data suggest that a composite regression analysis would be appropriate. Using a value of 9.65 in. for X_c, the data were fitted with the following result:

$$(11.26)$$
$$(11.27)$$
$$\hat{V} = \begin{cases} 24.23 - 2.675\,S + 0.2961\,S^2 & \text{for } S \leq 9.65 \text{ in.} \\ 24.23 - 2.675\,S + 0.2961(19.3S - 93.1) & \text{for } S > 9.65 \text{ in.} \end{cases}$$

in which $\hat{V}$ is the predicted value of V ($\times 10^3$ acre-feet) and S is the snow water equivalent (inches). Equations 11.26 and 11.27 appear to provide a reasonable fit to the data for both high and low values of S; it does not appear to result in biased values for years in which the April 1 snow water measurement is low. It is evident that the model is not entirely rational for values of S below 4.52 in. because the slope becomes negative. Equations 11.26 and 11.27 provide what appears to be a reasonable fit to the data, with a correlation coefficient of 0.902 and a standard error of estimate of 11.42×10^3 acre-feet. Thus the equations explain over 80% of the variation.

TABLE 11.15   Snowmelt Runoff Data

Year	S	V
1952	32.6	88.0
1953	6.9	16.7
1954	16.4	35.4
1955	11.6	17.5
1956	9.6	21.0
1957	16.4	40.2
1958	17.9	62.8
1959	7.4	13.8
1960	8.6	17.4
1961	9.0	18.2
1962	24.7	53.8
1963	1.7	18.9
1964	8.2	29.1
1965	13.4	51.9
1966	12.9	39.4
1967	9.7	55.7
1968	18.3	56.4
1969	34.4	107.1
1970	4.3	22.9
1971	8.6	24.3
1972	2.0	20.9
1973	23.8	91.3
1974	6.2	16.6
1975	14.1	37.4
1976	14.4	25.2

Note: S = snow water equivalent (in.) and V = snowmelt runoff volume ($\times 10^3$ ac-ft.).

The irrationality for low S values is due to the negative linear term of Equation 11.26. For low S values, this term subtracts from predicted estimates such that, for S values less than 4.52 in., the estimates decrease with increasing S values. To overcome this irrationality, a constraint was included in the numerical fitting of the composite model. Whenever the adjustment of numerical fitting wanted to force the coefficient of the linear term of Equation 11.26 to be negative, the program reset it to zero. The other coefficients were then optimized, with the following result:

(11.28)
(11.29)
$$\hat{V} = \begin{cases} 16.93 + 0.106\,S^2 & \text{for } S \le 15 \text{ in.} \\ 40.78 + 3.18(S-15) & \text{for } S > 15 \text{ in.} \end{cases}$$

For this version of the model, the best fit was obtained with an intersection point X_c of 15 in. The model yielded the following goodness-of-fit statistics: relative bias = 0; RMSE = 11.76 in.; $S_e/S_Y = 0.4535$; $R^2 = 0.8115$; $R = 0.9008$. These values are essentially the same as for Equations 11.26 and 11.27, but slightly poorer. The advantage of Equations 11.28 and 11.29 over Equations 11.26 and 11.27 is that including the constraint on the linear coefficient yielded rational estimates of the runoff for low values of S.

11.6.2 A MULTICOMPONENT TIME SERIES MODEL

The objectives of this application are to (1) demonstrate the calibration of a multicomponent model; (2) show the analysis with a temporally varying database; and (3) assess the effect of initial parameter estimates. The analysis will also provide additional illustrations of the importance of the step size D and the convergence tolerance T.

The model represents a watershed with two storage systems: surface water (S_1) and ground water (S_2). One input to the watershed model is a time sequence of daily rainfall depths (millimeters). Part of the rainfall drains on the surface (storage S_2) while the remaining portion enters the ground water storage (S_2). One fitting parameter PINF determines the fraction of each rainfall depth that goes to ground water storage S_2; 1 – PINF enters surface storage.

In each time increment, water drains from each storage unit. The total runoff, Q, is the sum of the two outflows. The surface outflow, SRO, is computed by

$$\text{(11.30)} \qquad\qquad \text{SRO} = S_1 \left(1 - e^{-\text{PSRO}}\right)$$

where PSRO is a fitting parameter. The ground water runoff GWO is computed by

$$\text{(11.31)} \qquad\qquad \text{GWO} = e^{\text{PGW}(S_2)} - 1$$

where PGW is the ground water fitting parameter. Once SRO and GWO are computed, they are summed to get the total flow TRO, which can then be compared to the measured depth of runoff, which is also a model input. The outflows from storages SRO and GWO are then subtracted from the storages S_1 and S_2, respectively. Then the algorithm is repeated for another time increment.

The objective of the fitting is to modify the values of PINF', PSRO, and PGW to get the best agreement between the computed flows TRO and the measured flows Q, where Q and TRO are time series. The initial values of the two storages (S_1 and S_2) could influence the degree of agreement between TRO and Q.

The same goodness-of-fit statistics used for fitting the functions of earlier examples can be used to assess the quality of the fitting. Specifically, the relative bias is a useful indicator of the systematic error; an absolute value less than 5% to 10% is generally considered acceptable. The standard error ratio, S_e/S_Y, is a good measure of the nonsystematic variation, with values less than 25% generally considered good and values above 75% considered poor. The R^2 can also be used, although its value is questionable for nonlinear models, especially those with autocorrelated measurements such as in time series.

A small data set of 18 days will be used to demonstrate issues with calibrating a multicomponent model that shows variation of predictions as a function of time. Table 11.16 includes the daily rainfall P and measured runoff Q. For all runs the initial storages S_1 and S_2 were assumed to be zero. Two analyses were made, each with four separate runs. The two analyses differed in the initial estimates of the three parameters, with the values of the first analysis being considerably different from the optimum values (PINF = 0.4, PSRO = 0.2, and PGW = 0.3). The results of the individual runs are given in Table 11.17. For the four trials of run 1, the initial parameters are not close to the true values. For trials 1B, 1C, and 1D, the initial parameters are the final values of the previous trial. Different step sizes and convergence tolerances were tried.

TABLE 11.16   Data for Application in Section 11.6.2

P	Q	TRO
0.8	0.1878	0.1847
2.4	0.7567	0.7289
0.0	0.5262	0.5222
0.0	0.3848	0.3814
0.0	0.2893	0.2840
0.0	0.2213	0.2154
0.8	0.3640	0.3614
2.6	0.9601	0.9501
1.2	0.9966	1.0104
0.0	0.6996	0.7203[a]
0.4	0.6199	0.6316
0.0	0.4650	0.4651
0.0	0.3566	0.3495
0.0	0.2773	0.2675
0.0	0.2177	0.2082
0.0	0.1721	0.1643[b]
0.0	0.1368	0.1313
0.0	0.1092	0.1060

[a] Largest error (0.0207).
[b] Largest relative error (−0.0453).

TABLE 11.17   Results for Application in Section 11.6.2

| Run | Initial Parameters | | | Inputs | | | | Final Parameters | | | Goodness of Fit | | |
	PINF	PSRO	PGW	S_1	S_2	Tol.	D	PINF	PSRO	PGW	$\bar{e}/\bar{Y}$	S_e/S_Y	R^2
1A	0.750	0.050	0.500	0	0	0.00050	0.050	0.657	0.440	0.124	−0.049	0.181	0.9710
B	0.657	0.440	0.124	0	0	0.00010	0.010	0.651	0.448	0.163	0.000	0.057	0.9970
C	0.651	0.448	0.163	0	0	0.00005	0.005	0.650	0.449	0.158	−0.004	0.045	0.9980
D	0.650	0.449	0.158	0	0	0.00010	0.001	0.650	0.449	0.154	−0.008	0.043	0.9984
2A	0.450	0.150	0.350	0	0	0.00001	0.00500	0.430	0.199	0.286	0.000	0.0067	0.99996
B	0.430	0.199	0.286	0	0	0.00001	0.00300	0.430	0.196	0.287	−0.001	0.0077	0.99995
C	0.430	0.199	0.286	0	0	0.00001	0.00030	0.431	0.196	0.288	−0.001	0.0044	0.99998
D	0.431	0.196	0.288	0	0	0.00001	0.00003	0.431	0.196	0.288	−0.001	0.0044	0.99998
3A	0.400	0.200	0.300	0	0	0.00001	0.00001	0.400	0.200	0.300	0	0.0001	1.0

In the second run (2A to 2D), the initial parameter estimates were closer to the true values, so the final estimates were in better agreement. The goodness-of-fit statistics showed better agreement than with the first run.

11.7 CONCLUDING REMARKS

It is evident from the examples that numerical optimization is considerably more difficult than analytical optimization. It is difficult to state rules that will always yield the best solution. The examples showed that finding the optimum solution depends on the initial values of the unknowns, the tolerance selected for convergence, and the step size D. The initial estimates determine where on the response surface movement starts. For a complex response surface, movement can be constrained if the starting point is located where movement is difficult.

The interaction between the tolerance T and the step size D confounds the fitting process. It is not safe to say that the two inputs can be made very small, as the examples have shown that this can prevent convergence. Instead, it is best to make several runs using different values of the two factors.

The method of steepest descent presented here is the most basic method of numerical optimization. More advanced numerical procedures are available and can contribute to better chances of convergence. For example, one procedure uses a practice that is referred to as acceleration/deceleration. With this option, the derivatives are not recomputed if movement in one iteration improves the objective function and the step size D is doubled with each iteration. When movement is not improved, the step size D is halved for several iterations in the hope that a smaller step size will allow movement in a narrow part of the response surface.

While numerical optimization can be difficult to apply, it has the important advantage of being able to be systematically fitted to complex functions. The complex models allow data sets with highly nonlinear trends and discontinuities to be fitted such that the models are more physically rational. The advantages clearly outweigh the disadvantages.

PROBLEMS

11.1   Create a least squares response surface for a linear model fitted to the following data set:

X	1.2	1.7	2.6	4.2	4.9	5.3	5.9	7.4
Y	2.2	3.4	3.8	3.6	3.5	4.9	6.0	5.1

11.2   Create a least squares response surface for a linear model fitted to the following data set:

X	0.9	1.4	2.3	3.6	3.9	4.6	5.5	6.0
Y	5.6	3.8	4.1	4.0	2.1	1.8	2.5	0.8

11.3   Create a least squares response surface for a linear model fitted to the following data set:

X	1.2	1.7	2.6	4.2	4.9	5.3	5.9	7.4
Y	11.9	8.1	8.4	5.3	4.1	3.4	3.8	2.1

11.4   Create a least squares response surface for a linear model fitted to the following data set:

X	0.9	1.4	2.3	3.6	3.9	4.6	5.5	6.0
Y	3.6	5.4	6.6	8.6	7.9	8.5	9.5	8.7

11.5   Graph a response surface for the model $Y = a + bX$ that suggests b to be more important than a.

11.6   Graph a response surface for the model $Y = a + bX$ that suggests a to be more important than b.

11.7   Assume that the response surface F for a problem is $F = X^2 - 7X + 10$. Compute the response surface such that the minimum is within the range of the values of X.

11.8   Assume that the response surface F for a problem is $F = 2X^2 - 5X + 7$. Compute the response surface such that the minimum is within the range of X.

11.9   For the objective function of Problem 11.7, compute the slope at $X = 3$ both analytically and numerically.

11.10   For the objective function of Problem 11.8, compute the slope at $X = 3.5$ both analytically and numerically.

11.11   For the response surface $F = -X^2 + 9X - 3.6$ compute the slope at (a) $X = 1.65$; (b) $X = 6.35$.

11.12   For the response surface $F = -3X^2 + 12X - 6$ compute the slope at (a) $X = -1.1$; (b) $X = 3.7$.

11.13   For the objective function of $F = X^2 - X + 5$, compute the slope at $X = 1.5$ both analytically and numerically.

11.14   For the objective function of Problem 11.13, compute the slope at $X = -1.5$ both analytically and numerically.

11.15   Compute the analytical and numerical derivative of the function $Y = 2e^{-0.3X}$. For the numerical analyses use step sizes of 0.005, 0.01, 0.05, and 0.10. Compute the errors and discuss the results.

11.16   Compute the analytical and numerical derivative of the function $Y = 2.0/(1.0 + e^{-0.4X})$. For the numerical analyses use step sizes of 0.005, 0.01, 0.05, and 0.10. Compute the errors and discuss the results.

11.17   Use the steepest descent method to find the minimum of the function of Problem 11.7. Assume a base point of $X_o = 6$ and a step size of 0.3.

11.18   Use the steepest descent method to find the minimum of the function of Problem 11.8. Assume a base point of $X_o = -1$ and a step size of 0.25.

11.19   Use the steepest descent method to find the minimum of the function of Problem 11.7. Assume a base point of $X_o = 6$ and a step size of 0.3.

11.20   Use the steepest descent method to find the minimum of the function of Problem 11.8. Assume a base point of $X_o = -1$ and a step size of 0.25.

11.21   Use the steepest ascent method to find the maximum of the function of $F = -1.5X^2 + 5.5X + 1$. Assume a base point of $X_o = 1.1$ and a step size of 0.05.

11.22   Use the steepest ascent method to find the maximum of the function of $F = -X^2 + 9X - 3.6$. Assume a base point of $X_o = 5.1$ and a step size of 0.08.

11.23   Use the steepest ascent method to find the maximum of the function of $F = -X^2/6 + 13X/6 + 2$. Assume a base point of $X_o = 5.5$ and a step size of 0.06.

11.24   Use the steepest ascent method to find the maximum of the function of $F = -0.2X_1^2 - 0.12X_2^2 + 3.6X_1 + 2.5X_2 + 3$. Assume a base point of $X_1 = 8$ and $X_2 = 11$ with a step size of 0.05.

11.25   Use the steepest descent method to find the minimum of the function of Problem 11.8. Assume a base point of $X_o = -1$ and a step size of 0.25.

11.26   Using the data of Problem 11.1 and the steepest descent method, determine the least squares solution for the linear model $Y = a + bX$. Assume the base point ($a_o = 3$ and $b_o = 0.5$). Compare the analytical and numerical solutions.

11.27   Using the data of Problem 11.2 and the steepest descent method, determine the least squares solution for the linear model $Y = a + bX$. Assume the base point ($a_o = 8$ and $b_o = -2$). Compare the numerical and analytical solutions.

11.28   Using the data of Problem 11.3 and the steepest descent method, determine the least squares solution for the linear model $Y = a + bX$. Assume the base point ($a_o = 3$ and $b_o = 0.5$). Compare the analytical and numerical solutions.

11.29   Using the data of Problem 11.4 and the steepest descent method, determine the least squares solution for the linear model $Y = a + bX$. Assume the base point ($a_o = 8$ and $b_o = -2$). Compare the numerical and analytical solutions.

11.30   Using the data of Problem 11.1 and the steepest descent method, determine the least squares solution for the power model $Y = aX^b$. Use the base point ($a_o = 1$, $b_o = 0.5$).

11.31   Using the data of Problem 11.2 and the steepest descent method, determine the least squares solution for the exponential model $Y = a\,e^{-bX}$. Use the base point ($a_o = 1$, $b_o = 0.5$).

11.32   Using the data of Problem 11.3 and the steepest descent method, determine the least squares solution for the power model $Y = aX^b$. Use the base point ($a_o = 1$, $b_o = 0.5$).

11.33   Using the data of Problem 11.4 and the steepest descent method, determine the least squares solution for the exponential model $Y = ae^{-bX}$. Use the base point ($a_o = 1$, $b_o = 0.5$).

11.34   Using the situation of Problem 11.10, show the effect of D on the final values by varying D with the values of 0.01, 0.05, and 0.10. Compare the error in the solution and the number of iterations required.

11.35   Using the situation of Problem 11.11, show the effect of D on the final solution by varying D with the values of 0.01, 0.05, and 0.10. Compare the error in the solution and the number of iterations required.

11.36   Using the situation of Problem 11.12, show the effect of D on the final values by varying D with the values of 0.01, 0.05, and 0.10. Compare the error in the solution and the number of iterations required.

11.37   Using the situation of Problem 11.13, show the effect of D on the final solution by varying D with values of 0.01, 0.05, and 0.10. Compare the error in the solution and the number of iterations required.

11.38   Calibrate the watershed model of Section 11.6.1 to the following rainfall, $P(t)$, and runoff, $Q(t)$, data:

$P(t)$	0.00	0.10	0.08	0.00	0.00	0.14	0.00	0.09	0.05	0.09	0.00	0.06	0.00	0.00
$Q(t)$	0.01	0.02	0.06	0.05	0.03	0.02	0.06	0.07	0.05	0.06	0.04	0.03	0.02	0.01

11.39   Calibrate the watershed model of Section 11.6.2 to the following rainfall, $P(t)$, and runoff, $Q(t)$, data:

$P(t)$	0.08	0.06	0.00	0.00	0.11	0.13	0.06	0.00	0.00	0.09	0.00	0.08	0.05	0.00
$Q(t)$	0.03	0.04	0.04	0.02	0.03	0.04	0.06	0.05	0.04	0.03	0.04	0.03	0.04	0.04

Index

Note: Page numbers ending in "f" refer to figures. Page numbers ending in "t" refer to tables.